HERMANN GÜNTHER GRAßMANN (1809–1877):
VISIONARY MATHEMATICIAN, SCIENTIST AND NEOHUMANIST SCHOLAR

BOSTON STUDIES IN THE PHILOSOPHY OF SCIENCE

VOLUME 187

HERMANN GÜNTHER GRAẞMANN (1809–1877): VISIONARY MATHEMATICIAN, SCIENTIST AND NEOHUMANIST SCHOLAR

Papers from a Sesquicentennial Conference

Edited by

GERT SCHUBRING
University of Bielefeld

KLUWER ACADEMIC PUBLISHERS
DORDRECHT / BOSTON / LONDON

A C.I.P. Catalogue record for this book is available from the Library of Congress.

ISBN 978-90-481-4758-8

Published by Kluwer Academic Publishers,
P.O. Box 17, 3300 AA Dordrecht, The Netherlands.

Kluwer Academic Publishers incorporates
the publishing programmes of
D. Reidel, Martinus Nijhoff, Dr W. Junk and MTP Press.

Sold and distributed in the U.S.A. and Canada
by Kluwer Academic Publishers,
101 Philip Drive, Norwell, MA 02061, U.S.A.

In all other countries, sold and distributed
by Kluwer Academic Publishers Group,
P.O. Box 322, 3300 AH Dordrecht, The Netherlands.

Printed on acid-free paper

TABLE OF CONTENTS

LIST OF ILLUSTRATIONS

Illustration 1: Portrait of Hermann G. Graßmann

GERT SCHUBRING

INTRODUCTION - REFLECTIONS ON THE COMPLEX HISTORY OF GRASSMANN'S RECEPTION

ACKNOWLEDGEMENTS

This volume is one of the results of the international conference on *150 years of 'Lineale Ausdehnungslehre' - Hermann G. Graßmann's work and impact.*[1] It took place from May 23rd to 28th 1994 in Lieschow (Rügen Island), near Stettin (today Szczecin), Grassmann's principal place of work, and of the university town Greifswald, with which Grassmann and his family had many close relationships. The volume contains a large number of contributions to this conference, supplemented by those provided by A. Brigaglia and A. Zaddach who were unable to participate in person. We regret that not all papers presented could be included in the present volume. Some of those missing have already been published in a complementary volume (Schreiber 1995).

Not only is Grassmann's work in mathematics and physics discussed in this volume, but here for the first time his work and impact in the fields of linguistics, Indo-European philology, language teaching, and technology receive a comprehensive analysis, in keeping with Grassmann's commitment to one central tenet of the particularly Prussian concept of neohumanism: the need to relate sciences and humanities.

The conference owes its success to its having been organized as a "collective endeavour": in order to encourage in-depth analyses of some major aspects of Grassmann's work, the conference was planned over an extended period by a committee composed of mathematicians, historians and educators of mathematics, and linguists: by Peter Bergau (Bielefeld), Albrecht Beutelspacher (Gießen), Erika Hültenschmidt (Bielefeld), Günter Pickert (Gießen), Karin Reich (Stuttgart), Erhard Scholz

---

1 According to the rules for the assimilation of foreign names to English typography, his name has to be typed as "Grassmann" throughout this book. Since the orthography for names was not standardized in Germany during the nineteenth century and since even the editors of his *Gesammelte Werke* did not pay much attention to it, it has to be emphasized here that he himself wrote his name as 'Graßmann' (see also the reproduction of his personal file, illustr. no. 4, and illustr. no. 1).

(Wuppertal), Peter Schreiber (Greifswald), Gert Schubring (chairperson, Bielefeld), Heinz Schwartze (Gießen).

Significant contributions to the preparations were made by A.C. Lewis (Canada) and A. Zaddach (Chile) as "foreign correspondents." An important share of the success is due to our local organizer P. Schreiber.

Editing this volume required an even more comprehensive and intense cooperation. As most contributions had been presented in German, they had to be reedited in English for this book. Many colleagues in England and North America helped in rendering texts not originally written by native speakers into good English. This process simultaneously provided opportunities for the discussion of the papers to continue, so that editing became a collective enterprise as well. For their active support, I should like to thank in particular I. Grattan-Guinness, A.C. Lewis, D. Rowe, A. Swimmer, and R.S. Turner. The "maieutic" work of Günter Seib (translator, Bielefeld), who prepared English versions of some papers for editing, was particularly valuable.

I am especially indebted to Günther Pickert, David Rowe and Erhard Scholz for their assistance in publishing this volume, as they were always prepared to discuss editorial questions and to give advice.

I am pleased to acknowledge that funding for the Rügen Conference was made possible through a grant from the *Deutsche Forschungsgemeinschaft*, and I am also grateful to the other institutions and organizations who helped to organize it: the Universities of Bielefeld and Greifswald and the *Fachsektion Geschichte* of the *Deutsche Mathematiker Vereinigung*.

I. THE ALLEGED "TRAGEDY" OF HERMANN GRASSMANN

The occasion for the 1994 conference was provided by the 150th anniversary of the publication of Hermann Grassmann's principal work in its first edition. The conference was the very first in which debate and exchange of views about Grassmann's work ever took place, except for one afternoon session of the *Berliner Mathematische Gesellschaft* (BMG) on 21st April, 1909: the occasion then was Grassmann's 100th birthday, and contributions were made by F. Engel, Eugen Jahnke (BMG chairman), and Hermann E. Grassmann (his son most involved with mathematics).[2]

Given this lack of previous analysis, one of the major themes of the 1994 conference had to be the reception of Grassmann's achievements. That reception was quite different in the various disciplines in which he was active. His contributions to physics were received rapidly and

---

[2] The papers are published in: *Sitzungsberichte der Berliner Mathematischen Gesellschaft*, 8. Jahrgang, 79-114 (*Beilage zu: Archiv der Mathematik und Physik*, 3. Reihe, Band 15, 1909).

effectively; his work in linguistics was immediately understood as seminal and has retained its importance down to the present. In contrast, Hermann Grassmann is held in the historiography of mathematics to be one of the classical cases illustrating the phenomenon that major scientific innovations are noticed by the expert public and are able to exercise an effect only after a substantial time-lag.

The belated recognition in mathematics is closely associated with his fame. The delay is a theme underlying almost all contributions at the Conference and in these Proceedings. In fact, the history of Grassmann's reception in mathematics reveals essential patterns even in the development of mathematics: his immediate adherents' restriction of their own work to traditional geometric space; the disdain and misgivings that F. Engel and E. Study (the editors of Grassmann's *Gesammelte Werke*) evinced about the abstract-algebraic dimensions in his work; the decisive switch to the algebraic conceptions brought about by French mathematicians. This is why it makes sense to explicate the pervasive problem of reception and to discuss some aspects of the complex but hitherto unexplored history in a separate chapter within the introduction to these Proceedings.

Among present-day mathematicians, Grassmann's belated impact is a well-recognized topic. J. Dieudonné, speaking for Bourbaki who had opened up entirely novel and promising fields of application for Grassmann's ideas, deplored the "tragedy" of the latter's delayed recognition as follows:

"In the whole gallery of prominent mathematicians who, since the time of the Greeks, have left their mark on science, Hermann Grassmann certainly stands out as the most exceptional in many respects, when compared with other mathematicians, his career is an uninterrupted succession of oddities: unusual were his studies; unusual his mathemtical style; [...]; unusual and unfortunate the total lack of understanding of his ideas, not only during his lifetime but long after his death; deplorable the neglect which compelled him to remain all his life professor in a high-school." (Dieudonné 1979, 1).

In a similar vein, Barnabei, Brini and Rota, in a much-read article on exterior calculus of invariant theory, alluded to

"the tragedy of Hermann Grassmann which has been unfolding since his death, by a succession of misadventures and misunderstandings of his work unique in the history of modern mathematics." (Barnabei et al. 1985, 120).

Undoubtedly, a tragic life story makes an historical figure even more fascinating and pathetic. The extent of the "misadventures," however, should not be overestimated in Grassmann's case, as Rowe shows in his paper. It is a phenomenon well known in the history of science that incisive innovations require a lapse of time to overcome inertia and to exercise their effects. Compared to this general tendency, the impact of Grassmann's work was not so extraordinarily belated after all.

This perspective is confirmed by another fact which is generally not taken into account by those who lament the time-lag which Grassmann suffered before his work was recognized. A mere seventeen years after his death, the publication of his *Gesammelte Werke* on mathematics and physics began. In view of the many editions of "collected works" now available for mathematicians of all periods, we perhaps do not fully appreciate that towards the end of the 19th century such an effort at publication represented exceptional veneration - and a strong impetus towards enhancing a reception which had already begun to be established. The exceptional character of this venture becomes obvious if we consider other German mathematicians and the comparable efforts to publish their collected papers. In 1892, the year Felix Klein had the project of publishing Hermann Grassmann's collected works adopted, only a small minority of mathematicians had received this honor: Gauß (since 1870), Riemann (1876), J. Steiner (1881), C. G. J. Jacobi (since 1881), Möbius (1885), Borchardt (1888), Dirichlet (1889) and H.A. Schwarz (1890). Weierstraß, who like his disciple Schwarz initiated publication of his own works himself, saw his first volume printed in 1894. Plücker's works, too, began to be published only after those of Grassmann (since 1895). The works of two of the most important mathematicians who had been active in Germany began to be published only in the 20th century - Leibniz (since 1923) and Euler (since 1911). On an international scale, only a small number of authors had been thus honored at the time, most of them in France: Condorcet, Lagrange, Laplace, Cauchy. It is notable that Grassmann was the only mathematician not active at a university or academy who was honored by such a publication. That the project ran to an exceptional length of six volumes underlines the extent of Grassmann's recognition even more. The publication of his collected works was not only an expression of Grassmann's renown within the mathematical community - it also laid the foundations for a much extended reception.

If there were nevertheless obstacles to the reception of Grassmann's work - and it is of course problematical to say that the actual reception was less profound or pervasive than how we might imagine it potentially to have occurred - then those specific obstacles are of structural interest for the history of science.

I. 1 The "Grassmannianer"

It was Felix Klein who most actively promoted the dissemination of Hermann Grassmann's work (see Tobies' contribution). Nevertheless, Klein held rather negative views of the fervent adherents of Grassmann whom he called *Grassmannianer*. The mathematicians around Klein sided with him in this matter. In his work on the history of 19th-century mathematics, produced in his old age, Klein devoted an entire chapter to the *Grassmannianer* that characterized them as a particularly reductionist school who showed a one-sided loyalty to their master and were therefore prevented from developing their field further:

"There are two things in Grassmann's nature and fate that, more and more effectively as time went on, made him the head of a school, or better said, a sect affected with all the fanaticism usual in such cases. The first is his pronounced sense for special algorithms, with which the initiate becomes so intimately familiar that they acquire for him an obligatory status and become a distinguishing mark of the narrow solidarity of the adepts. Because of their interest in the correctness of their style from the point of view of their orthodocy, these sectarians go off the track, neglecting what is really mathematically essential, a penetrating investigation of the problem." (Klein 1979, p. 169)

On the other hand, according to Klein, the very "tragedy" of Grassmann's neglect by his contemporaries generated the cohesion between the *Grassmannianer*:

"The second influential thing is the fact that Grassmann did not receive the recognition that was his due during his lifetime, so that his partisans now see in him the martyr, whom they must surround by glory in order that he may come into his own. It is appropriate to this glory for the partisan to choose all literary and computational forms of expression so as to contrast the master and himself from everything that is usual and thereby effectively to withdraw from a comparative competition" (ibid.).

As a case in point, Klein discussed the hermetic character of the peculiar terminology developed and used by Grassmann's son Hermann Ernst (Jr.) in his own book on projective geometry (loc. cit). Friedrich Engel, who otherwise disagreed with Klein's views, considered it necessary to distance himself from the *Grassmannianer* in the introduction of the Collected Works he had edited:

"I have never been a one-sided adherent of Grassmann and will never become one, but for this very reason I could at least claim to be impartial." (Engel 1894, p. VI).

And in dedecating the last volume to Study, Engel reminded Study of his conversion from a pure Grassmannian to a less fervent advocate:

"As you know well, we got somtimes into dispute there [in Leipzig]: I knew little of Grassmann then and opposed your exaggerated, as it seemed to me, enthusiasm for Grassmann's methods. Since then, there has been a change in both of us. Your judgement of these methods has become much more sober. And I have become what I would never have dreamt of at the time, the editor of Grassmann's collected works, and now even his biographer. While I have not become a partisan of the Master from Stettin, I have come to estimate and love him in doing so" (Engel 1911, p. VI).

In his correspondence with Study, Engel voiced even harsher criticism of the *Grassmannianer*. Thus he complained that particularly young mathematicians in particular allowed themselves to be drawn into this system.

This may have been an allusion to Alfred Lotze (1882-1964). In the *Enzyklopädie der mathematischen Wissenschaften*, in the part of geometry and the analysis of geometrical systems, Lotze was the author of the article on Grassmann's "Ausdehnungslehre": this in 1923, when Klein no longer took an active part in editing the encyclopedia. Lotze deviated somewhat from the original plan sketched by the first author Hermann Rothe,[3] in that he represented Grassmann's work reduced to the conceptions of his school, rather than in an open manner that mediated between Grassmann's approaches and the development mathematics had taken in the meantime.

The Grassmannians' way of presenting themselves displayed two themes which became relevant for the reception accorded to Grassmann's ideas. Both suggest an aspect of the concept of "scientific school" (which Klein had termed "sect") that has been too-little noted in the contemporary discussions of scientific schools that go on in the field of science studies: "schools" do not serve invariably to promote scientific progress. By a too rigid adherence to the master's theories, they may on the contrary bring about ossification in a particular field of science.[4]

The first theme concerns the tension between the Grassmannians loyalty towards the master's concepts and their reinterpretation and transformation of those concepts on the basis of the continuing development of the field. As D. Rowe's contribution shows, this theme becomes particularly salient in the controversy between F. Klein and V. Schlegel of 1875.

Schlegel insisted on the superiority of Grassmann's *Ausdehnungslehre* in the first volume of his own "System der Raumlehre," stating that Grassmann's "new analysis is the only one which is appropriate to" arithmetic and geometry (and to mathematical physics as well) (Schlegel 1875, p. V), Klein who was involved in the discussion and was himself not unbiased, maintained in his review published in 1875 that it would have been more productive for Schlegel to have presented Grassmann's conceptions and accomplishments in the light of the "directions that research has taken subsequently" (Klein 1875, p. 235). Schlegel replied in the preface to the second volume, persisting in his view that Grassmann's methods offered "the shortest and easiest approach to the results of ancient and modern geometry and algebra" (Schlegel 1875, p. VIII).

Curiously enough, the *Grassmannianer* did not form the only such school at the end of the 19th and the beginning of the 20th centuries: There was another, competing group within the same field: the Hamiltonians. F. Klein placed them on the same level as the *Grassmannianer*: "Alle these characteristics of the convinced sectarians now recur among the Quaternionists, Hamilton's pupils [...] It hardly

3 For instance, he has renamed the chapter "Begründung der Ausdehnungslehre durch G. Peano" as "Begründung der Punktrechnung durch G. Peano" and transformed it, hence into a "Grassmannianer" presentation.

4 A well known example is given by the orthodox transmission of Newton's doctrines in the England of the eighteenth century.

need be mentioned that Grassmannians and Quaternionists oppose each other strongly, and that each of the schools has separated into wildly warring camps" (Klein 1979, p. 169f.).

As M. Crowe's study shows, the Hamilton School evinced an even more marked sectarian character: it even boasted of a "pope," P. G. Tait, who watched over the integrity of the master's theory and called deviants to order (see Crowe 1967, pp. 117 ff.). In the fight over a standardized vector notation the disputes between the two schools that Klein had mentioned reached an intensity that made it visible to the entire mathematical community. Even a committee established by the International Congress of Mathematicians of 1908 proved unable to settle the matter (see Reich 1989, 1992).

How the Grassmannians themselves affected the reception of Grassmann's work can also be seen from the second theme of their self-presentation: the conceptual restrictiveness with which they took up his ideas and elaborated them. It must be noted that the prominent members of this group were without exception not important, productive mathematicians, and that their fields of work were just as traditional as their fields of action were restricted. Victor Schlegel (1843-1905), Grassmann's first biographer, worked as a teacher for all his life, first at a *Gymnasium*, later at a *Gewerbeschule* (vocational school) and finally at a *höhere Maschinenbauschule* (technical college). In his work, he concentrated on traditional elementary geometry, confining the presentation of Grassmann's ideas to three-dimensional geometry in his own "system of spatial theory." His reviews in the yearbook *Jahrbuch über die Fortschritte der Mathematik*, monopolized discussion of all publications on the *Ausdehnungslehre* so that no widening of scope was possible here either. Rudolf Mehmke (1857-1944), professor at technical colleges, published mainly on geometry and mechanics. His comprehensive textbook on point and vector calculus also remained three-dimensional, with an emphasis on projective geometry. Grassmann's son Hermann E. Grassmann jr. (1857-1922), first a teacher and since 1902 professor at universities, strove all his life to systematically elaborate his father's *Ausdehnungslehre* and to get it recognized. Further representatives of the group of *Grassmannianer* were Ferdinand Kraft (1844-1924), teacher at technical colleges, later engineer and lecturer at the renowned technical university of Zurich, Ferdinand Caspary (1853-1901), teacher, engineer and professor at a technical university, Eugen Jahnke (1863-1921), teacher and professor at a technical university.

The conceptual restricitiveness in the approach of many *Grassmannianer* to the work of the master was particularly salient in the case of "point calculus."[5] In the *Enzyklopädie der mathematischen*

5 The term "Punktrechnung" seems to have been introduced by Mehmke in 1884.

Wissenschaften, Rothe presented Möbius' barycentric calculus as a method of point calculus which "was the foundation for Grassmann's *Punktrechnung*" (Rothe 1916, p. 1293). Lotze in his own encyclopedia article on Grassmann, also conceived the *Ausdehnungslehre* as a point calculus and hence concluded that it agreed with Möbius' methods (Lotze 1923, pp. 1432 and 1486). While Lotze conceded that Grassmann had founded his *Ausdehnungslehre* on the "abstract concept" of an *n*-dimensional extensive quantity, he declared this irrelevant, as Grassmann had "undoubtedly originally" been "guided by geometric aspects," and because the concrete meaning of his quantities and connections could only be derived within the frame of geometry (ibid., p. 1543). Because of that, Lotze said, it made more sense to confine oneself to Euclidean space from the outset:

"If one confines oneself, however, to point calculus in Euclidean space, one may conversely start from the concrete meaning of the (outer) product of four simple or multiple points established by definition, deriving from that, in connection with an appropriate definition of the equality of geometric quantities and with the requirement that the geometric (outer) multiplication is to be distributive with respect to addition, the meaning of the other geometric quantities and of their conncections" (ibid., p. 1544).

Lotze interpreted Peano's contribution to be his realization of this confinement to Euclidean space as point calculus (ibid.) Conversely, he criticized the vector analysis advocated by the Hamiltonians: it already assumed, he said, the Euclidean metric. Lotze advanced this criticism in a way that excluded critical debate: There was no organic connection, he said, between vector analysis and a comprehensive system of geometric analysis, and hence it was not so important for applications in geometry and mechanics (ibid. p. 1550).

This restricted interpretation was certainly of great practical interest for the contemporary applications in geometry and mathematical physics.

The restricted understanding of the *Ausdehnungslehre* as a "point (and segment) calculus" in Euclidean space was publicly criticized by Engel. Sharp protest prompted Engel to elaborate his position. For the *Fortschritte* yearbook, Engel reviewed the first volume of H. E. Grassmann's principal work *Projektive Geometrie der Ebene unter Benutzung der Punktrechnung dargestellt* which had been published in 1909. To his positive review, Engel added a more general remark:

"At least I cannot conceal that the calculus with points and line segments does not appear to me to be the true instrument to treat projective geometry as the latter distinguishes the infinitely distant. Indeed we possess a perfect, even the only appropriate instrument in the shape of homogenous coordinates, and the designations of the theory of invariants incidentally permit the realization of the ideal that G r a s s m a n n senior had in mind, namely to conduct all calculation independent of the choice of coordinate system. At the same time, this method makes the extensive quotients superfluous" (Engel 1912, p. 591).

The protests that these comments provoked compelled Engel to publish a "Berichtigung" in the DMV periodical, in which he declared his regrets about "an error." With regard to the first sentence, he wrote:

"This creates the impression that in his volume Grassmann founded projective geometry on the calculus of points and line-segments, thus distinguishing infinite distance. This, however, is by no means the case; rather, the role of line-segments in Grassmann's book is quite peripheral, and projective geometry is represented by means of point calculus alone without deploying the infinitely far. This is a point I insist on making. However, I remain convinced as before, as I have expressed in that review as well, namely that the designations and methods of the theory of invariants are superior to those developed by Grassmann, not only to those originating from Grassmann senior but also those perfected in the meantime. In particular, I cannot make friends with the extensive quotients, for, ingeniously as they have been conceived, the arbitrary coordinate systems which Grassmann intended to ban reappear precisely there" (Engel 1913, p. 103).

While Engel excepted Grassmann's son from his criticism, he stood by his critique of the limited scope of Grassmann's concepts and did not retreat from his claims about the latter's inferiority compared to more modern developments of the theory of invariants. Debates of this kind, however, obviously remained exceptional, all the more so within the camp of the *Grassmannianer*.

I. 2 Reception is always contemporary

Beyond the special problems that the intervention of the *Grassmannianer* raised for the impact of Grassmann's work, a more general one has to be taken into account: reception can only take place within the associated contemporary conceptual horizon. It is typical that the two mathematicians most involved in editing Grassmann's collected works - Friedrich Engel and Eduard Study - neglected the fundamental aspects, and received and interpreted him from a geometric point of view. The two editors' principal field of activity was geometric research, and they took insufficient notice of the foundational and algebraic parts.

Typical is Engel's negative assessment of the foundational aspects in a letter to Study dated October 6th, 1893, one year after he had agreed to be the editor of the collected works. In that letter he agrees with Study's critique of the A_1:

"It is indeed inconceivable that Grassmann was satisfied with so empty fundamental concepts like *Element*, *Grundänderung*, etc. The §§ 14 ff. [of the A_1] are partly a bit of a hoax."[6]

Engel also published this kind of assessment in his edition of the collected works. One of his notes on the edition of A_1 says it could not

6 In the Friedrich Engel Archives, Library of the Gießen Mathematics Department figures not only the letters from Study to Engel but also the Letters from Engel to Study which Engel had got back after Study's death. Unfortunately, the collection with its enormous number of letters is only summarily classified.

be denied "that Grassmann, in his striving for generality, sometimes lost his footing. This is particularly true for the fundamental concepts on which the *Ausdehnungslehre* has been established in the present work. These fundamental concepts are far too indetermined, too devoid of content to lend themselves to the kind of conclusions Grassmann drew from them" (Grassmann 1894, p. 404).

As an example, he quoted the definition of the *Element* in § 13 and in his edition of the A_2,[7] Engel repeated his critique, claiming that the fundamental concepts of A_1 are "so general and hence so empty of content that they do not suffice to establish a *real* (my emphasis. G.S.) system." This was the reason why "he later had to interpret more into his own fundamental concepts." This drawback was not longer present in the A_2, Engel said, as Grassmann here assumed much from elementary mathematics and analysis (Grassmann 1896, p. V).

Study, too, voiced harsh criticism of Grassmann's objectives with regard to foundations, but even more radically than Engel. In discussing Grassmann's concept of "innere Grössen" (inner magnitudes) in the prize essay on geometric analysis characterized them as "aschgraue Abstraction," as "Sprung ins Nebelhafte" and commented:

"since his inner magnitudes ["innere Grössen"] are, in my view, nothing else than arbitrary signs on the paper" (letter no. 99, 28 June 1894).

In a letter dated May 14th, 1895, after discussing Grassmann's concepts of line complexes, he exclaimed:

"Yes, my veneration for G. has suffered a heavy jolt, from which it will not recover. He is indeed so full-mouthed about everything. All is fundamental, even the tritest things."

Given the contemporary limitations of Engel's and Study's judgements, one may doubt whether Engel was correct when he claimed that "of purely mathematical matter, the *Nachlaß* contains nothing which would merit publication," in his letter to Study on May 4th, 1906. Engel believed that the *Nachlaß* contained interesting and publishable things only on mathematical physics. As the *Nachlaß* has obviously been destroyed subsequently (see the remarks in this volume) possible losses resulting from these restrictive contemporary judgements unfortunately cannot be assessed or made up.

The relationship between algebra/analysis and geometry usually provides a prime indicator for the respective reception of Grassmann's ideas within a given mathematical culture. Thus, in the 1860s and 1870s Grassmann was studied most intensely by the school established by Clebsch and Klein, which was situated within the Königsberg tradition. He was largely ignored by the so-called Berlin school with its focus on

---

7 A_1 resp. A_2 are the traditional abbreviations for the first, 1844, edition of the *Ausdehnungslehre* and for the second one of 1862.

pure mathematics and analysis.[8] As Tobies stresses in her contribution, the encyclopedia contains a large number of references to Grassmann's work. It must be added, however, that these references are concentrated on the volumes on geometry and mechanics, while Grassmann remains practically unquoted in the volumes on algebra and analysis.

It is highly revealing that awareness of Grassmann's work extended to algebra and analysis as well in another mathematical culture, the French. In his contribution, Schlote points out that E. Cartan was the first to clearly elaborate the algebraic aspects of Grassmann's work - in the French issue of the encyclopedia. The French version, indeed, was no mere translation, but rather a version which had been worked on, sometimes extensively. Lotze's article was not translated at the time, because it appeared after the World War I which brought this joint venture of German and French mathematicians to a halt.

Ironically, Cartan's contribution was a re-edition of Study's article in the German issue: "Theorie der gemeinen und komplexen höheren Größen" of the first volume (on arithmetics and algebra). Cartan did not only decisively extended the article's historical part, but also restructered its main body. Engel and Study observed this revision with misgivings. On April 25th, 1908, Engel wrote to Study: "Just now I received the part of the encyclopedia's French issue containing your cartanized article."

The sharp contrast between Grassmann's reception in Germany and in France is shown in the fact that Study, in his own contribution, mentioned Grassmann only briefly in two notes, while Cartan added a substantial essay on "Le calcul extensif de Grassmann" (Cartan 1908, pp. 378-393). In 1922, Cartan developed his theory of differential forms on the basis of his own reception of Grassmann. The later establishment of multilinear algebra by Bourbaki can be directly tied to Cartan and thus to a line of development of Grassmann's ideas in France which is so quite different from that in Germany.

In the context of the British school of symbolic algebra, Grassmann was received in yet a different way, one which proved in the long run to be of decisive importance for the understanding of Grassmann's foundational intentions. A.N. Whitehead claimed that his seminal work, the *Treatise on Universal Algebra* (1898), was based on three systems of Symbolic Reasoning: Hamilton's, Grassmann's and Boole's (Whitehead 1898, p. v). Still more, this entire work proves the profound impact which Grassmann's calculus of extension had on Whitehead's conceptions. In full accord with Grasmann's own ideas, Whitehead develops the logical foundations and shows their universal applicability in mathematics (and physics). Whitehead himself emphasized "the greatness of my obligations in this volume to Grassmann" and his two

---

[8] On the relation between the two schools see my study: "Königsberger vs. Berliner Schule - Kämpfe um Gauß' Lehrstuhl in Göttingen," its publication is in preparation.

versions of the *Ausdehnungslehre* (ibid., p. x).[9] Several of the contributions in this volume on the modern reception of Grassmann reveal the important role of Whitehead in this.

While comparing mathematical cultures, it might be useful to add a remark about developments at the "periphery"[10] where there innovations often go beyond the state of the art attained in the "metropoles," even though these innovations may be noticed indirectly at best. While I was participating in a conference on the history of science in Istanbul in 1991, a 1988 reedition of a textbook *Linear Algebra* caught my eye which had originally been published in 1882 by Hüseyin Tevfik Pascha (in English!).[11] When I tried to find out later when the term of *linear algebra* - which was so decisive for the development of mathematics in the 20th century - was first used, I was informed by Gregory Moore that van der Waerden had been the first to use this term (i. e., without the suffix "associative" with which it already occurs in B. Peirce in 1881).[12]

Hüseyin Tevfik Pascha (1832-1901), educated in the Ottoman Military Academy, was active there and in private endeavours of teaching mathematics and the sciences. As military attacheé to France from 1869 to 1870, he improved his knowledge in mathematics. Another stay abroad, from 1872 to 1880 in the United States, was decisive for introducing him into mathematical research. It seems that he was in contact with P.G. Tait, but he achieved results independently. What he published in his 1882 textbook was the exposition of a three-dimensional linear algebra (and of its appplications to elementary geometry): the transposition of the quaternions to three dimensional space - as a non associative algebra. The presentation abounds with reverence for Hamilton; Grassmann is not mentioned. One can, therefore, understand the notion "linear algebra" as originating from an approach aiming at generalizing both the associative and the non-associative cases.[13]

On the future development of the reception of Grassmann's ideas one can be quite optimistic. This is evidenced not only by the many contributions at the Grassmann Conference and by the broad range of

---

9 As an appendix, Whitehead's book even contains a "Note on Grassmann" and a bibliography of his publications.

10 For the concept of relation between "metropole" and "periphery" see Lew Pyenson 1989.

11 and a second edition of 1892. The book and its author, entirely unknown until the reedition even in Turkey, were rediscovered when a copy appeared in the catalogue of a German antiquarian bookseller.

12 See also Moore (1995, p. 294) where he mentions an earlier use of the term by H. Weyl in 1918.

13 Whitehead mentioned the term "linear algebra" in 1898 when he announced it as the subject of the intended second volume of his *Treatise* (Whitehead 1898, p. v).

applications of his ideas but also by the decisively improved accessibility of his major work to the international public. The first translation of the A_1 (into Spanish) went almost unnoticed, probably since it occurred at the "periphery" - in Argentina, in 1947[14]. But two more translations published in the wake of the 150th anniversary of the A_1 - into French by D. Flament (1994) and into English by L. Kannenberg (1995)[15] - will obviously attract the attention of an even larger public.

II. ON THE CONTRIBUTIONS IN THIS VOLUME

The **first** part is devoted to aspects of Grassmann's life and work.

Given the still limited understanding of the *Ausdehnungslehre*, Beutelspacher surveys the main innovations and mathematical ideas developed in this key work.

H. Schwartze reviews Grassmann's life and describes his teaching activity, including his successes and failures and the contemporary conceptions of teaching methodology. From the perspective of a modern textbook author, he presents the two mathematics textbooks Grassmann published.

A report on the fate of Grassmann's *Nachlaß* follows. Grassmann's family in the past followed no coherent policy in collecting and preserving his papers, so that it seems that no part of the *Nachlaß* is still extant.

The contributions in the **second** part examine how some of Grassmann's characteristic ideas emerged and how they can be related to the larger intellectual context of his time. A.C. Lewis tackles the problem of exploring how Grassmann developed mathematical concepts of his *Ausdehnungslehre* by analyzing how Grassmann's *Theorie der Ebbe und Flut*, his teacher examination dissertation of 1840, paved the way for his new theories. Lewis focuses on differences in the presentation of the operations and shows how notational complications and inconsistencies might have led Grassmann to more foundational abstraction.

This part of the volume focuses naturally on the relationship of philosophical concepts to Grassmann's work, especially the intriguing impact of *Naturphilosophie*. Since mineralogy was a favorite subject of *Naturphilosophie*, E. Scholz discusses the significance which the extensive crystallographic work of Grassmann's father Justus, markedly influenced by romantic *Naturphilosophie*, had on the emergence of Hermann's vectorial ideas. This is particularly relevant since dynamist

---

14 I am grateful to J. Bosco Pitombeira de Carvalho (Rio de Janeiro) who informed me of this Spanish translation and provided me with a copy.

15 Kannenberg's translation even contains more of Grassmann's publications: his prize essay on geometrical analysis and several articles with applications of the *Ausdehnungslehre* in mathematics and physics.

crystallography established the notation of a three-dimensional vectorial system of forces. Scholz's paper demonstrates the richness of dynamist principles in crystallography and their traces in Hermann Grassmann's early works.

It is highly revealing to assess Justus Grassmann's adherence to the broad movement of *Naturphilosophie* in early nineteenth century Germany, but difficult to do so, since he, like his son Hermann, was reticent about the roots of his thinking. This attitude is characteristic of *Gymnasium* teachers of this period in Prussian history, who were eager to show and defend their originality. M.-L. Heuser tries to identify corresponding concepts in Schelling's work, one of the major *Naturphilosophen*, and the probability of his influence on Justus.

Given the critical philosophical strand in Grassmann's mathematical thinking, it has been common to emphasize the influence of Schleiermacher's *Dialektik*. Schubring's paper proposes a broader notion of reception. He shows that the years immediately after 1839 - the decisive period of formation for Hermann's mathematical conceptions - were marked by a close scientific cooperation between Hermann and his brother Robert, as they jointly prepared the theses required by the state teacher examinations. Since Robert prepared a voluminous dissertation on the history of philosophy of mathematics during this period, Hermann became acquainted both with traditional and recent conceptions in this field. In particular, the paper shows congruencies between Hermann Grassmann's views and those of J. F. Fries, who was the most knowledgeable of contemporary philosopher reflecting on mathematics, and who proposed a "Syntaktik" as foundational discipline for all parts of mathematics. Moreover, Hermann's switch to a more formalistic style can be attributed to the cooperation with his brother Robert, who demonstrated a penchant for abstract and formalist systematization.

Turner's paper discusses Grassmann's direct interaction with the famous scientist Hermann Helmholtz - universalist researcher in science, mainly in physics and physiology. By strictly applying the principles of his *Ausdehnungslehre*, Grassmann was able to bring a decisive innovation to color theory. Showing that color space is geometrically representable as a three-dimensional space, Grassmann was able to end the traditional debate on "primary" colors and to open, by his barycentric mix-law, the way for modern color theory. As Turner suggests, Grassmann was able to do this partly because he was an outsider and not hampered by the traditional ways of thinking on color vision. On the other hand, this marks his limitations, too: he could not keep abreast of the on-going research in this field.

The greatest number of Grassmann's books concern **language** teaching. In view of how often some of them appeared in multiple editions, Grassmann seems to have enjoyed his greatest success in language teaching! There have been no studies of his ideas in this field.

Hültenschmidt's paper is the first to describe Grassmann's work on language teaching and to analyze his conceptions. The chapter focuses on a scientific school grammar published in 1842. Ostensibly a grammar text for the German language, the book proves to be actually a concretization of theories on general grammar, and Hültenschmidt shows how Grassmann transposes innovations in grammatical theory established in France, to Germany, in a more modern form. This transposition is carried out, however, within the framework of a romantic nationalism claiming the existence of a "Deutsche Kulturnation." These views were even more fanatically propagated by Hermann's brother Robert, whose later writings on language teaching serve to complement the analysis.

Hültenschmidt also shows that Grassmann applied structural patterns to linguistic analysis which were analogous to those in mathematics: structuring in form of symmetrical oppositions. One of these oppositions, between "Form" and "Begriff," proves to unite two essential dimensions in linguistics which had traditionally been treated separately. In an appendix, the historical and the systematical importance of Grassmann's famous "Aspiraten" law is briefly introduced, not only for Sanskrit studies, but for linguistics in general.[16]

The **third** part of the volume examines the influence of Grassmann's work in the nineteenth century and demonstrates the extraordinary scope of that influence. Grassmann's influence was earliest and strongest in Germany and Italy.

Tobies gives an overview of the reception in Germany and shows that it is mainly seen in members of the Clebsch school. As it makes clear, most initiatives for dissemination were taken by Felix Klein. Rowe's paper has already been alluded to in this Introduction. It analyzes in particular the reception in Germany in the 1870s and shows that Grassmann's alleged "tragedy" has been grossly exaggerated. The key questions of Grassmann's reception become visible in the controversy between F. Klein and V. Schlegel in 1875. Nadeník's paper gives an overview on the work of Czech mathematicians at the end of the nineteenth century. It shows how Grassmann's work was studied and disseminated there beginning in the 1880s and '90s, mostly in the context of geometry and of physics, after the mathematical community had become familiar with Bellavitis's and Hamilton's publications. The author emphasizes parallel patterns in the lives of Grassmann and the Czech Bolzano.

After these more "regional" studies, the other papers in this chapter focus on particular conceptual developments. Peano's importance for the

---

16 Jean-Claude Muller's contribution at the Conference, "Grassmanns Beschäftigung mit dem Rigveda," discussing the Aspiraten Law from the point of view of Indo-European studies was not delivered for publication.

dissemination of Grassmann's ideas in Italy is well known. Here, Brigaglia studies another feature: how Grassmann's theory of *n*-dimensional spaces was immediately taken up by C. Segre and used as early as in his doctoral thesis of 1883. Later, studying Grassmann's work more intensively, Segre extended his own work to projective *n*-dimensional geometry and influenced a number of mathematicians.

Schlote demonstrates those of Grassmann achievements in algebra which are often concealed within a geometrical context: his extensive studies on the nature of connections and on compositions of connections. Particularly innovative were Grassmann's studies on the multiplicative connections. Schlote shows how Grassmann provided the basis for a theory of hypercomplex number systems and broke with traditional conceptions. The paper also shows simultaneously the limitations in Grassmann's approach and its flaws.

Dorier explores another line of development of Grassmann's work affecting algebra: the concepts of basis and dimension, which later led - together with advances in the study of linear equations - to linear algebra, a key theory of modern mathematics. Dorier reconstructs how the notions of basis and dimension were already present in the theory of extension, in particular in the notion of a system of *n*-th order. A remarkable side effect of the reconstruction is that the theorem known as the Steinitz Exchange Theorem originates from the *Ausdehnungslehre*. Dorier studies the contributions of followers like Peano and Burali-Forti and Marcolongo to a theory of linear vector spaces. He complements this by presenting an overview on the development of the theory of linear equations from Cramer to Frobenius, and the development of the concepts of basis and dimension within field theory, all this eventually merging in the modern algebra as so masterfully elaborated by van der Waerden in 1930.

While Grassmann's reception in mathematics proceeded slowly, it was highly effective and broad in mechanics. There an urgent demand developed for the new calculus, since it permitted physics to operate with forces and other directed quantities. Reich shows the speed of dissemination and acceptance of the extension theory within mechanics - in Germany as well as in other countries. At the same time, she elaborates its limitations, mainly those caused by the competition which Grassmann's theory had to face from Hamilton's disciples. She discusses both the terminological and the conceptual differences between the respective versions of vector calculus. While neither of the two factions dominated in mechanics originally, the Hamiltonians prevailed in electrodynamics and spread their approach from there to other parts of physics, as this branch waxed in importance.

The next two papers deal with logic, and in particular with mathematical logic, which is especially relevant here since Grassmann aimed at establishing a foundational branch common to all other areas of mathematics. Grattan-Guinness gives a brief systematic exposition of the development of logic in the nineteenth century, notably of algebraic and mathematical logic. Since Robert's work in logic is so closely intertwined with that of his brother Hermann, the author analyzes the reception of both into the contemporary mainstreams. Although both were read quite soon (and Robert even more frequently) by important representatives of modern developments (in particular by C. S. Peirce, Frege, Peano, and Whitehead), an immediate impact of their ideas cannot be traced, in this period. Peckhaus also analyzes the influence of the two brothers on Ernst Schröder's programme for an "algebra of logic." Though not very influential in his own time, Schröder's logical work received a major stimulus from Robert Grassmann's logic.

The **fourth** and last part of this volume is devoted to more recent developments in the sciences influenced by Grassmann.

In the algebra section, Brini and Teolis introduce Grassmann's two essential multiplications - the progressive and the regressive product - and show how one can construct the exterior or Grassmann algebra $G(\mathcal{V})$ over an n-dimensional vector space $\mathcal{V}$ with them. Endowed with the Hodge operator $*$, $G(\mathcal{V})$ turns out to be a CG-algebra $G(\mathcal{V}, \vee, \wedge, *)$ which can be seen as the linear analog of the Boolean algebra of sets with union, intersection and complement. Brini and Teolis develop the correlations between CG-algebra and the geometric or Clifford algebra and discuss Bourbaki's method of introducing the duality between $G(\mathcal{V})$ and $G(\mathcal{V}^*)$. The paper concludes with a perspective on recent research on the Cayley factorization problem.

Hestenes gives a general reappraisal of Grassmann's seminal work which has provided indispensable tools for modern mathematics. Grassmann achieved decisive success in realizing his ultimate goal, to establish a universal instrument for geometric research. Hestenes asks why - despite the fertility of Grassmann's vision - due recognition was so belated, pointing out that there is, besides the reasons usually discussed, an internal reason, namely missing elements in the theory. One such element is that Grassmann did not complete his theory to deal with general rotations in space. It was the English mathematician W. K. Clifford (1845-1879) who did so, and it is the algebraic system known as *Clifford algebra* which integrates both Grassmann's algebra and quaternions; its unifying character stems from reducing all multiplicative aspects to one central product in the geometric algebra. Hestenes vividly emphasizes the perspectives which this unifying approach opens for mathematics and physics.

It should be noted that both Brini/Teolis and Hestenes use the signs $\vee$ and $\wedge$ for the two products according to conventions in recent work on CG-algebras, i. e. in the reverse way as introduced by Bourbaki and used here by Zaddach in his paper.

Lawvere emphasizes the foundational aspects in Grassmann's

Ausdehnungslehre and their formation by his philosophical perspective, understood as a dialectical one. By interpreting some of Grassmann's basic notions within the conceptual framework of category theory, Lawvere achieves a reconstruction of Grassmannian conceptions in terms of the category of graded algebras, endowed with a boundary operation. He is, thus, able to show shortcomings in Engel's and Study's interpretation of Grassmann's ideas.

The geometry section also contains three papers. Starting from the perspective of geometrical applications, Swimmer discusses the competing basic concepts 'vector' and 'point', maintaining that Grassmann's preference for 'Strecke' (vector) deterred him from achieving his ultimate methodological goal. The author undertakes to propose a "point calculus"-interpretation instead of a vector calculus interpretation of Grassmann's mathematical symbolism - in the spirit of Möbius's barycentric calculus and in the interest of applications to geometry and physics. It is shown how addition and multiplication can be effected by developing the algebra of weighted point systems. By discussing some examples from geometry, Swimmer shows how to interpret algebraic operations in terms of geometrical meaning. Applications to physics are demonstrated by using line-bound vectors.

A further important field of modern mathematics where Grassmann's concepts are being applied and developed is differential geometry, based on Elie Cartan's work. A basic concept there is given by the *Grassmann manifold* (or simply the *Grassmannian*). Klingenberg examines this concept and shows how extraordinarily modern Grassmann's ideas were, explaining its applications in contemporary geometry, in particular in the Fundamental Theorem of Vector Bundles. Zaddach's contribution complements that of Brini and Teolis. It shows the problems and fallacies arising in what he calls the "old Grassmannian style," i.e. the works of the *Grassmannianer*, who did not differentiate between progressive and regressive multiplications and merged them into one sole operation. Zaddach presents the modern concepts of progressive and regressive multiplication and shows how Bourbaki succeeded in clearly defining them. By analyzing Bourbaki's related work on exterior algebras, the author develops important applications in projective three-space, giving as a particularly impressing example of Grassmann's ideas the modern representation of the ruled hyperboloid.

The last section provides an impression of the scope of applications within which Grassmann's work is being developed today in the traditional field of physics as well as in new fields like robotics and theoretical informatics and technology. Further applications have been shown during the Conference on "Invariant Methods in Discrete and Computational Geometry" (Curaçâo, June 1994). Grassmann algebra was the central focus there, since symbolic operations in computer aided geometric reasoning are based on it (see N. White 1995).

Braunss first gives a personal account on the qualitative change which has taken place within mathematical physics since the 1950s and which is characterized by including algebras, products, manifolds, bundles into its theoretical framework. Grassmann's work has exercised an essential foundational impact here as well. The author then shows how concepts like Grassmann algebra and Grassmann product are applied in quantum physics.

Franksen's paper is an entirely novel contribution to array-based logic so that it also has to provide the necessary background for the proposed concept of *nesting*. The author's research in logic and theoretical informatics is stimulated by the technological task to ensure maximum speed and minimum storage requirements in industrial computer implementations. Although first working on *non-nested arrays*, he goes on to show the richness of a nested array approach. Here, too, Grassmann's philosophical-dialectical stance has inspired its reconstruction in array-based logic. His "Stufen" are interpreted in a hierarchical sense - as levels of nesting. The author gives an introduction to array-based logic and its underlying array-theoretic foundation. The reconstruction of Grassmann's philosophical ideas is based upon the understanding of algebraic form and connection as nesting by *pairing*. The author gives explanations on technical conventions for the computer display of arrays and then develops his concept of invariance under nesting, largely by exploiting Ch.S. Peirce's work on associative algebras. The new concepts and propositions are applied to interpret Grassmann's concept of "Stufen." The author maintains that Grassmann's key idea in his work on the distributive law and the element-by-element addition can be understood as the array-theoretic operation of interchanging two adjacent levels of nesting.

Pfalzgraf is able to show how Grassmann's concepts play a key role in mathematically modelizing a robot. In modelling, a particularly important case is given by singularities, since such configurations give rise to non-reachable points for robots. With the help of results from projective geometry and exterior algebra, considerable advances can be made towards a general solution.

Hopefully, this volume will contribute to further deepening the study of Grassmann's multi-faceted work and to its application to even more fields of science. The cooperative atmosphere at the Rügen Conference between representatives of different disciplines was an encouraging sign and a stimulus for such endeavours.

Institut für Didaktik der Mathematik, Universität Bielefeld

REFERENCES

Nachlaß Friedrich Engel. Wissenschaftliche Korrespondenz. In: Mathematisches Institut, Bibliothek. Universität Gießen

M. Barnabei, A. Brini, G.-C. Rota, "On the exterior calculus of invariant theory," *Journal of Algebra*, 1985, *96*: 120-160.

Elie Cartan, "Nombres complexes," *Encyclopédie des sciences mathématiques pures et appliquées. Edition Française. tome I, vol. 1: Arithmétique* (Paris: Gauthier-Villars, 1908), 329-468.

Michael J. Crowe, *A History of Vector Analysis* (Notre Dame: University of Notre Dame Press, 1967).

Jean Dieudonné, "The tragedy of Grassmann," *Linear and Multilinear Algebra*, 1979, *8*: 1-14.

Friedrich Engel, *Grassmanns Leben, nebst einem Verzeichnisse der von Grassmann veröffentlichten Schriften und einer Übersicht des handschriftlichen Nachlasses.* Hermann Grassmanns *Gesammelte Werke*, Band 3.2. (Leipzig: Teubner, 1911).

Friedrich Engel, "Review of: H. Grassmann. Projektive Geometrie der Ebene unter Benutzung der Punktrechnung dargestellt. Erster Band: Binäres. Leipzig und Berlin 1909," *Jahrbuch über die Fortschritte der Mathematik*, 1909, *40* [1911], 589-590.

Friedrich Engel, "Berichtigung," *Jahresbericht der Deutschen Mathematiker-Vereinigung*, 2. Abtheilung: Mitteilungen, 1913, *22*: 102-103.

Hermann G. Grassmann, *Gesammelte mathematische und physikalische Werke.* Band 1.1.: *Die Ausdehnungslehre von 1844 und die geometrische Analyse.* Hrsg. Friedrich Engel. (Leipzig: Teubner, 1894).

Hermann G. Grassmann, Band 1.2.: *Die Ausdehnungslehre von 1862.* Hrsg. Friedrich Engel (Leipzig: Teubner, 1896.

Hermann G. Grassmann, *Teoría de la Extensión. Nueva disciplina matemática expuesta y aclarada mediante aplicaciones.* Trad. Emilio Oscar Roxin (Buenos Aires: Espasa-Calpe Argentina, 1947).

Hermann G. Grassmann, *La Science de la Grandeur Extensive. La «Lineale Ausdehnungslehre».* Traduction et Préface de Dominique Flament et Bernd Bekemeier (Paris: Blanchard, 1994).

Hermann G. Grassmann, *A New Branch of Mathematics. The* Ausdehnungslehre *of 1844 and Other Works.* Transl. Lloyd C. Kannenberg (Chicago, La Salle: Open Court 1995).

Hüseyin Tevfik Pascha, *Linear Algebra*, reedition of the first edition of 1882 and the second of 1892 by Kâzim Çeçen (Istanbul: Istanbul Teknik Üniversitesi, 1988).

Felix Klein, "Review of Victor Schlegel, System der Raumlehre, Erster Teil," *Jahrbuch über die Fortschritte der Mathematik*, Jahrgang 1872, (Berlin: Georg Reimer, 1875), pp. 231-235.

Felix Klein, *Vorlesungen über die Entwicklung der Mathematik im 19. Jahrhundert*, (Berlin: Julius Springer, Part I 1926, Part II 1927). [Felix Klein, *Development of Mathematics in the 19th Century*. Part I. Translated by M. Ackermann. (Brookline, Mass.: Math. Sci. Press, 1979.)]

Alfred Lotze, "Die Grassmannsche Ausdehnungslehre," *Encyklopädie der mathematischen Wissenschaften mit Einschluß ihrer Anwendungen*, Bd. III, 1.2.

(Leipzig: B.G. Teubner, 1923), 1425-1550.

Gregory H. Moore, "The Axiomatization of Linear Algebra: 1875-1940," *Historia Mathematica*, 1995, *22*: 262-303.

Lew Pyenson, "Pure Learning and Political Economy: Science and European Expansion in the Age of Imperialism," *New Trends in the History of Science*, eds. R. P. W. Visser et al. (Amsterdam: Rodolpi 1989), 209-278.

Karin Reich, "Das Eindringen des Vektorkalküls in die Differentialgeometrie," *Archive for the History of Exact Sciences*, 1989, *40*: 275-303.

Karin Reich, "Who needs vectors? Discussion of calculus in history", *Learn from the Masters!*, Proceedings of the Kristiansand Conference on the History of Mathematics and its Place in Teaching (Kristiansand, Norway. August 1988) Otto Bekken, John Fauvel, Bengt Johansson, Frank Swetz (eds.) (Pennsylvania State University 1992), 214-228.

Hermann Rothe, "Systeme geometrischer Analyse," *Encyklopädie der mathematischen Wissenschaften mit Einschluß ihrer Anwendungen*, Bd. III, 1.2. (Leipzig: B.G. Teubner, 1916), 1277-1423.

Victor Schlegel, *System der Raumlehre, nach den Prinzipien der Grassmannschen Ausdehnungslehre und als Einleitung in dieselbe dargestellt, Zweiter Teil: Die Elemente der modernen Geometrie und Algebra* (Leipzig: Teubner, 1875).

Peter Schreiber (ed.), *Hermann Graßmann - Werk und Wirkung. Internationale Fachtagung anläßlich des 150. Jahrestages des ersten Erscheinens der "linealen Ausdehnungslehre"* (Greifswald: Ernst-Moritz-Arndt-Universität, Fachrichtungen Mathematik/Informatik, 1995)

Neil L. White (ed.), *Invariant Methods in Discrete and Computational Geometry* (Dordrecht: Kluwer, 1995).

Alfred North Whitehead, *A Treatise on Universal Algebra, with Applications* (Cambridge: Cambridge University Press, 1898).

Illustration 2: Participants in the Grassmann Conference 1994

PART I

BIOGRAPHY

Illustration 3: The building of the Friedrich-Wilhelms-Schule in 1840

ALBRECHT BEUTELSPACHER

A SURVEY OF GRASSMANN'S *LINEALE AUSDEHNUNGSLEHRE*

It is hardly conceivable that a mathematical book (a) was so much ahead of its time and (b) received so little attention from contemporary scholars as Herrmann Günther Grassmann's *Die lineale Ausdehnungslehre* from 1844.

In this book Grassmann studies from a very general point of view the properties of compositions ("products"). In particular, he has no problem introducing non-commutative products, which were absolutely new for his contemporaries (cf. Hamilton's invention of the quaternions.)

The underlying concept of the *Ausdehnungslehre* is the idea of a geometric calculus which combines the advantages of synthetic and analytic geometry: Grassmann's approach takes from synthetic geometry the concept that one deals with geometric objects, such as points, and not with numbers; from analytic geometry he takes the idea of calculating with these objects.

The *lineale Ausdehnungslehre* introduces two revolutionary concepts. First, and most importantly, the notion of a general n-dimensional vector space. Grassmann introduces the notion of linearly dependent vectors and develops the "elementary" theory of finite-dimensional vector spaces, as can be found in all today's books on linear algebra.

The second idea is the introduction of multilinear algebra, that is the product of vectors or "multivectors". This leads to a complete theory of the "exterior algebra," a part of mathematics whose importance could not have been seen at that time.

One must imagine that this took place at a time when most mathematicians considered only the real 3-dimensional vector space $\mathbf{R}^3$ given by its coordinates. Hamilton proceeded to the 4-dimensional vector space $\mathbf{R}^4$, but arbitrarily many dimensions and coordinate-free vector spaces were unheard of. It is still not completely clear how Grassmann could conceive of these until then unimaginable ideas.

Grassmann got little, if any acknowledgement. Plücker, whose ideas pointed in a similar direction, refused to review the book.

This was certainly due to (a) the extremely abstract mathematics, which none of Grassmann's colleagues used and (b) the very philosophical language, which most mathematicians disliked.

G. Schubring (ed.), Hermann Günther Graßmann (1809–1877): Visionary Mathematician, Scientist and Neohumanist Scholar, 3–6.

The *Ausdehnungslehre* from 1844, the so-called A_1 was by no means a success. Grassmann therefore decided to write a new edition A_2, which appeared in 1862. For this edition Grassmann changed his style completely: Whereas A_1 is written in a philosophical style, A_2 is in the usual definition-theorem-proof style. In A_2 products are treated in greater generality and with more care than in A_1. Grassmann introduces a very general "combinatorial product" and studies the "inner product" also for quantities of different levels. Today, a mathematician can read A_2, once he has got used to the notions, without great difficulties. Also A_2 contains a lot more results, so it is a completely new book. The book is an admirable example of a completely developed theory – which had nearly the same success as A_1.

We shall now discuss the content of the *Ausdehnungslehre* in greater detail.

First we shall deal with linear algebra. Grassmann introduces the concepts of a vector ("extensive Größe"), linearly dependent vectors ("stehen in einer Zahlbeziehung"), basis ("System von Einheiten"), and the defining properties of a vector space; he calls an n-dimensional vector space a "Gebiet n-ter Stufe." Then he proceeds to state and verify the theorems of linear algebra: the exchange lemma, the exchange theorem. He introduces the dimension ("Stufenzahl eines Gebiets") and proves the dimension formula for the span of two subspaces.

Now we come to the multilinear part, for which Grassmann has become famous. For a fixed basis $\{e_1, e_2,..., e_n\}$ of a vector space V he introduces "bivectors" $e_i \wedge e_j$ in such a way that the set $\{e_i \wedge e_j \mid i < j\}$ forms a basis of a vector space, which has dimension $\binom{n}{2}$. This product is called the outer product, since it assigns to any two vectors of the original vector space an element of a different vector space.

Grassmann then proceeds to multivectors. Here, to any r-tuple $(v_1, v_2, ..., v_r)$ of vectors he assigns a new symbol $v_1 \wedge v_2 \wedge ... \wedge v_r$ (multivector of rank r, "Multivektor r-ter Stufe"). These multivectors also form a vector space; they satisfy the following properties:

$$v_1 \wedge v_2 \wedge ... \wedge v_r = o,$$

if the v_i are linearly dependent.
If the v_i are linearly independent, then

$$v_1 \wedge v_2 \wedge ... \wedge v_r = v_{\pi(1)} \wedge v_{\pi(2)} \wedge ... \wedge v_{\pi(r)}$$

if π is an even permutation and

$$v_1 \wedge v_2 \wedge \ldots \wedge v_r = -v_{\pi(1)} \wedge v_{\pi(2)} \wedge \ldots \wedge v_{\pi(r)},$$

if π is an odd permutation.

Now let $\{e_1, e_2, \ldots, e_n\}$ be a basis of the vector space. It follows that the set of all multivectors of rank r of the form

$$e_{s_1} \wedge e_{s_2} \wedge \ldots \wedge e_{s_r},$$

where $\{s_1, s_2, \ldots, s_r\}$ runs over all r-subsets of $\{1, 2, \ldots, n\}$, forms a basis of a new vector space of dimension $\binom{n}{r}$.

The exterior algebra is the vector space of dimension 2^n that has the vectors $e_{s_1} \wedge e_{s_2} \wedge \ldots \wedge e_{s_r}$ with $1 \leq r \leq n$ as a basis.

The $\wedge$-composition is called a "progressive product," since at any stage $r<n$ it produces vectors of higher stage.

The exterior algebra, also called Grassmann algebra, has proved to be an extremely important structure; it plays a central role in Bourbaki's *Algèbre Multilinéaire*.

Grassmann also introduced the "regressive product" (in A_1 this is called "einwendiges Produkt") as a kind of converse of the progressive product.

The inner product is very important. The inner product of two quantities of the same level is a scalar; this product has all properties of the usual inner product: the inner product of a unit (element of a basis) is 1, whereas the inner product of two distinct units of the same level is 0. This enables Grassmann to develop the whole theory of orthogonality up to the theorem that any (real) vector space has an orthonormal basis.

Let us also mention that, apart from these theoretical milestones, the *Ausdehnungslehre* contains three "gems" (Zaddach 1994, p. 72), which are applications of his theory to concrete geometric questions:

- Without coordinates Grassmann could represent the line through two points in a space of arbitrary dimension.
- Without using coordinates he could represent the equation of a conic section through five points in general position in a plane.
- He could generalize the cross ratio in such a way that he could apply this to sets of four mutually skew lines in 3-space.

Finally we would like to point out the significance of the word "lineal." (I would like to thank Gert Schubring for this hint.) This is not, as is often assumed, a misprinted version of "linear." It is the adjective to the noun "Lineal", which means ruler.[1] There is something deeper in that. It is very likely that the "lineal" refers to the first part of the

---

1 and in American English: straight-edge.

Ausdehnungslehre as opposed to a (never published) second part of a "circuläre Ausdehnungslehre" which should deal with circular measures, such as angles, rotation etc.

This means that Grassmann is also based upon the traditional ruler-and-compass approach and is much more concrete than might seem at first glance.

Mathematisches Institut, Universität Gießen

REFERENCES

N. Bourbaki, *Éléments de Mathématique. Algèbre, Chapitre 1 à 3* (Hermann: Paris 1970).

Werner Burau, Christoph J. Scriba, "Grassmann, Hermann Günther," *Dictionary of Scientific Biography*, Volume XV, Supplement I, Charles C. Gillispie (ed.) (New York: Scribner, 1978).

Arno Zaddach, *Grassmanns Algebra in der Geometrie* (B.I. Wissenschaftsverlag: Mannheim, 1994).

HEINZ SCHWARTZE

ON HERMANN GRASSMANN'S LIFE AND HIS WORK AS A MATHEMATICS TEACHER

INTRODUCTION

Hermann Günther Grassmann (1809-1877) is known as a scientist versed in many fields, in particular in mathematics, physics, and linguistics. This is analysed in a number of contributions in this book.

He was even more universal in his activity as a teacher. This contribution is intended to provide a view of his teaching, in particular in mathematics.

Grassmann owes much to his father Justus Günther Grassmann (1779-1852) who had been scientifically prominent himself, among other things through a textbook on trigonometry. Certain mathematical results obtained by Hermann Grassmann in his "Ausdehnungslehre" go back to ideas his father had. Grassmann came from a family of Protestant pastors. According to his own statement - in a twenty-five page curriculum written in Latin - he was what we would call a "late developer" today. He was able to unfold his excellent gifts only as a student. He made his decision to focus on mathematics even later in life, after having finished his studies of theology and languages, after he had already become a teacher.

He thus never studied mathematics and physics at a university, but attained the teaching *facultas* in these subjects belatedly, as an autodidact.

1. GRASSMANN AS A STUDENT AND AS A TEACHER

He began his studies of languages and theology in Berlin. All his later studies were autodidactical. In 1831, he passed his state examination for teaching both languages and mathematics - the latter for the lower grades of secondary schools. In 1834, he qualified as a theologian, and shortly afterwards began working as a teacher of mathematical subjects at the *Gewerbeschule* in Berlin. In 1836, he became teacher for mathematics, German and other subjects at a "Bürgerschule" (Otto-Schule) in Stettin.

G. Schubring (ed.), Hermann Günther Graßmann (1809–1877): Visionary Mathematician, Scientist and Neohumanist Scholar, 7–18.

In 1839/40, he prepared for his examination to become a full mathematics teacher (these years were decisive for the maturing of his *Ausdehnungslehre*). After half a year's interlude at the *Gymnasium*, where he had been a student himself, he changed to the Friedrich-Wilhelms-Schule, an important *Realschule*, in 1842. In 1852, he became "Oberlehrer" (senior teacher) for mathematics and physics at a grammar school, the *Vereinigtes Königliches Stadtgymnasium*, later *Marienstiftsgymnasium* in Stettin, where he taught for 25 years, as successor to his father Justus Günther Grassmann. He died there in 1877, still active as teacher at this renowned school. Thus Grassmann remained a schoolteacher for all his life.

1.1 Studies

Grassmann's university studies extended mainly to languages, besides theology. While he took a mathematics examination in his first state examination in 1831, his interest in this subject developed only many years later.

It is interesting to note the variety of subjects in his teacher examination report. The document admits him "with full confidence" to teaching languages in lower and upper grades of secondary schools and prognosticates that "he will educate himself to be a competent teacher of classical languages for all classes" (Engel 1911, p. 40).

The document goes on to evaluate his examination lesson in teaching history, and he is given a "limited" *facultas* to teach this subject. Then follows the report on his teaching elementary mathematics which attests his ability to teach this subject in lower grades. The document also contains reports on natural science, philosophy and pedagogy, on teaching German literature and theology; he is also given the right to teach religion in lower grades. The report closes with a general remark: "His teaching skills already acquired entitle to the best expectations in all respects" (Engel 1911, p. 41).

1.2 His teaching career

In 1834, shortly after his examination in theology, Grassmann held his first teaching post as a teacher for mathematical subjects at a vocational school (*Gewerbeschule*) in Berlin. His predecessor had been the famous Jacob Steiner, the renowned geometer. Steiner had been famed for possessing a particularly good teaching method, and this was of course the yardstick applied to the teaching of other teachers at this school. This is undoubtedly one of the reasons why Grassmann's evaluation as a teacher was not so favorable at this school (Schubring 1991, p. 169):

"...Grassmann left this school after only half a year, the director gave the reasons as lack of classroom discipline, describing the consequences this had for his teaching: '...Discipline, in our school, has always been considered flawless. During the last half year, it was shaken in the mathematics lessons... By energetic intervention and appropriate measures, these disturbances have been mended... One of the natural consequences of those disturbances is that the pupils did not learn as much mathematical knowledge as before and as they could have learned under different circumstances, and they have lagged behind particularly in calculus.'"

Grassmann did not feel happy in Berlin, because he was lonely and as he had not yet recognized his true vocation as a mathematician. He submerged himself in scientific studies and seems to have attempted everything possible to return to his native town of Stettin. In this he succeeded in 1836 by attaining a teaching post at a Stettin "Bürgerschule." In this school, he had to give many lessons in various subjects such as religion, German, mathematics, physics, and Latin (in those times, such a varied occupation was nothing unusual, even in grammar schools).

The return to Stettin seems to have made Grassmann happy. As Engel says in his biography, he devoted himself with utmost seriousness to teaching, at the same time pursuing intensive scientific studies. Besides mathematics, physics, and theology, he also got involved with chemistry. As to his true vocation, he was then still undecided between mathematics and theology. With an extreme show of energy, he submitted to two extensive examinations at once: to his second theological examination, and to a second teacher examination. The latter was to entitle him to a full teaching post in mathematics and physics.

First, he very successfully passed his theological examination. The decisive turn in his life, however, was brought about by his second teacher examination, a fact which can be seen from the extensive report. It was mainly his essay on the tides which showed his excellent mathematical skills. The report says:

"In mathematics, the aforementioned possesses knowledge which is as comprehensive as it is profound and which extends all across natural science, including mechanics and the entire higher analysis. His test essay treated the theory of the tides very methodically and strictly, and he had even the good fortune to choose his own particular method which deviates in some parts from Laplacean theory.""...and he showed, in general, a mathematical education so profound that he is perfectly qualified to teach all the classes of a *Gymnasium* and of a higher *Bürgerschule*" (Engel 1911, p. 70).

Further very favorable judgements were recorded for other disciplines as well, such as physics, mineralogy, theoretical and analytical chemistry, and he obtained permission to teach these subjects as well. It is most admirable that Grassmann passed both examinations almost simultaneously with such excellence, given the fact that he had to meet his teaching obligations in full over the entire period. After this examination, he had become fully aware that mathematics was to become his lifework. (It must be noted, however, that he succeeded in doing pioneer work in linguistics and physics after this decisive turn.)

Grassmann's essay about the tides, by the way, already sketched elements of his *Ausdehnungslehre*, that is "Vektorbegriffe," "geometrische Summe," and "geometrisches Produkt." Grassmann himself noted in this essay that he had abandoned the traditional method "because nature is subjected to constraint by introducing arbitrary coordinates" (Engel 1909, p. 8). He wanted to "calculate with the geometrical objects themselves." In any case, Grassmann was now moved to give up all thought of religious office. This did not cause him to give up promoting Christian faith, however.

Grassmann must have been very content to be allowed to succeed to his father's teaching post at the (later) *Marienstiftgymnasium* in 1852. On the occasion of this change of position, he received a very favorable assessment of his pedagogical skills. To do Grassmann justice, however, we must try to give a more precise description of the quality of his teaching (in the next section), as the literature is undecided in this aspect.

During the first years of his new teaching activity, Grassmann had much work in preparing his lessons which were now confined in principle to mathematics and physics. Engel's biography gives details about his curriculum which was mainly confined to senior classes.

From that time, Grassmann did intensive work in mathematical research, among other things on the second part of his *Ausdehnungslehre*. During that period, he also had numerous contacts with leading mathematicians of the time, i.e. his correspondence with Möbius. The latter urged Grassmann to bring his Ausdehungslehre "into a shape better suited to the mathematician," an intention Grassmann himself had announced.

His heartfelt wish, to become an **academic** teacher, was not fulfilled, however. Two direct applications for a university position to the *Kultusministerium*, of 1847 and of 1862, were not successful.: the first on the basis of a negative expertise by E. E. Kummer (Engel 1911, pp. 126-129). Regarding the second, the *Ministerium* advised him that professorial salaries were lower than his salary at the well-endowed Stettin *Gymnasium* (ibid., pp. 231-232). An initiative by the Greifswald mathematician Grunert in 1868 to call Grassmann to the second mathematical chair was rejected by his colleagues of the *Fakultät* (ibid., pp. 278-287).

2. GRASSMANN'S QUALITIES AS A TEACHER

The question whether Grassmann was a good teacher is of interest for the fact alone that it is frequently treated in the literature, with varying answers. It is many-faceted besides: Did Grassmann like to be a teacher, or did he serve pedagogy only reluctantly besides his scientific ambitions? Was he satisfied with his nonacademic teaching post? Were his lessons rather more scientific, or rather more pedagogical? The unequivocal answer to the latter question is that Grassmann was always intent on taking "modern paths." This can already be seen from his own textbook on arithmetic.

A further question is whether he was a good teacher in the sense of being attentive to students and whether he treated them with friendliness and understanding. It is impossible, of course, to answer all these questions exhaustively, and this is why we just give some quotes from the literature.

Without doubt, Grassmann was a model with regard to performance of duties, diligence, friendliness and other teacher virtues. His teaching success, however, seems to be marred by the fact that the majority of this grammar school's students were inattentive and undisciplined (Schubring 1991, 168). Many students thus presumably did not learn as much in his lessons as would have been possible with a better method. It must be noted, of course, that "there were no methods specific for mathematics at the time, the methods being taken over from linguistic subjects" (ibid., 172).

Generally, teaching at that time was a "lecture" given by the teacher, the students remaining more or less inactive. Only towards the end of the century did the authorities see to it that progressive educational methods prevailed in the grammar schools as well, which consisted mainly in permitting the students to work out mathematical problems for themselves. It is interesting to note that Grassmann's father also had difficulties with classroom discipline. The two Grassmanns were so productive as mathematicians that for them the systematics of mathematics was the main thing. This can be seen from the textbooks written by Hermann and by Justus Grassmann. Besides, strict systematics were considered to be the best condition for learning performance in the methodological views of contemporaries.

Engel's biography probably contains the largest number of both positive and negative statements about the quality of Grassmann's teaching made in later years by his former students. There is uniform praise of his benevolent, good-natured, and easy going behavior towards students. Those who were interested in mathematics or knew they would have to make use of the subject in later life, were full of praise that Grassmann's teaching required attention and independent work. Several former students, however, pointed out that Grassmann's methods was not elementary enough for the less gifted. These abused Grassmann's good nature by disturbing the lessons as much as possible. On the whole, the level on which Grassmann taught mathematics must have been too high by far.

Just as contradictory are the statements concerning the role of exercise and drill in Grassmann's mathematics lessons. There seems to be agreement on the opinion, that repetition and drill were neglected in order to be able to develop the mathematical results as generally as possible. There was perhaps such a chasm between Grassmann's genius and the students' ability to grasp that Grassmann's goodwill could not bring about the teaching success he so ardently desired. It should not be overlooked, however, that he was more successful in other subjects, and that his personality alone had a good effect on his students. Engel says:

"The claim frequently raised in these reports that Grassmann only cared for the gifted students is certainly exaggerated. In any case, he honestly tried to pull the feebler minds along, but was frequently thwarted by the latters' indifference, all the more so because mathematics was of little weight then for moving up a grade." (Engel 1911, p. 266).

It is possible that Engel's closing judgement of Grassmann is too favorable, as he might have wished to save the latter's honor out of veneration. Well known is Felix Klein's harsh judgement, who said "that Grassmann failed in at least one aspect: he was a bad teacher" (Klein 1926, p. 175).

3. THE GRASSMANNS AS TEXTBOOK AUTHORS

Fortunately, Hermann Grassmann's and his father's textbooks are still accessible today (see the three in the references). They inform us both about the Grassmanns' teaching and about their mathematico-scientific work. Hermann Grassmann's "Arithmetik" shall be used to name some general points which are particularly salient as compared to other schoolbooks of the time and to present-day ones. In fact, not only the natural numbers, but also the whole numbers are treated from the very outset (Grassmann using the term of "Größen" <quantities> for numbers resp. number areas). The volume appears "modern" by its feature of introducing fractions at once in connection with division.

The volume "Arithmetik" is for several grades, just like the other books. It also introduces rational numbers and "irrational numbers." A disadvantage seemes to be that variables (according to the conventions of the time) always meant natural numbers, a fact which made the abstract representation of whole and fractioned numbers rather complicated.

There is a marked absence of exercises; for these, the author points to well-known problem collections "in particular to Heis' excellent collection" (Grassmann 1861, introduction), whereas the representation in the book itself is as a rule abstract. Axiomatics is a program which, while not yet being consistently axiomatic, is at least purely theoretical. This shows the concept Grassmann intentionally used in his town teaching. To quote his "Arithmetik": "It remains to comment on the pedagogical side of the book and on its practical use. There will be few to contest that the strictest mathematical method ought to be preferred to all others even for the first scientific teaching of mathematics." With regard to "demonstrations" containing fallacies or circular reasoning, he remarks (also in his introduction), that "the objectionable way of so-called demonstrations is prevalent in present-day textbooks of arithmetics wherever the point is to found and establish the system." Finally, Grassmann opposes the notion that the "strict demonstration" could be too difficult for the student's grasp.

Pedagogically more elaborate is the book "Trigonometrie" in Hermann Grassmann's textbook series, and a contemporary reviewer, the well-known school mathematician Schlömilch, praised it "for belonging to the rarer species of textbooks which graciously show that they have sprung from extended pedagogical practice..." (Engel 1911, p. 268).

The book's intellectual demand, however, is more extensive as it exclusively uses formulae in variables to derive all formulae, and to "dissolve" triangles and quadrangles, a method otherwise not in use with textbooks. In general, there are no exercises with certain number quantities. Trigonometry, as "volume 2," was to complete the calculatory part of school mathematics; the intended volume 3 "Geometrie" was never published.

Hermann Grassmann's textbooks neither refer to certain school forms nor to school types. Against that, Justus Grassmann's textbook on trigonometry is intended "for the senior grades of the Gymnasien." As to its style, the volume differs by its strongly "verbal" diction from Hermann Grassmann's textbook which is written in a more abstract language of formulae. The "introduction" just gives a general presentation of the advantages teachers and students will draw from a book accompanying the lessons. Against that, Hermann Grassmann, in his own introduction, already communicates conceptional matters, the hint to teaching methodology which says guidelines of a kind not contained in the arithmetic volume will be given to permit using a heuristic method. These guidelines are then given as required in footnotes. Furthermore, the author justifies his decision to start the introduction of the trigonometrical functions with cosine. The reason given, namely that the cosine relates the two legs of the argument angle seems external; more deeply, it practically leads to introducing the scalar product ("explanation" Nr. 38) which of course does not appear under this name. It may be noted that both Grassmann senior and junior introduce the concept of the *gerichtete Winkel* (oriented angle).

It would be erroneous to believe Hermann Grassmann's trigonometry to be an "easy" textbook because of its didactical methodology. In fact, the "Additionssätze" (theorems of addition) for trigonometric functions, the "half-angle theorems", are derived in rapid succession abstractly; and one has to keep in mind that the author had placed the emphasis on correct demonstrations.

From a present-day pedagogical perspective, one would probably think that Grassmann goes too far with his demand that young students should learn strict demonstrations. His father, by the way, was also in favor of rather "strict" conceptions with regard to the requirements asked of students. Thus, he wrote: "All methodological efforts which are intended to facilitate learning a science thus would seem to be of extremely problematical and in any case quite inferior value if they do not at the same time enhance the science to which they are devoted by more extensive clarifications of the elements..." (quoted after Engel 1911, p. 2). Engel remarks rightly that Justus Grassmann's *Schulprogramm* "Begriff und Umfang der reinen Zahlenlehre" is reminiscent of his son's introduction to the latter's *Ausdehnungslehre* in several respects.

CONCLUSION

Grassmann was still scientifically productive in his last years of life; he even produced a particularly rich list of publications in this period. Hence, he met with increasing signs of recognition, from mathematicians as well. This may have comforted him a bit over not having succeeded in obtaining a teaching post at a university.

Grassmann's work in mathematics has had extremely strong repercussions in the sciences which are reported in the other contributions to this volume. If we enquire into his effect on present-day school mathematics, the introduction of vector calculus must be mentioned at least. Many methodologists of mathematics instruction were initially quite opposed to vector calculus, and the number of its advocates at the Gymnasium grew very slowly. In present-day Germany, vector calculus still has a sadly subordinate role in secondary school.

In his 1909 speech, Engel laments that Grassmann was forced to use up his strength for all his life in school service, and that the time allotted to him for his scientific work was "small and piece-meal."

Upon Grassmann's death, his significance for the *Marienstifts-Gymnasium*, but also for Stettin and science became obvious. It was seen that he was one of the most important citizens of his home town. Special mention was made of the fact that he always remained modest and disinterested despite the excellence of his achievements.

It has often been said that Grassmann's significance for mathematics has been recognized too late, particularly in Germany. But the versatility of his personality was seen. Thus, there is an extensive obituary in the *Schulprogramm* of the Stettin *Marienstiftsgymnasium* of 1878, that is one year after his death. It says that "numerous works on the objects of teaching sprang from his productive mind." Among those named are a textbook on the German language (1841), a German reader (1843), a textbook on the first instruction in Latin, a program on crystal shapes, a survey of acoustics and elementary optics, the textbooks already mentioned on arithmetic, and trigonometry, a textbook on mechanics (1867), a textbook of German plant names (1870), etc. "All this a lifework already for a man under the stress of school life, for Grassmann the smallest part of his literary work..." the obituary goes on to praise his two major work fields, mathematics and lingustics, and its conclusion that "only a quite independent spirit could dare to break his own paths in mathematics, on which others followed him only after decades" shows rather clearly that Grassmann's school did see his importance for future mathematics. The fact that he did not meet with immediate recognition among his peers is criticized, and that he was only later in life honored by an *Ehrendoktor* conferred by Tübingen University, by membership in several academies, and by obituaries written by famous colleagues in linguistics and mathematics.

Fachbereich Mathematik, Universität Gießen

REFERENCES

Joachim Buhrow, "Hermann Grassmann - Späte Anerkennung eines originellen Mathematikers," *Der Mathematikunterricht*, 1993, *39*, 6: S. 14 - 24.

Friedrich Engel, *Vortrag anläßlich des 100. Geburtstages von Hermann Grassmann vor der Universität Greifswald* (27. Januar 1909). (Greifswald: 1909).

Friedrich Engel, *Grassmanns Leben (= H. Grassmanns gesammelte mathematische und physikalische Werke*, Band 3. Teil 2) (Leipzig: Teubner, 1911).

Hermann G. Grassmann, *Lehrbuch der Mathematik für höhere Lehranstalten. Erster Teil: Arithmetik* (Berlin: Enslin, 1861).

Hermann G. Grassmann, ...*Zweiter Teil: Trigonometrie* (Berlin: Enslin, 1865).

Justus G. Grassmann, *Lehrbuch der ebenen und sphärischen Trigonometrie. Für die oberen Klassen der Gymnasien* (Berlin: Reimer, 1835).

Felix Klein, *Vorlesungen über die Entwicklung der Mathematik im 19. Jahrhundert*. Teil 1. (Berlin: Springer, 1926).

o. Verf.-Angabe. "Hermann Grassmann. Sein Leben und seine mathematisch-physikalischen Arbeiten," *Mathematische Annalen*, 1879, *14*, 1-45.

o. Verf.-Angabe, *Schulprogramm des Kgl. Marienstiftsgymnasiums Ostern 1878*. (Stettin: 1878).

Victor Schlegel, *Hermann Grassmann. Sein Leben und seine Werke* (Leipzig: Brockhaus, 1878).

Gert Schubring, *Die Entstehung des Mathematiklehrerberufs im 19. Jahrhundert* (Weinheim: Deutscher Studienverlag, 2. Aufl. 1991).

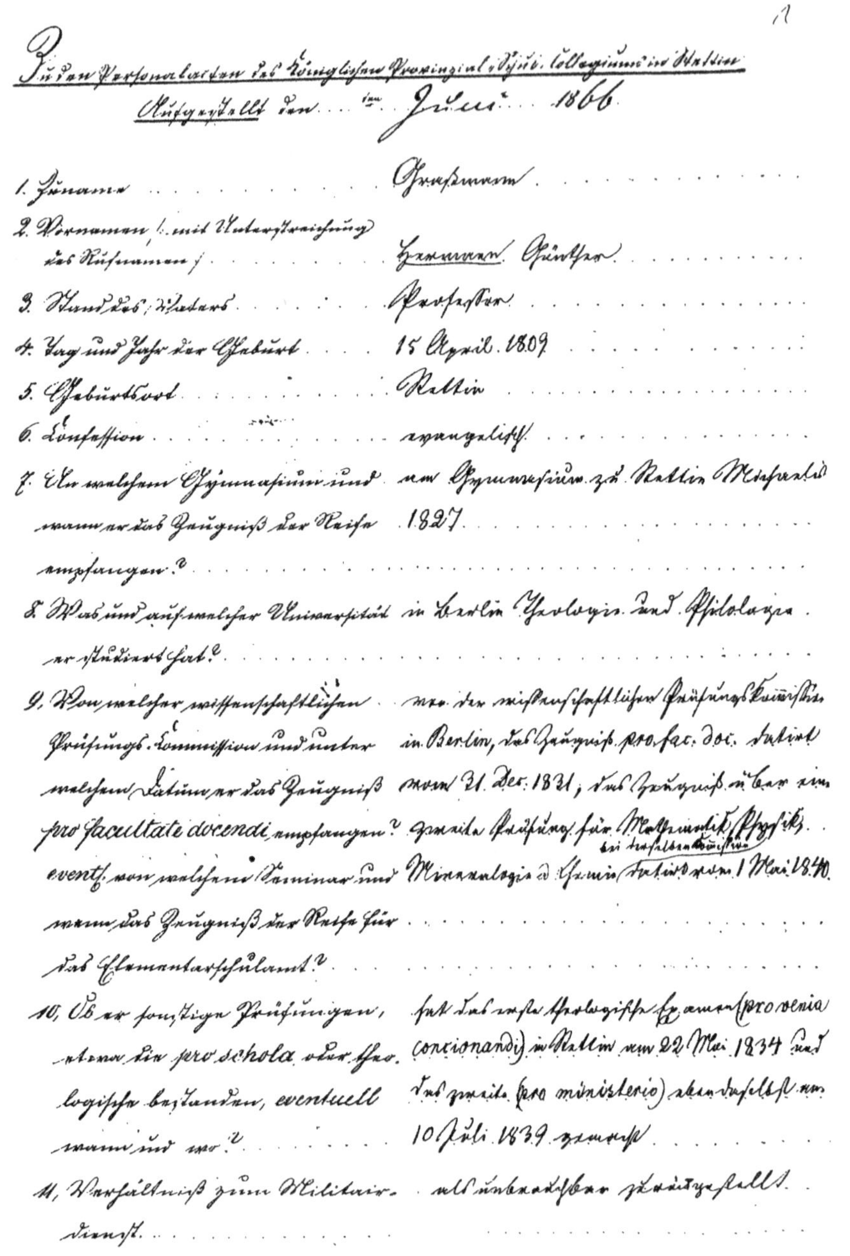

Illustration 4a: The personal file form filled in by Graßmann (1866) for the *Provinzialschulkollegium* Stettin, first page
The transscription is given on page 350-351.

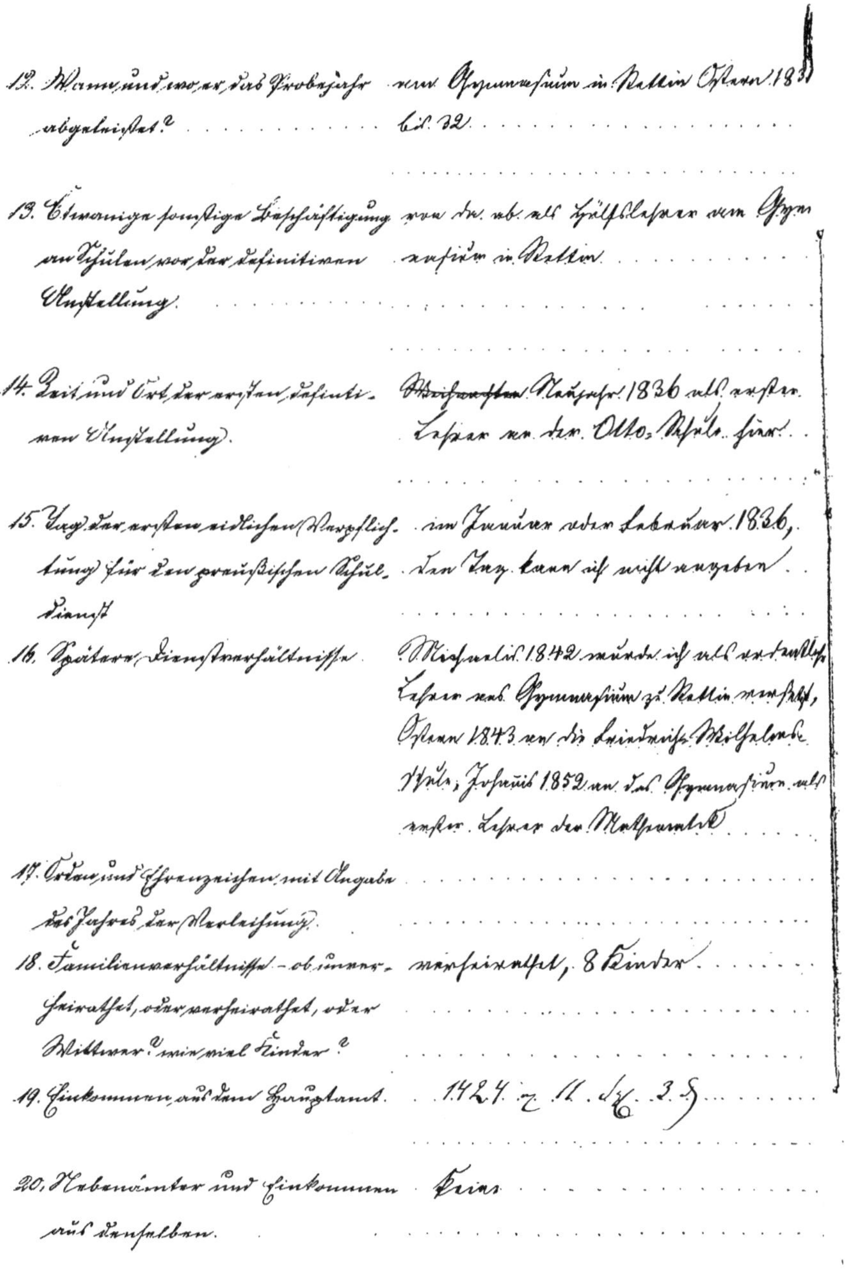

12. Wann und wo er das Probejahr abgeleistet?	am Gymnasium in Stettin Ostern 1831 bis 32.
13. Etwaige sonstige Beschäftigung an Schulen vor der definitiven Anstellung.	von da ab als Hülfslehrer am Gymnasium in Stettin.
14. Zeit und Ort der ersten definitiven Anstellung.	~~Weihnachten~~ Neujahr 1836 als erster Lehrer an der Otto-Schule hier.
15. Tag der ersten eidlichen Verpflichtung für den preußischen Schuldienst	im Januar oder Februar 1836, den Tag kann ich nicht angeben.
16. Spätere Dienstverhältnisse	Michaelis 1842 wurde ich als wissenschaftl. Lehrer ans Gymnasium zu Stettin versetzt, Ostern 1843 an die Friedrich-Wilhelm-Schule, Johannis 1852 an das Gymnasium als erster Lehrer der Mathematik
17. Orden und Ehrenzeichen mit Angabe des Jahres der Verleihung.	
18. Familienverhältnisse - ob unverheirathet, oder verheirathet, oder Wittwer? wie viel Kinder?	verheirathet, 8 Kinder
19. Einkommen aus dem Hauptamt.	1424 Thlr 11 Sgr 3 Pf
20. Nebenämter und Einkommen aus denselben.	Keins

Illustration 4b: The personal file form filled in by Graßmann (1866) for the *Provinzialschulkollegium* Stettin, second page
The transscription is given on page 350-351.

GERT SCHUBRING

REMARKS ON THE FATE OF GRASSMANN'S *NACHLAß*

Grassmann's collection of private papers (his *Nachlaß)* must have been of considerable size at the time of his death. At the end of his biography, Friedrich Engel (1861-1941) published a list of the Grassmann papers and manuscripts concerning mathematics and physics (as usual, Engel only considered these two disciplines and no other material). This list, systematically arranged, is impressively long (Engel 1911, pp. 373-395), and had the manuscripts on Grassmann's other fields of work been included, it would have been much longer. Ever since I began with research on German mathematicians of the nineteenth century in the late 1970s, I have tried to locate this voluminous *Nachlaß*.

Engel himself was enigmatic about the papers:

"It has not yet been established where the manuscripts will be stored in future. In the *Jahresbericht der Deutschen Mathematiker-Vereinigung* it will be stated where they shall be kept" (ibid., p. 373).

Since a great part of the *Nachlaß* had been in Engel's care for almost twenty years, the vagueness of this announcement is remarkable. It points to conflicts about the long-term plans for the papers, either between Engel and members of the Grassmann family or within the family. The promised notice about the definite storage never appeared.

Probably, Engel had intended to have Grassmann's *Nachlaß* deposited at the *Handschriftenabteilung* of the Göttingen University Library, since there existed a recent agreement between the *Deutsche Mathematiker-Vereinigung* that this Library would function as the central depository for the *Nachlässe* of mathematicians.

On the other hand, both Engel and Study were eager to collect autographs. For instance, Study sent back to Engel a letter by Gibbs which Engel had shown him and expressed his regrets that he had to return it:

"den ich Dir gerne für meine Autographensammlung ausgespannt hätte" (letter no. 81, 4 May 1893, Engel-Nachlaß).[1]

---

1 "Which I should have liked to wheedle from you for my collection of autographs."

G. Schubring (ed.), Hermann Günther Graßmann (1809–1877): Visionary Mathematician, Scientist and Neohumanist Scholar, 19–26.

Detailed study showed me that Engel had had much of the Grassmann material in his hands since 1893, in order to evaluate it for the *Gesammelte Werke* project; he made the necessary parts available to the respective collaborators for the individual *Werke*-volumes. Did Engel simply keep the Grassmann papers, so that they became a part of his own literary estate at the time of his death? Since T. Hawkins had published an article about Engel's *Nachlaß* in Gießen, I explored these archives. In fact, a considerable part of Engel's scientific correspondence is extant in the library of the Mathematical Department there, but it contains no manuscripts or letters from Grassmann's *Nachlaß*.

However, that evidence is not decisive. It became clear to me that the Engel *Nachlaß* in the Gießen collection is not complete; important parts are missing, like his correspondence with F. Klein and with S. Lie. None of the experts whom I consulted could explain these gaps. Egon Ullrich (1902 - 1957) who had taken care of Engel's papers after his death and deposited them in the Library (he even published Engel's last volume of the Lie-*Werke* after the War) had already deceased; his daughter Friedl, herself a mathematician and married to Martin Kneser's brother, had no information. The only child of the Engels had died in 1922 so that there are no descendants who might still be in possession of the Grassmann-Engel papers or have knowledge of them. In an allied bombing raid on Gießen during December 1944, Engel's house was destroyed (while the library suffered only minor damages); his widow Caroline moved in with relatives in Oldenburg the same month. Although the possibility cannot definitively be excluded that material from the Grassmann-*Nachlaß* was burnt with Engel's house or was part of the small losses from the library, neither possibility is probable.

The main focus of my search, therefore, has been within the Grassmann family. To a degree this search has been productive. I first tried to locate Grassmann's descendants. This search was quite complicated, but eventually I succeeded in finding a direct descendant, a great-grandson: and he in Bielefeld, in the mathematics department! He is Peter Bergau (born 1926), grandson of Karl *Justus* Grassmann (1851-1926), Hermann's first son. He brought me in contact with another direct descendant, the chemist Dietrich Grassmann, and made inquiries with other family members. Although we were able to make inquiries with practically all lines of Hermann's descendants, none of them was in the possession of Hermann's *Nachlaß,* or of parts of it, or had information about its fate. There was one exception, however. Peter Bergau himself was not only happy to have kept some of Hermann's and his father's (Justus) books (some with handwritten notes), he also was able to show me a letter written to him on December 18th, 1952, by an aunt, Margarete Grassmann (born 1875, deceased in the late 1950s).

"I do not know whether you know that I was the guardian of all the valuable old Grassmann books, bound in pig leather, dating from many generations ago? The director of the Stettin municipal library had taken them into his custody according to my own wish, and I believed at that time [i.e. at the end of World War II] to be able to leave Stettin with a calm mind; but things were to develop differently. Everything fell into Polish hands, and Dr. Ackerknecht, whose address I recieved later, also knew nothing of the whereabouts of all this valuable material."

Although Margarete Grassmann alluded only to books and not to papers or manuscripts, this was a very promising clue since this aunt was the widow of Hermann Ernst Grassmann (1857-1922), mathematically the most active of Hermann's sons and Engel's main partner in utilizing the *Nachlaß* of his father. I thus went to Szczecin in 1985. The staff of the Municipal Library (*Książnica Szczecińska Wojewodzka i Miejska Biblioteka Publiczna*) was very forthcoming in assisting me. In the library's storage I not only found a substantial part of the former school library of the Friedrich-Wilhelms-Schule, later named Friedrich-Wilhelm-Realgymnasium (where Justus Grassmann had been director from 1901 to 1911), but also numerous books bearing the signature of members of the Grassmann family. The personal library mentioned by aunt Margarete, or at least a part of it, has thus survived. Unfortunately the Grassmann *Nachlaß* proper was not to be found in Szczecin, except for two documents related to it. These were typewritten copies of the two famous *curriculum vitae* by Hermann Grassmann (see my contribution on the cooperation between Hermann and Robert in this volume).

I was later able to discover more clues to the fate of the *Nachlaß* thanks to a hint from Jean-Claude Muller (Luxembourg), who reported on extant correspondence of Hermann Grassmann at the Rügen Conference. One of the letters he showed was Gauß's reply of December 14, 1844, to Grassmann's sending him the A_1. This text was a transcription of an original, handwritten letter; Engel had made the transcription and offered it to the *Handschriftenabteilung* of the Göttingen University Library for its Gauß *Nachlaß* collection. The importance of the transcription lies in a series of attached notes that Engel had made referring to the the original and its location. Engel's note, dated "25.1.99," gives clear information:

"Das Original ist nach H. Grassmanns Tode in ~~den Besitz~~ [replaced by:] die Verwahrung von dessen Sohn H. Grassmann übergegangen. ~~und befindet sich augenblicklich nebst andern Briefen von Grassmann hier~~ [i.e.: in the hands of F. Engel]" (Nachlaß Gauß, Briefe B: Grassmann)[2]

---

2 "The original has gone, after H. Grassmann's death, into the possession (replaced by: the custody) of the latter's son, H. Grassmann, and is at present here, together with other letters by Grassmann [i.e. in Engel's hands]."

The phrase later cancelled by Engel proves that he had the *Nachlaß* material at his disposal during the editing process and that he returned it later on to Hermann Grassmann, Jr. Moreover, his replacing of "possession" by "custody" hints to some dissent among the Grassmann sons about the distribution and location of their father's *Nachlaß*.

More can be learned about the fate of the *Nachlaß* by studying how the idea of publishing Grassmann's Collected Works emerged and became established. As M. Crowe reports, this idea was originally due to J. W. Gibbs (1839-1903), (Crowe 1967, p. 161). Actually, Gibbs's interest was narrower: he wanted to obtaining Grassmann's 1840 examination thesis on the theory of tides. Toward this end, he had urged Hermann Grassmann Jr. since 1887 to make this thesis available for the public by allowing it to be published. Hermann Jr.'s first response to this proposal, dated 21 June 1888, shows the scattered situation of the *Nachlaß* at that point. Hermann Jr., then living in Halle, reported:

"Upon a scrutiny of papers of my father in my hands, I found only preparatory work for that paper; just as unsuccessful was the search in my mother's house in Stettin. I succeeded in the end, however, with a brother of mine who is teacher of mathematics in Brandenburg [i.e. Karl Justus Graßmann], in finding the draft of the treatise on tides" (J. W. Gibbs Papers, no. 140).

Hermann Jr. expressed agreement with Gibbs that an edition of the 1840 thesis would be very valuable from an historical point of view, but he objected that he felt the time not ripe for such an historical endeavour: In his judgement, the number of people who worked with Grassmann's methods or appreciated them was still so small, at least in Germany

"that it seems for the present more necessary to me to enhance interest and understanding for this study and to encourage larger circles to cooperate by publishing further applications. Only when the understanding of the multiplication calculus has progressed further, will the interest in the history of the development of these ideas appear." (ibid.).[3]

Gibbs urged the son to change his mind, and in the meantime he tried to find a journal in the United States that would publish the thesis (cf. Crowe 1967, p. 161). After another dilatory letter from Hermann Jr. of 6th April 1892, Gibbs was desperate to come to terms with him. Hermann Jr. had written that his brother Justus, had examined the extant copy of the thesis (actually, it was the copy which their father had made in 1841) and had judged at least its last third to be unfit to print (ibid., no.143). Gibbs decided to overcome this resistance by appealing to a higher instance. On 15 July 1892, he wrote to Felix Klein and proposed that Klein publish the Grassmann essay in the *Mathematische Annalen*. He did not fail to mention the fears and hesitations of the Grassmann

---

3 Quoted with kind permission of the Beinecke Rare Book and Manuscript Library, Yale University.

sons (Nachlaß Klein, 318). Immediately, Klein began to look for a way to realize Gibbs's proposal. He, too, learned first-hand how "undecided and fearfull" the sons were and found that they judged the essay unpublishable. But, as Klein was able to report to Gibbs by September 22, he had found a solution to overcome the obstacle:

"I have given the matter a turn now with which I hope that you will be satisfied. That is, I shall try to find out whether I will not succeed in publishing a complete edition of Grassmann's works, or at least of his treatises, which will then be supplemented by as much of Grassmann's *Nachlaß* as will prove useful upon progress of its examination" (Gibbs papers, no. 146).

In the event, this proposal to publish the entire Collected Works and to insert the essay there met with the unanimous agreement of the Grassmann family. Klein, in a circumspect manner, had proposed Hermann Jr. to be one of the principal editors of the collected works. Modestly, he declined and restricted himself to participating in the edition of the A_2 (letter of 18 September 1892: Klein *Nachlaß*, 9, 486). Klein's subsequent proposal to commission Engel and Study as editors was happily accepted (letter of September 26th: ibid., no. 487) and shortly thereafter the brothers and sisters agreed to the project. Engel agreed to become the chief editor (ibid., 8, no. 875).

The state of my research on the Grassmann *Nachlaß* can be summarized as follows.

Grassmann had lived in an official residence for the professors of the Stettin *Gymnasium*. Upon his death, his family had to leave it and to move to a smaller flat (Augustastrasse 49). Probably at this time, the *Nachlaß* was separated: the mathematical and physical manuscripts were given to the sons, all of whom were active in mathematics and physics. The bulk of the papers went to Hermann Jr. The eldest, Justus, obtained at least the copy of the 1840 thesis. The family correspondence and the manuscripts regarding linguistics, language teaching, etc. remained with the mother (perhaps the manuscripts and correspondence of Hermann's father Justus did as well). Upon the death of the mother on April 5, 1889, this material, as well as the family library remained with the two daughters (Agnes and Emma) who continued to live there (later on joined by their brother Max) but moved several times within Stettin. Some part of this material was probably sold some time later: A very personal letter from Hermann to his wife Therese of 1858 now belongs to the *Sammlung Darmstädter* (*Staatsbibliothek Preußischer Kulturbesitz*), and was acquired by Darmstädter before he donated his collection in 1909 to the *Preußische Staatsbibliothek*.

Hermann Jr. lent his share to Engel for the editing process of the Collected Works, while Justus never let other people touch his manuscripts. Upon finishing the edition, Engel gave the mathematics and physics manuscripts and the scientific correspondence back to Hermann Jr. (they now worked together at the same university, in Gießen). After Hermann Jr.'s death on 21 January 1922, his widow Margarete was forced by the economic crisis that followed the war to make room in her flat for borders. She donated her husband's library to the Gießen *Mathematisches Seminar* and his *Nachlaß* to the Gießen University Library. Since this *Nachlaß* comprises only approximately one and a half meters on the shelfs[4] while the entire original *Nachlaß* had been much more voluminous,[5] one can assume that she retained the *Nachlaß* of her husband's father and transferred it to Stettin where she moved in the end of 1923 in order to live in the house of her brother and sister (family Holste, Gartenstraße). Justus, after first teaching in several Prussian towns, returned to Stettin in 1901 as director of the Friedrich-Wilhelm-Gymnasium and in 1911 became *Provinzialschulrat*. He lived in another flat (Kantstrasse 9) than that of his brother and his sisters (last residence: Derfflingerstrasse). When he died in 1926, Margarete was the representative of the older generation in Stettin. Thus, the family library was given to her. Perhaps some of the *Nachlaß* material which had been in the hands of Justus was moved to her house, as well. The fate of the part of the *Nachlaß* which had remained with the mother is entirely unclear. It has never been classified and registered as had been the case with the other part already by 1878 - a fact patent in the obituary by Sturm, Schröder, et al. of 1879 in the *Mathematische Annalen* which contained as appendix an inventory for the mathematics and physics part (conforming in its structure to Engel's inventory of 1911).

When Margarete Grassmann fled from Stettin before the end of World War II, she deposited the family library in the Municipal Library.[6] She went to the Sudetenland, and it is not likely that she took *Nachlaß* materials with her. If such materials had stayed with her in Stettin during the war, then one must assume - since only two related documents are extant - that they were destroyed there at the end of the war. Another

---

4 Communication by Bernd Bader, Universitätsbibliothek Gießen. Hermann E. Grassmann's *Nachlaß* mainly consists of manuscripts of his lectures and of his research regarding the *Ausdehnungslehre*. There is also a small collection of professional correspondance. The donation of the *Nachlaß* was effected in November 1922.

5 In a letter of 6 May 1922, directed to Georg Wolff, one of the few disciples of her husband, Margarete Grassmann urged Wolff to inspect "den wissenschaftlichen Nachlaß" since the high rack ["Gestell"] on which all the cases are set up which contain the manuscripts is going to be sold and she did not knew where to store the cases otherwise (correspondence of Georg Wolff concerning the publication of H.E. Grassmann's last volume of his *Projektive Geometrie*; in my autograph collection).

6 There are no files extant in the Municipal Library about its direction during the war, and hence no documents about this transfer either.

possibility is that the papers were given to other descendants, but since none of the surviving descendants has any information, it would seem that the *Nachlaß* is definitively lost.

Institut für Didaktik der Mathematik, Universität Bielefeld

REFERENCES

Sources:

Niedersächsische Staats- und Universitätsbibliothek, Handschriftenabteilung:

Cod. Ms. Carl Friedrich Gauß (Nachlaß Gauß)

Cod. Ms. Felix Klein (Nachlaß Klein)

Universität Gießen, Bibliothek des Mathematischen Instituts:

Nachlaß Friedrich Engel

Yale University Library, Beinecke Rare Books and Manuscripts Library:

Scientific Papers of Josiah Willard Gibbs

Publications:

Lebrecht Beyer, *Graßmannsches Familien-Buch. Zweiter Teil. Aus Mitteilungen der Familie zusammengestellt* (Sondershausen 1925).

Michael J. Crowe, *A History of Vector Analysis* (Notre Dame: University of Notre Dame Press, 1967).

Friedrich Engel, *Hermann Grassmanns gesammelte mathematische und physikalische Werke*. Dritten Bandes zweiter Teil: *Grassmanns Leben* (Leipzig: Teubner, 1911).

Robert Graßmann, *Graßmannsches Familienbuch, aus alten Urkunden zusammengetragen* (Stettin: R. Graßmann, 1876).

Thomas Hawkins, "The Friedrich Engel Archive in Giessen," *Historia Mathematica*, 1976, *3*: 213-214.

Karl Scheibert, *Geschichte des Geschlechts Graßmann und seiner Nebenlinien* (Görlitz: Starke, 1937).

Gert Schubring, *Die Entstehung des Mathematiklehrerberufs im 19. Jahrhundert. Studien und Materialien zum Prozeß der Professionalisierung in Preußen (1810-1870)* [1983], 2., korrigierte und ergänzte Auflage (Weinheim: Deutscher Studien Verlag, 1991).

Illustration 5: The official residences of the professors of the *Marienstiftsgymnasium* (reconstructed)

PART II

EMERGENCE OF GRASSMANN'S IDEAS AND THEIR CONTEXT

Das Marienstifts-Gymnasium 1832—1884.

Illustration 6: The building of the *Marienstiftsgymnasium* during Grassmann's professorship

ALBERT C. LEWIS

THE INFLUENCE OF GRASSMANN'S THEORY OF TIDES ON THE *AUSDEHNUNGSLEHRE*

Michael Crowe has shown in his *History of Vector Analysis* (Crowe 1985) that Grassmann's earliest mathematical work, the *Theory of Tides*, contains almost all of the key vectorial notions that appeared four years later in the *Ausdehnungslehre*. On the other hand, Grassmann never published the *Theory of Tides*—it first appeared in 1911 in Grassmann's collected works (Grassmann 1840). Many of the physical applications, however, did appear in 1877, the last year of Grassmann's life, as "Die Mechanik nach den Prinzipien der Ausdehnungslehre" in the *Mathematische Annalen*. The only essential difference between this later version and the 1840 appearance appears to me to be ostensibly minor changes of notation. I believe, however, that it is just this difference that points to the contribution that the *Theory of Tides* can make to our understanding of how the *Ausdehnungslehre* came to be.

HISTORY OF THE TEXT

The collected works editor for the theory of tides appears to have been Justus Grassmann,[1] one of Hermann's eleven children and at this time mathematics teacher and director of the Friedrich-Wilhelm-Realgymnasium in Stettin. Justus had available the original manuscript which is described in the list of Hermann's *Nachlaß* as being in the archive of the *Wissenschaftliche Prüfungskommission* in Berlin (Grassmann 1894–1911, III, 2, 387). Is there a chance it still exists?[2] My attention to the notation of the *Theory of Tides* would have been impossible without the carefully edited version in the collected papers. I suspect that few mathematicians have been so well served by their collected works. Fortunately the standards were such that even exacting historians do not feel the need to verify the text against the originals.

1 Justus Grassmann is joint editor with Friedrich Engel for the relevant volume of the *Werke* (III, 1) and his is the sole name at the end of the introduction to the *Theory of Tides* text.

2 See Gert Schubring's extensive searches for such official documents (Schubring 1991, pp. 23-24, 222-223).

G. Schubring (ed.), Hermann Günther Graßmann (1809–1877): Visionary Mathematician, Scientist and Neohumanist Scholar, 29–36.

This is doubly fortunate since many of the originals are either difficult or impossible to obtain today.

Not only has the collected papers text been carefully treated, with all changes duly noted, but it is commented upon rather extensively. The commentary analyzes the mathematics where the editor sees something noteworthy or doubtful, and it compares the text with other related writings of Grassmann or with sources used by Grassmann. Friedrich Engel in *Grassmanns Leben* (Grassmann 1894–1911, III, 2) devotes a chapter to the *Theory of Tides* and demonstrates how the work was a turning point in Grassmann's career: Grassmann was led by this work to take up mathematics instead of theology.

THE HISTORICAL AND MATHEMATICAL CONTENT

The 1844 *Ausdehnungslehre* is the fullest flowering of Grassmann's mathematical and philosophical program. The *Theory of Tides* concerns only the mathematical strand, for Grassmann states that he will not present the principles of the geometrical analysis from which he derived his coordinate-free expressions; rather, limiting himself to the subject of his treatise, he will "borrow" the needed results from the related algebraic expressions (Grassmann 1894–1911, III, 1, 17).

Though Grassmann did provide some new mathematical points in his attempt to propagate the geometrical analysis, these were either not sufficiently important or not closely enough related to his new notation to attract any special attention from the official evaluators of this examination treatise. The examiner's report, according to Justus Grassmann, was probably written by a Professor Conrad, who refers to Grassmann's work as a sound and rigorous treatment of the theory of tides that "uses his own distinctive method that, not without success, departs in places from the Laplacean theory" (Grassmann 1894–1911, III, 1, 208). The *Werke* commentary confirms what histories of the theory of tides tells us: it seems that Grassmann contributed nothing essentially new to the theory but, quite apart from his new method of analysis, did exhibit an amazing grasp of the subject, especially considering that he was largely self-taught.

However good it is to have such a scholarly version of the text, the historian always wishes for more. Justus, as editor, had available some notes from the Grassmann *Nachlaß* that were preliminary to the *Theory of Tides*. These included notes on Lagrange and Laplace and also on topics to be covered in the *Theory of Tides*. Justus mentions that in one of these manuscripts one can see some of Grassmann's difficulties through some of the scattered remarks he made: the only example Justus gives is Grassmann's remark "The geometrical method is not yet ready to use"—"ist noch nicht reif" (Grassmann 1894–1911, III, 1, 209). I don't believe that Grassmann would have judged the geometrical analysis as ready until the *Ausdehnungslehre*. Thus the *Theory of Tides* is the only extant source that provides some evidence of why and how Hermann developed the calculus of extension.

NOTATIONAL ADVANTAGES AND DISADVANTAGES

Grassmann's main purpose in the *Theory of Tides* is to exhibit a new approach to an old topic. He points out, in case the reader misses it, that his single equations, such as

$$\frac{d^2}{t^2} p \doteq P \dot{+} Q \dot{+} R \dot{+} \cdots$$

can only be completely expressed algebraically by three equations:

$$\frac{d^2}{t^2} x = X + X_1 + X_2 + \cdots$$

$$\frac{d^2}{t^2} y = Y + Y_1 + Y_2 + \cdots$$

$$\frac{d^2}{t^2} z = Z + Z_1 + Z_2 + \cdots$$

In the above geometrical equation Grassmann has put dots above the + and = . Grassmann explains that $\doteq$ designates equality in magnitude and direction of two displacements ("Strecke") and is to be read "geometrically equal." Similarly in the first occurrence of $\dot{+}$ he states simply that this is the geometrical addition symbol. Where there is no ambiguity, he states, the dot may be dropped from above the equal sign. It will also be generally dropped from above the *d* in the differential quotient.

In Nr. 21 the geometric product of two displacements is defined as the surface area of the parallelogram formed by the two. Two such products are deemed to be equal if they have equal area (taking their sign, + or – , into account) and lie in parallel planes. As the editor points out, there is no further elaboration, anywhere in the work, of how the positive or negative signs arise. Grassmann states that the multiplication sign will be the same cross as has been used for algebraic multiplication but with a dot above it. As the editor points out, however, the dot is never subsequently used over the cross; sometimes a dot alone is used because juxtaposition of two letters has been earlier used for a displacement when the letters represent the end points. The basic properties of the geometric product are given, albeit without the rigor of his later presentations in the *Ausdehnungslehre* or the *Lehrbuch der Arithmetik* (1860) (Nr. 21):

$$a \cdot a = 0$$

$$\begin{cases} a \cdot (b \dotplus c) = a \cdot b + a \cdot c \\ (b \dotplus c) \cdot a = b \cdot a + c \cdot a \end{cases} \quad a, b, c \text{ in the same plane or}$$

$$a \text{ a product of two factors}$$

$$b \cdot a = -a \cdot b$$

$$a \cdot b = ab \times \sin(ab)$$

A geometric addition of two surface areas is defined so that if the displacements are not all in the same plane the above distributive property can be written:

$$a \cdot (b \dotplus c) = a \cdot b \dotplus a \cdot c$$

Further multiplications are introduced in the first chapter: Grassmann in Nr. 22 writes $na_r a_s$ to represent the force (the "inner force" as he terms it) acting between the two points a_r and a_s, "where $a_r a_s$ represents the connecting displacement ("Verbindungsstrecke") and n is a number." No further explanation or example is given of just what n means or what the properties of this apparently new multiplication are.

This allows such notational (and, as Grassmann would add, conceptual) simplifications as:

$$\dot{\sum} \frac{d^2}{t^2} p_r \cdot p_r = \dot{\sum} P_r \cdot p_r$$

for

$$\left.\begin{aligned} \sum\left(\frac{d^2}{t^2} x_r \times y_r - \frac{d^2}{t^2} y_r \times x_r\right) &= \sum (X_r y_r - Y_r x_r) \\ \sum\left(\frac{d^2}{t^2} y_r \times z_r - \frac{d^2}{t^2} z_r \times y_r\right) &= \sum (Y_r z_r - Z_r y_r) \\ \sum\left(\frac{d^2}{t^2} z_r \times x_r - \frac{d^2}{t^2} x_r \times z_r\right) &= \sum (Z_r x_r - X_r z_r) \end{aligned}\right\}$$

Lagrange, Mec. anal., part. II, sect. III, 7

Laplace, Mec. cel., sect. 25, Book 1.

"The geometrical sum of all the accelerations, each geometrically multiplied by the corresponding carrier displacement, is equal to the geometrical sum of the (statical) moments {of all the outer forces}, taken with respect to the origin of the carrier displacement as the center of revolution" (Grassmann 1894–1911, III, 1, 32).

Another significant product is the linear product:

$$a \mathbin{\hat{\cdot}} b = ab\cos(ab)$$

$$a \mathbin{\hat{\cdot}} b = b \mathbin{\hat{\cdot}} a$$

$$a \mathbin{\hat{\cdot}} (b \dot{+} c) = a \mathbin{\hat{\cdot}} b + a \mathbin{\hat{\cdot}} c$$

So we have a multiplicity of multiplications. Some of them were foreshadowed in Justus Günther's crystallographic work but much has probably been developed in the *Theory of Tides* especially to handle the expressions in Lagrange and Laplace. Before mentioning the advantages as Grassmann saw them for these now familiar looking notations it may be good at this point to remind ourselves of the modern importance of vectors.

VALUE OF VECTORS

In the wisdom of retrospection we can see fairly clearly the value of the geometrical analysis. Saunders MacLane, for example, has pointed to the time when

"three dimensional space, where points are determined by three coordinates, was reduced to a space of vectors. At first this seemed to be just a technique to write one vector equation instead of three equations in coordinates—but it is actually much more. Vector spaces visualize the solutions of simultaneous linear equations, they formulate electrodynamic fields and other fields in physics, they allow geometry in more dimensions, even infinite ones, and they provide a setting for the functional analysis of linear operators in analysis." (Mac Lane 1992, pp. 11–12).

Indeed, "a technique to write one vector equation instead of three equations" might *appear* to be Grassmann's sole motivation in the *Theory of Tides*. (MacLane does not refer to Grassmann or to anyone else.) Certainly none of the modern advantages listed appear to me to have been exploited to any degree in the *Theory of Tides*. (A possible exception is the last, linear operators, which might be seen in the differential operators of the *Theory of Tides*.) But of course we know that in the *Ausdehnungslehre* four years later, *all* of the modern motives listed by MacLane are at least hinted at if not fully exploited.

Grassmann, with his usual assurance about the larger view, would have nothing to do with a mere notational "technique." He claimed in the *Theory of Tides* that the coordinate method was more removed from the course of nature: instead of dealing with ideas they were more involved with transformations and developments that were "only a symbolic representation of the idea" and had nothing to do with the idea itself. The formulaic development of his geometrical method, on the other hand, allows us to proceed "step by step from one concept to another and thus, by translating term by term into the conceptual language, we immediately obtain a true and natural method of proof" (Grassmann 1894–1911, III, 1, 17).

SIGNIFICANCE FOR THE *AUSDEHNUNGSLEHRE*

The focus of this paper is not the origins of vector spaces but the narrower one of how the *Theory of Tides* contributed to the *Ausdehnungslehre*. To address this question I looked for differences in the presentation of the geometrical analysis between the two works. The most interesting one, I believe, is the evident difficulty Grassmann had in symbolizing operations between his geometrical entities. I conjecture that it is in trying to overcome these symbolizing difficulties that Hermann made one of his major contributions to this Grassmann-family subject, namely the foundational contribution that would try to establish the geometrical analysis as a branch of mathematics.

One major problem with Grassmann's presentation is the variety of different notations and their inconsistent use (sometimes dots, sometimes not, depending on the context). I believe that Grassmann wanted strongly-typed operations but in the *Ausdehnungslehre* changed the notation from the dotted forms to an in-line form consisting of brackets. This is easier from a typesetting point of view at least but loses the visible reminder of the analogies between algebraic operations and geometric operations. That is, one always has a centered dot or cross involved for both types of multiplication and a plus sign for both types of addition with only the raised dots (sometimes) to distinguish the types.

We also noted above a hint of the problem that would result if, as in the *Ausdehnungslehre*, one is not limited to three dimensions. In the *Theory of Tides* Grassmann distinguished between displacements and oriented areas with respect to the operations between them.[3] Is it necessary to introduce at least one new operation for each new higher order entity? I think that it is at this point that Grassmann would look rather closely at just what "new entity" means: to what extent is a directed line segment a "different entity" from an oriented surface area? I imagine Grassmann thinking, in effect, that in a complete development of the pure geometrical analysis (as opposed to this hybrid treatment in the *Theory of Tides*) it should be possible to give a clearer picture of these operations. Otherwise their complications threaten the simplicity of the geometrical notation. This need for attention to operations is surely, then, one of the factors that led him to that greater abstraction that may have doomed the *Ausdehnungslehre* to its initial miserable reception, but that makes the work appear so modern to us. This work introduces the *Formenlehre* or general theory of forms—a kind of universal algebra that would have the geometrical analysis as one possible model—in order to deal with these issues.

---

3 The two notions are closely related. I maintained in (Lewis 1993) that Grassmann's move to higher dimensions was motivated by the definition of multiplication.

In the *Ausdehnungslehre* there is not only a classification of generalized geometrical objects into steps or spaces of different dimension, but also a classification (or typing) of operations. Since the objects are generated by the operations there is a duality which we might express in terms of mappings today and which Grassmann expressed dialectically.[4] He has expressed the relationship as a genus–species one: the *Formenlehre* determines a genus, multiplication for example, as having certain properties (the key one being distributivity over a lower-step operation), and included in this genus would be the inner and outer products (the geometrical and linear) as species of multiplication. This schema leads to a complicated presentation when it comes to the regressive product. Nevertheless, the general *Formenlehre* has the same type of abstracting, unifying role that set theory or, I think more akin, category theory has had in this century. His recasting of the theory of tides helped Grassmann to reach this point.

Bertrand Russell Editorial Project, McMaster University, Hamilton

REFERENCES

Michael J. Crowe, *A History of Vector Analysis: The Evolution of the Idea of a Vectorial System* (New York: Dover Publications, 1985).

Hermann G. Grassmann, *Theorie der Ebbe und Flut. Prüfungsarbeit 1840.* In (Grassmann 1894–1911, III, 1).

Hermann G. Grassmann, *Hermann Grassmanns gesammelte mathematische und physikalische Werke*, 3 vols. (Leipzig: B. G. Teubner, 1894–1911).

Robert Grassmann, *Die Formenlehre oder Mathematik* (Stettin, 1872). Reprinted Hildesheim: Georg Olms, 1966.

Albert C. Lewis, "H. Grassmann's 1844 Ausdehnungslehre and Schleiermacher's Dialektik," *Annals of Science*, 1977, *34*: 103–162.

Albert C. Lewis, "Hermann Grassmann's discovery of the n-dimensional vector," *Proceedings of Rencontre international d'histoire des mathématiques: nombre complexe et vecteur*, Paris, 24–26 November 1993. To be published by Maison des Sciences de l'Homme, Paris.

Saunders MacLane, "The Protean Character of Mathematics," *The Space of Mathematics: Philosophical, Epistemological, and Historical Explorations*, ed. Javier Echeverria, Andoni Ibarra and Thomas Mormann (Berlin/New York: Walter de Gruyter, 1992), 3-13.

Gert Schubring, *Die Entstehung des Mathematiklehrerberufs im 19. Jahrhundert. Studien und Materialien zum Prozeß der Professionalisierung in Preußen (1810-1870)* [1983], 2., korrigierte und ergänzte Auflage (Weinheim: Deutscher Studien Verlag, 1991).

---

4 Hermann and his brother Robert read Schleiermacher's *Dialektik* in 1840, see (Lewis 1977). Manuscript notes on the general theory of forms in the Nachlaß, some dated 1846, are also by his brother. Robert, rather than Hermann, took up this topic in later writings, notably in (R. Grassmann 1872).

Justus Günther Graßmann,
* 19. 6. 1779 in Sinzlow,
† 9. 5. 1852 in Stettin,
Professor in Stettin.

Illustration 7: Portrait of Hermann's father Justus G. Graßmann

ERHARD SCHOLZ

THE INFLUENCE OF JUSTUS GRASSMANN'S CRYSTALLOGRAPHIC WORKS ON HERMANN GRASSMANN

HERMANN'S *PROGRAMMSCHRIFT* OF 1839

The philosophical background of Hermann Grassmann's *Ausdehnungslehre,* as well as the basic features of the early formulation of vectorial ideas in the context of the theory of tides in his *Staatsarbeit,* are outlined in the contributions of M.-L. Heuser and A. Lewis, both in this volume. However, both aspects -- philosophically motivated conceptual design and the physical context -- also came together in another early treatment of vectorial ideas, namely the contributions to crystallography made by Hermann's father, Justus Günther Grassmann (1799 - 1852).

There is no doubt that Hermann became acquainted with his father's ideas about crystallography early enough for them to have influenced his early elaboration of a vectorial symbolism of forces and motions in his *Staatsarbeit* on the theory of tides (1840). This is documented by his essay, *Ableitung der Krystallgestalten aus dem allgemeinen Gesetze der Krystallbildung,* published as a *Programmschrift* (H. Grassmann 1839) of the *Otto-Schule* in Stettin where Hermann was working as a teacher. F. Engel remarked on the occasion of the publication of this *Programmschrift* in H. Grassmann's *Collected Works* that it is

"... quite remarkable ... that H. Grassmann mentions neither the paper nor the name of his father, although he not only uses his father's notions ["Vorstellungen"] but even his peculiar terminology ["Benennungen"] ..." (Grassmann *Werke* 2.2, 244).

For this reason it may be useful to look more closely at what *notions* Hermann adopted from his father's work on crystallography and to investigate the possible influence of that work on the genesis of the *Ausdehnungslehre.* Here I should like to distinguish between the general principles of crystal formation that Justus Grassmann adopted and the specific mathematical contributions that he made toward developing a symbol system adapted to the representation of internal structure (according to the dynamist hypothesis) and external shapes of crystals, both of which he pursued in his *geometric theory of*

G. Schubring (ed.), Hermann Günther Graßmann (1809–1877): Visionary Mathematician, Scientist and Neohumanist Scholar, 37–45.

combinations ("Geometrische Combinationslehre"). Both aspects left distinct traces in Hermann Grassmann's program essay from 1839.

DYNAMIST PRINCIPLES OF CRYSTAL FORMATION

Already at the beginning of the century, *Christian Samuel Weiss* (1780 - 1856) had started a research program in crystallography based on dynamist principles. In a critical evaluation of *René Just Haüy's* geometric atomistic theory of crystal structure, Weiss in 1804 proclaimed his intention to establish a theory which interpreted crystal formation as governed by a regular system of forces working in the crystal. These forces were not assumed to act between presupposed centres of matter like atoms or molecules, but were rather assumed to constitute the generative principles of matter itself. Hence, they were not directly observable, but had to be inferred from their effects. Their most visible effect was the production of the external surfaces of the crystal figure and the establishment of planes along which the crystal was most easily split by external mechanical forces. In the following decade, Weiss's investigations led him to conclude that the whole system of crystal forces always contains three spatially extended *main forces,* from which all the other internal forces can be derived by composition. Weiss classified the force-arrangements and the ensuing crystal figures into seven systems according to symmetry considerations. For two decades, and particularly in the Germanic countries, this classification and its underlying principles acted as a seminal influence on a tradition of crystallography that followed dynamist principles. Without going into detail here,[1] I should like to outline the fundamentals of dynamist crystallography on the basis of the presentation given in Hermann Grassmann's essay of 1839. This will serve two purposes: to briefly present the dynamist conception of crystal structure and to show how the young Hermann Grassmann assimilated essential features of dynamist crystallography into his own thought. It should be remembered that the principles outlined by him were not his own invention, but in fact had been adapted from his father's work.

Grassmann began with some very general considerations on the role of crystal figure as an expression of the *essence* of bodies in inorganic nature; these considerations followed very much along the lines of speculative *Naturphilosophie*. Then Hermann Grassmann came directly to the central point of dynamist crystallography:

---

1 Compare Scholz 1989, 1994 and Heuser, this volume.

"If we look at any individual crystal, we are at once struck by the perfectly plane boundaries, and we are led immediately to the conception that each boundary plane must be generated by some force directed orthogonal to the plane." (H. Grassmann 1839, 116)

This "immediately" formed conception provided the methodological link that dynamist crystallography assumed to exist between the matter-generating force-system inside the crystal (the *essence* of crystal formation) and the empirical *appearance of the faces* in the crystal figures. Grassmann extended this idea by clearly stating an equivalence principle: between two crystals, equal physical behaviour with respect to (optical) reflection or (mechanical) cleavage allows us to infer equality, both with respect to quantity and kind (*gleich grosse und gleicharthige Kräfte*) of the internal forces. From this, we may deduce the physical behaviour of a crystal in parallel faces, since these faces express the *opposite sides* of the same force.

Now the appearance of a crystal figure does not necessarily offer a one-to-one correlation with its essence, as not every plane which may *possibly* be formed by the internal force system *actually* is formed. Two real crystal figures formed by the same force system may differ. The reason for this, in Grassmann's view, is that any plane may be "overgrown" ("*überwachsen*") by some limiting plane; then only the latter is directly observable.[2] Thus, Grassmann concluded,

"... the real appearance (*die wirkliche Erscheinung*) or suppression of a boundary plane is to be considered as something contingent, whereas the essential (*das eigentlich Wesentliche*) is nothing but the interior structure of the crystal." (ibid.)

Thus we find in Grassmann's version a beautiful description of the *essence* of the crystal as its *interior structure*. And this interior structure is to be understood as the assumed internal system of forces which is equivalent to (i.e. defines and is characterized by) a system of possible boundary planes.

The force systems obey two fundamental principles. The first was considered by Grassmann as the *natural law of crystal formation*, which he stated in the following manner:

"... that when two forces of known direction and quantity appear as plane-forming, then also the intermediate force formed from these two may appear as plane-forming, where under intermediate force one has to understand the diagonal in the parallelogram made up from the two forces ...." (ibid. 117)

Therefore the "first law" of crystal formation was nothing but the parallelogram law of composition. The second law was Weiss's conception that the entire system of forces can be derived from *elementary forces*.

---

2 In this case the "hidden" plane direction may be observed indirectly by cleavage experiments.

These were, in brief, the basic principles of dynamist crystallography that H. Grassmann expressed in a clear-cut form. We can recapitulate these ideas by saying that dynamist crystallography established the notion of a three-dimensional vectorial system of forces with integer (or rational) coefficients that represent the interior structure of a crystal and that regulate the latter's formation, external shape, and physical behaviour. As already stated, these principles were not at all new in Grassmann's essay. They were rather the common property of the dynamist crystallographers of the early nineteenth century, going back essentially to C. S. Weiss. Weiss had even already begun to look for a mathematical symbolism to represent the force and plane arrangements in his crystal systems. He had found support from other authors of the dynamist tradition, like J. C. Hessel, M.-L. Frankenheim and Justus Günther Grassmann.

GEOMETRIC THEORY OF COMBINATIONS

Justus Günther Grassmann developed his *geometric theory of combinations* at the close of the 1820s in order to establish a set of symbols adapted to the representation of the force systems of crystals and their proper classification. The geometric "theory of combinations" dealt with directed line segments as the *elements of combinations*. For Grassmann these elements represented the "direction and measure of motions" that constituted the crystal and were thus to be combined according to the parallelogramm law (J. Grassmann 1830, 5). Each elementary or combined directed line segment was considered by Grassmann as a *support* ("Träger") of an orthogonal plane which could appear in the outer shape of a crystal.

Grassmann started his construction of combinations with three "elements" (or "elementary supports") *b, c, d* and with directionally-inverted "negative elements" b', c', d'. He then built up combinations of these elements by means of a repetition-procedure which he called (in the terminology of the combinatorial mathematics of his time) "*complexions*" ("Complexionen"). For such a "complexion" as *bbbccddd* he introduced as an abbreviated notation the expression $b^3c^2d^3$ or in general $b^\beta c^\gamma d^\delta$ or $b'^\beta c'^\gamma d'^\delta$ etc. with *exponents of repetition* ("Wiederholungsexponenten") $\beta, \gamma, \delta \in \mathrm{N}$. Then he introduced the *combination of complexions* in a rather obvious way:

"One complexion is combined with another by separation of the elements and recombination of those of equal kind, where the negative elements have to be treated as such." (J. Grassmann 1830, 4)

Transcribing Grassmann's accent notation for "negative Elements" by using exponents in Z this meant

$$b^{\beta}c^{\gamma}d^{\delta} + b^{\tilde{\beta}}c^{\tilde{\gamma}}d^{\tilde{\delta}} = b^{\beta+\tilde{\beta}}c^{\gamma+\tilde{\gamma}}d^{\delta+\tilde{\delta}}$$

In modern terminology, Justus Grassmann introduced into his "calculus of complexions" a three-dimensional free Z-module, whose elements (that is, Grassmann's "complexions") represented directed lines of Euclidean space in an algebraic symbolism. He thus introduced with great clarity a three-dimensional vectorial calculus with integer coefficients.

This was only the basic structure of Grassmann's "geometric theory of combinations", the structure on which he built an algebraic representation of the Weissian crystal systems and their internal structure.[3] Crucial for this development was a symbolical characterization of the Weissian symmetries of crystal systems. In the language of contemporary crystallography, Grassmann's goal was to characterize in his calculus of complexions "equivalent supports ["gleichwerthige Träger"]" with respect to a crystal system. Once the question had been put into this framework, it was not too difficult for Justus G. Grassmann, after an appropriate choice of "elementary supports" *b, c, d* adapted to the crystal system, to characterize the Weissian holoedric symmetries for six of the seven crystal systems by permutations and change of sign of the "exponents of repetition."

The simplest example is the *regular system* in which all elementary supports are equal and pairwise orthogonal. Then two complexions $b^{\beta}c^{\gamma}d^{\delta}$ and $b^{\tilde{\beta}}c^{\tilde{\gamma}}d^{\tilde{\delta}}$ are obviously equivalent if their exponents of repetition are interchanged by any permutation or change of sign (symmetry of the extended octahedral group O^*) :

$$b^{\beta}c^{\gamma}d^{\delta} \sim b^{\tilde{\beta}}c^{\tilde{\gamma}}d^{\tilde{\delta}}$$
$$\Leftrightarrow \left(|\tilde{\beta}|, |\tilde{\gamma}|, |\tilde{\delta}|\right) = \sigma\left(|\beta|, |\gamma|, |\delta|\right), \textit{where } \sigma \in S_3$$

On the other hand, in the *monoclinic system* with $b \perp < c,d >$ not orthogonal to *d* and *b*, with *c*, *d* of general lengths, supports are only equivalent if transformed by reflections at the *c,d*-plane and/or point

---

3 Justus Grassmann could not directly refer to Weiss's original publications, but had to rely on mineralogical literature in Mohs' tradition, accessible to him in Stettin (J. Grassmann 1829, p.97). His own explanation of methodological principles and of underlying ideas about nature, however, show strikingly that he did not adhere to Mohs' morphological approach in continuation of natural history, but was rather strongly inspired by speculative *Naturphilosophie* (cf. Heuser's contribution, this volume).

inversion (i.e. symmetry of type C_{2h}). For the "complexions", that means

$$b^{\beta}c^{\gamma}d^{\delta} \sim b^{\tilde\beta}c^{\tilde\gamma}d^{\tilde\delta}$$
$$\Leftrightarrow \left(\tilde\beta,\tilde\gamma,\tilde\delta\right) = \pm(\pm\beta,\gamma,\delta)$$

To give a last, slightly more involved example, in the "rhombohedric" *(trigonal) system* (of symmetry D_{3d})[4] Grassmann chose three edges meeting in the summit of a rhombohedron of rotational symmetry of order three as elementary supports *b, c, d*. That led to

$$b^{\beta}c^{\gamma}d^{\delta} \sim b^{\tilde\beta}c^{\tilde\gamma}d^{\tilde\delta}$$
$$\Leftrightarrow \left(\tilde\beta,\tilde\gamma,\tilde\delta\right) = \pm\sigma(\beta,\gamma,\delta), \sigma \in S_3$$

Grassmann proceeded in an analogous way for the other Weissian systems, giving an elegant characterization of symmetries by permutations and change of sign. The only omission in comparison to Weiss was that he did not discuss the hexagonal system with symmetry D_{6h} since that system does not admit such a simple permutation representation.

In a last step, Grassmann listed the simple crystal figures for all the six systems which he had covered. He derived them from one support by forming all equivalent ones in one system. That leads to a full and simple system of bounding planes (in modern terminology a plane orbit of the symmetry group) of a convex polyhedron. Depending on the choice of the original support, combinatorially different polyhedra may arise. A similar procedure, including complete lists of simple crystal figures, had already been used by Weiss. Grassmann's geometric combinatorial calculus, therefore, did not yield new results in this respect, but it led nevertheless to an algebraic symbolism well adapted to the problem context.

In her contribution to this volume M.-L. Heuser discusses the philosophical debts and affiliations of Justus G. Grassmann. There is no doubt that Justus' philosophical commitments heavily influenced the

---

4 This example sheds light on J. Grassmann's position between Mohs and Weiss. Grassmann applied Mohs' terminology ("rhombohedric system"), but used it conceptually, from the point of view of structural characterization and of shape classification, in the Weissian sense as the trigonal ("3- und 3-gliedriges") system. In Mohs' tradition (even in the refined version of K. F. Naumann, 1824) the Weissian hexagonal ("6- und 6-gliedriges") and trigonal systems (symmetries D_{6h} and D_{3d}) were considered as one, the "rhombohedric system" leading to a morphological classification coextensional with the six crystal families in the modern sense (Wondratschek 1983, pp.721f.).

development of his calculus of complexions. Thus he was, for example, quite explicit about the background of his calculus in the dynamist philosophy:

"In our opinion atomism and dynamism stand in the same relationship to each other as does the exterior to the interior. ... Both can lead to identical results as has been shown by Haüy's crystallography. What position, number, and shape of the atoms and the integrating molecules [a technical term in Haüy's crystal structure theory, E.S.] allow the atomist to derive, the dynamist deduces from the forms of the complexions (Complexionen) in their elements, the exponents of repetition and their relationships. ... Wherever where one has started from atoms for the constructions of the phenomena it will be of much more advantage to start from forces and derive their combinations ...." (J. Grassmann 1829, 177f.)

Grassmann did not envision the superiority of his approach as lying on the technical level alone. With his proposal of a structural crystallography he also wished to contribute to a fundamental transformation of mathematics, from being a classical "theory of quantities" (*Größenlehre*) into a science of conceptual structures like the one in the geometric theory of combinations. As the result of such a transformation he hoped for a new unity of mathematics with natural sciences. This unity would result in a "... conscious representation of the accordance of nature and mind ... (and a) reconciliation and unification of the separated spheres ... ," which would lead in the end to a "... salvation ["Erlösung"] of the human mind from the torture of empty abstraction *and* of brute empiricism" (ibid.). These wide-ranging expectations reflected, of course, some of the central motivations of *Naturphilosophie* and German idealism, both of which were strongly present in his thought.

BACK TO HERMANN GRASSMANN

Similar reverberations of *Naturphilosophie*, although slightly less emphatic, are still to be found in Hermann Grassmann's early works, as in his essay on crystallography (1839) and his examination thesis (1840). In his *Ausdehnungslehre* of 1844, the logification of his approach had already developed so far that the traces of *Naturphilosophie* had become much weaker and even difficult to identify. Nevertheless, Grassmann gave a short application of the new theory of extensions to crystal shapes near the end of the book (section 174). The content of this section is not spectacular, however, from a purely mathematical point of view. Hermann Grassmann merely presented the "law of rationality" of crystallography in different versions, claiming their equivalence. To give two examples:

Given three edges of a crystal figure, not in one plane, then every other edge can be represented by a rational linear combination of these (with small coefficients).

The same holds for three boundary planes of a crystal figure, using Grassmann's representation of plane segments in the calculus of extension.

In fact, Grassmann gave different formulations for both statements, in vectorial form and in a barycentric geometrical form. Considering the less-than-fascinating content of this section, one is struck by Grassmann's opening description of this application:

'I shall conclude this presentation [of the *Ausdehnungslehre*, E.S.] by one of the most beautiful applications which can be made of the science treated, i.e. the application to crystal figures." (H. Grassmann 1844, 28)

What follows does not at all, at first glance, justify such an enthusiastic introduction. But Grassmann could expect, with some justification, that readers of his time might understand that the "beauty of the application" was rooted in the significance of crystallography for the program of *Naturphilosophie*. The *Ausdehnunglehre* itself gave clear indication of that significance, even though it had been less strongly stated in comparison to the initial forms that it had taken in his father's contributions and in his own earlier ones. Thus the "beauty of the application" lay rather in the presenting a good conceptual insight into the system of forces which constituted the "interior structure" of crystal matter in keeping with the approach of dynamist crystallography, rather than in the more technical aspects of the remarks that followed. Moreover, for the author of these lines, a quite personal pleasure may have reinforced this emphasis on beauty: the satisfaction of knowing that this reference alluded to the earliest concrete shape which the idea of extended magnitude had taken in his mind during the process of its formation.

Fachbereich Mathematik, Universität - Gesamthochschule Wuppertal

REFERENCES

Hermann G. Grassmann, *Ableitung der Krystallgestalten aus dem allgemeinen Gesetze der Krystallbildung*. Programm der Ottoschule Stettin 1839. Reprint in: Gesammelte mathematische und physikalische Werke 2.2. (Leipzig: Teubner 1902), 115--146.

Hermann G. Grassmann, *Theorie der Ebbe und Flut* (Prüfungsarbeit 1840). Werke 3.1, (Leipzig: Teubner 1911), 1--353.

Hermann G. Grassmann, *Die lineale Ausdehnungslehre, ein neuer Zweig der Mathematik, dargestellt und durch neue Anwendungen auf die übrigen Zweige der Mathematik, die Lehre vom Magnetismus und die Krystallonomie erläutert.*[1844] Werke 1.1 (Leipzig: Teubner 1894).

Hermann G. Grassmann, *Gesammelte mathematische und physikalische Werke*. Bände in 6 Teilbänden (Leipzig: Teubner 1894-1911).

Justus Günther Grassmann, *Zur Physischen Krystallonomie und geometrischen Combinationslehre* (Stettin 1829).

Justus Günther Grassmann, "Combinatorische Entwicklung der Krystallgestalten" *Annalen der Physik und Chemie,* 1830, *33*, Ergänzungsheft, 1--43, Tafeln I, II.

Erhard Scholz, *Symmetrie -- Gruppe -- Dualität. Zur Beziehung zwischen theoretischer Mathematik und Anwendungen in Kristallographie und Baustatik des 19. Jahrhunderts* (Basel -- Boston -- Stuttgart: Birkhäuser und Berlin: Deutscher Verlag der Wissenschaften 1989).

Erhard Scholz, "Schelling und die dynamistische Kristallographie", *Selbstorganisation. Jahrbuch für Komplexität in den Natur- Sozial- und Geisteswissenschaften*, 1994, *5*, 219--230.

Hans Wondratschek, "Introduction to space group symmetry", *International Tables for Crystallography*, vol. *A*., ed. T. Hahn, (Dordrecht: Reidel 1983), 712-731

MARIE-LUISE HEUSER

GEOMETRICAL PRODUCT - EXPONENTIATION - EVOLUTION. JUSTUS GÜNTHER GRASSMANN AND DYNAMIST *NATURPHILOSOPHIE*

Hermann Grassmann wrote his *Ausdehnungslehre* (1844) in partly philosophical language. If we are to understand his original ideas, we will have to concern ourselves with the philosophical implications of and the background to his works. What, then, was the exact nature of his philosophy? Since Grassmann did not present us with a work of philosophy in the narrower sense of the word we will only be able to understand indirectly what he has to say to us. In addition we will have to examine the cultural context in which his work was carried out.

Hermann's father Justus Günther Grassmann was also a mathematician; as is well-known, he had a decisive influence on his son's mathematical training. He composed a philosophical treatise of his own that he published as an addition to his work on crystallography in 1829. Both the basic ideas and the terminology of this treatise can definitely be viewed as offshoots of the Schelling school of dynamist *Naturphilosophie* which was the dominating force in philosophical thinking in Germany from 1797 to 1830 (see also Caneva 1974). In particular, Schelling's essay on dynamic processes in physics (Schelling 1800) is of central importance during this period.

The central topoi of the dynamists could be regarded as:

- the notion of unity of natural history and cognitive processes of creation or the "self-construction" processes of nature and the mind
- the method of construction in "intellectual *Anschauung*"
- the principles of expansion and contraction as basic principles of construction
- the production of space as a result of processes of expansion and contraction, i.e. as a result of vectorial processes
- the integration of space and force
- the theory of (mathematical) exponentiation as a theory of evolution
- the concept of qualitative morphogenetic mathematics.

I am indebted to Mr. Alun H. Davies for preparing the English version of my paper.

G. Schubring (ed.), Hermann Günther Graßmann (1809–1877): Visionary Mathematician, Scientist and Neohumanist Scholar, 47–58.

Before elaborating on these central ideas in context with Justus Grassmann's comments on the matter, I would like to sketch the main details of his biography and thus place my essay in a socio-historical context.

THE BIOGRAPHY OF JUSTUS GRASSMANN

According to Engel (1911, p.2) Justus studied theology at Halle University from 1799 to 1801.[1] These two years saw the blossoming of *Naturphilosophie*, since it was within this brief period that the most important systems of *Naturphilosophie* came into being and enjoyed great popularity at the universities, at least in the Thüringen area. The centre of this movement was in Jena and it was here that not only Fichte, Schelling and Hegel were all teaching philosophy, but also the brothers Schlegel were actively pursuing questions of philosophy and Ritter was carrying out experiments in galvanic theory. *Naturphilosophen* and scientists such as Troxler, Schubert, Heinroth, Feuerstein, Krause, Wagner, Oken et al. had all journeyed to Jena as disciples of Schelling.

Halle, too, lay within the sphere of influence of Jena university and at the end of the eighteenth century boasted one of the most important scientific faculties in Germany. At the same time as Justus Grassmann was a student at Halle, Achim von Arnim was studying theology, physics and mathematics in Halle. Achim von Arnim was, as was Novalis, not only an important Romantic poet of that epoch but also a physicist and *Naturphilosoph*. While still studying under Gilbert, Achim von Arnim published several essays on magnetism and electricity in, among other periodicals, the illustrious *Annalen der Physik*. It is likely that Justus Grassmann was acquainted with Achim von Arnim, whose thinking and ideas were akin to those of Schelling: after all, they were both studying the same subjects and were pursuing similar philosophical quests.

One of Achim von Arnim's interests at Halle University was the formation of crystals and combinatorial theory - areas which also fascinated Justus Grassmann. In Achim von Arnim's essay on a theory of electrical phenomena (Arnim 1799) we also find a chapter on the dynamist construction of crystallisation, and in his paper on magnetic materials (Arnim 1800) he concerns himself with Schelling's Dimension Theory which became relevant for Justus Grassmann, too (Grassmann 1829, p. 163).

We may safely assume that both Justus and Achim von Arnim attended lectures given by Klügel, who was an adherent of the Hindenburg school of combination theory (which orientates on Leibniz's

---

[1] The university archives at Halle record his immatriculation on April 6th, 1799, under the number 163.

characteristica universalia) and who lectured on mathematics in Halle at that time. It seems likely that both students attended lectures given by Gilbert, who was later to publish the *Annalen der Physik* and whose assistant Achim von Arnim later became. Perhaps Justus and Achim also studied the theological ideas of Stubenrauch, since he was the most important theologian in Halle at the time.[2] Three years later, in 1804, Schleiermacher was made professor of theology in Halle. One of Schelling's disciples - Henrik Steffens - was offered a professorship in *Naturphilosophie* at Halle; in addition, Reil was made professor. Reil was a medical scientist and a disciple of Schelling, who like Schelling advocated an antivitalistic position in his essay "Von der Lebenskraft" (Reil 1796).

Justus Grassmann's years as a student were influenced significantly by the dynamist *Naturphilosophie* movement which, among other things, received great impetus from the writings of Achim von Arnim who published his works while still a student at Halle University.

THE MAIN IDEAS OF DYNAMIST NATURPHILOSOPHIE

On the basis of Kant's *Metaphysische Anfangsgründe der Naturwissenschaft* (1794) it was assumed that matter not only possesses forces but in fact consists of forces. Thus what we perceive empirically as matter is merely the surface or the shell of nature. Inside nature, forces are at work which produce matter and which themselves are not empirically perceptible. Kant's idea was that matter was the product of the attracting and repelling motion of forces.

This dynamist *Naturphilosophie* was most convenient for the purposes of the *Naturphilosophen* of around 1800, since it was much easier to achieve the aim of conceiving a unity of nature and reason than with an atomistic theory of matter. The essence of nature was activity and the productive processes of nature could now be construed analogously to the constructive activity of human beings.

This motif is to be found as the mainstay of Justus Grassmann's essay on physical crystallonomy and the geometrical theory of combination (1829) in which he writes referring to his new mathematical approach: "... it justifies the most significant expectations that in a particular region (by which he means crystallonomy in particular, M.H.) the formations of reason present themselves to us as formations of nature" (Grassmann 1829, p. VI). He states further: "... these are the realized constructions of our reasoning that we retrieve in the form of crystals from them. And where is there to be found a single truly significant construction of the

---

[2] Stubenrauch was godfather to Friedrich Schleiermacher, whose students in Berlin included Hermann Grassmann and to whom the latter frequently made reference.

human mind that was not to be found through a construction of nature?" (ibid, p. 173). For "every significant internal synthesis will some day reveal itself as a synthesis of nature" (ibid, p. 176).

Grassmann adopted a monistic stance that Schelling had postulated thus: "Nature is visible mind and mind is invisible nature" (Schelling 1803, p. 56). This identist notion that the productive processes of reason function in principle in the same way as the productive processes of nature was one of the ideas central to the philosophical thinking of Schelling - among other things as a critical reaction to Kant's epistemology theory. According to Schelling, things are not to be construed the way Kant had meant in his *Kritik der reinen Vernunft* - namely that external natural phenomena are accessible to us only in the form of affectations of our perceptive faculties which we would merely subjectively structure using our forms of *Anschauung* and our categories of intelligibility, while the 'an-sich' - the very essence of nature - remains a closed book. In Schelling's world there is more to nature than superficial phenomena. Nature possesses an inner being which reveals itself as nature's productive and self-organizing forces. (See Heuser-Keßler 1986)

The inner, constructive activity of nature is not empirically accessible, it is perceivable only through intellectual *Anschauung*. Nature must be reconstructed on the basis of theory, which is why Schelling accorded mathematics such an important function in the perception of nature. Thus he wrote in 1800 that: "up to now, mathematicians have not learnt the mechanism by means of which dynamic construction can be subjected to calculus." (Schelling, 1800a, p. 533) And in his methodological essays on the process of construction he accords mathematics no little esteem in the words:

> "Geometry and mathematics altogether are the only sciences that have been able to present a model of that mode of perceiving things which we may also term the demonstrative mode." (Schelling 1802, p. 345; see also Heuser-Keßler 1995)

The reference to a non-empirical view of things is also to be found in the thinking of Grassmann. In his geometry textbook he writes: "It is by no means the logical moment that constitutes the mathematician but the force of construction." (Grassmann 1824, p. VIII) "This constructive activity," Grassmann states, "requires the application of our perceptive faculties," and, he continues, "I hardly need to add that by '*Anschauung*' I do not mean external, sensual forms of *Anschauung* but an internal form of *Anschauung*, an *Anschauung* of the activity of the constructing mind or an *Anschauung* of the inner construction itself" (ibid.).

Logical reflection cannot commence before the actual product of the construction exists. By 'construction' Grassmann means something different from what the intuitionists were later to propose: Grassmann did not regard this thinking as a continuation of the phenomenalist view of *Anschauung*. Rather, he postulated an internal, intellectual form of perception whose basic point of reference is the human mind. It was only thus, he opined, that scientific reasoning would be capable of a "less inhibited mode of flight." But since every significant "internal synthesis" would at some stage reveal itself to be a "synthesis of nature," we may safely rely on our purely intellectual faculties of construction. This quite optimistic approach to a true perception of nature was of great encouragement to those engaged in theoretical science and to those whose concepts of pure mathematics transcended empirical evidence.

THE PRODUCTIVE MECHANISMS OF MIND AND NATURE

Though cognition was Fichte's special area of research, we will not be able to enter into it in great detail at this stage. At any rate, it was Fichte who propounded the basic principles of expansion and contraction which were not only adopted by Schelling but also by Achim von Arnim, Justus Grassmann and other dynamists. Schelling did not merely consider these principles to be the basic productive principles of intellectual production - as Fichte did: Schelling regarded them as the basic productive principles of all nature.

Justus Grassmann's approach was the same: he imagined the formation of matter to be the 'alternation of contraction and expansion' (Grassmann 1829, p. 160). Grassmann, therefore, does not use the words 'attraction and repulsion' as Kant does, instead he talks of 'contraction and expansion' in the manner of thinking of the speculative *Naturphilosophen*. This difference in the formulation of the concept represents a difference in meaning. Kant assumed that there was a static, mechanical context of attracting and repelling forces that exerted an arithmetical influence one upon the other within a given space. He used the parallelogram as a mathematical model of force-addition. What Schelling meant when speaking of 'expansion and contraction' was, by contrast, a unity of space and force. Schelling thought that space was produced as a result of this very dynamic process - not as space that was the result of an *a priori* positing, i. e. as passive receptacle or "Behälterraum."

According to Schelling, the universe comes into being from a zero point as the result of a differentiation between plus and minus, or expansion and contraction. Were the expansion to continue unabated and infinitely, the result would be a universe expanding at an infinitely fast rate and producing an infinite vacuous extension. If the opposite were to happen, the result would be a point that contained infinity within itself.

This notion of expansion and contraction is also the basis of Justus Grassmann's space construction. He writes:

"We may now imagine that initially all planes intersect the focal point M, so that two of each coincide and begin to move away evenly on their supports ["Trägern"] with the result that at any given moment all the supports will be equal in size as measured from point M. We thus obtain a cube that is constantly increasing in magnitude, and this construction can be restricted at any given point in order to present a perceivable cube of a defined magnitude. To a certain extent shapes thus cease to be rigid, immutable products and they appear - at least in size - to be caught up in a never-ending development, flowing, as it were, visible to us only for a moment. It is thus through this idea that we have a clearer picture in our minds, namely, that it is not the finite product but the act of production that constitutes the preeminent moment of this development." (Grassmann 1829, p. 8f)

The decisive moment in Schelling's dynamism that transcends that of Kant was that the forces were not imagined to interact locally, within a given space. They were to be construed as forces that structured all space on a cosmic scale. Schelling asserts confidently: "It is not merely the competitive concurrence of the two forces - the attracting and the repelling - at all: it is a particular relation of the two forces one to the other with reference to space that makes matter possible, ..." (Schelling 1800, p. 27). It follows that an arithmetical approach to the effect of the forces simply will not do: these forces must be understood with reference to space, i.e. geometrically (Schelling 1800, p. 100).

THE EVOLUTION OF NATURE AS A PROCESS OF EXPONENTIATION AND THE (GEOMETRICAL) PRODUCT OF FORCES

Schelling understands the construction of the dimensions of space in a way that is markedly similar to the depiction of the geometrical product in Justus Grassmann's geometry textbook. Grassmann states:

"If we in fact take the concept of the product in its purest form, it will designate the result of a construction which transpires from a product that has already been produced (constructed) in the same way as this product was produced, and thus multiplication is merely a construction upwards to a higher (mathematical) power. In geometry the point is the original producing element; from it through construction the line emerges; if we take as the mere basis of a new construction the finite line constructed from the point, and if we treat it in the same manner as we formerly treated the point, then the rectangle emerges" (Grassmann 1824, p. 194).

Grassmann applies a self-reflexive principle in this instance, that is, a self-application of the act of construction to the act of constructing to the first degree or power, by which an act of constructing to the second power arises, i.e. a line becomes an area.

In his reflections on dynamic processes in physics (1800) Schelling talks of processes of exponentiation as self-reflexive processes, by which "processes of the first order" proceed to become "processes of the second order" and "processes of the second order" become those "of the third order." Thus the constructive activity to the second degree, which both for Schelling and for Grassmann produces an area, is a "constructing of the constructing" (Schelling 1800, p. 45), i.e. what was previously the productive process of the line becomes - in an act of self-application - the object of the producing agent to the second degree.

What is really sensational about the Schelling-Grassmann connection is that the mathematical discovery for which in the historiography of science (see Crowe 1967, p. 58-60) Justus Grassmann usually takes the honours had at least in qualitative terms already been formulated by Schelling. Schelling was already aware of the so-called "geometrical product" of forces. In his view, the product of these processes of multiplication was - matter. In his *Deduction des dynamischen Processes* he writes:

"Both forces are to be viewed as being set in opposition to one another in one and the same *Anschauung*. If both are opposed and separate, each of these forces will bring forth for itself an area, just as in the course of the preceding moment. (Paragraph 18.19) But in separation both are to be posited once again identically for *Anschauung*. This is - since the opposition of the forces themselves is to be conserved - possible only if their productions are depicted in a third production that is common to both previous ones. And since, as we have said, each of the forces produces the area for itself, that which is common to both (which must not be thought of as a mere addition to the first two but as arising out of a real penetration of or the multiplication of the products with each other) must be the second power of the area - the cube. As a consequence of this reciprocal exponentiation of mutual production the construction abstracts itself from the purely geometrical construction, a third dimension is associated with the first two and the real connecting link through which the two forces are posited for *Anschauung* as being non-identical and yet unified is (not the line or the area) but space itself, i.e. the mathematical quantity extended in three dimensions." (Schelling 1800, p. 31).

Schelling's view of things is that matter comes into being as the product of this process of multiplication of forces that are in the first instance to be considered as being without substance. The most basal operation is the separation of positive and negative forces from a zero point ("creatio ex zero"). For empirical *Anschauung* however this zero point is merely a void. For intellectual *Anschauung* by constrast primordial existence is a chaotic, directionless infinity from which structured three-dimensional space is to be deduced in unity with matter.

At the first stage of deduction the positive pole separates from the negative pole, but the two poles are still firmly related to one another, like the dipole of the magnet. According to Schelling, the first dimension of space is constructed using the dipole character of the magnet.

At the second stage of deduction the poles separate from each other - as is the case with electricity - and positive and negative electricity each produce an area for itself. At this point Schelling refers to Coulomb, whose experiments had shown that electricity is only ever to be found on the surface of solid bodies and for that reason is to be regarded as a "force of area."

Matter is constructed at the third stage of deduction as the product of the two two-dimensional areas of electricity. It is only at this stage of multiplication that something material is produced from that configuration of forces previously considered to be without substance, a configuration that creates one-dimensional and two-dimensional subspaces within primordial infinity. Compact mass and thus chemical processes become feasible.

Although this construction of matter was a hyothesis which, together with the results of contemporary experiments which Schelling included in his writings, was intended to inspire further research, Schelling's idea that matter was not something that is absolutely existent is, indeed, a modern one. Also, his notion that magnetism and electricity were to be considered as being without substance - as fields, so to speak - and as being constitutive of all matter and not only for the magnetic and electric 'fluida' postulated at that time, was preemptive of modern thinking. And Schelling's hypothesis that the dimensionality of physical space has something to do with basic forces does not turn out to be abstruse when viewed within the context of present-day thinking, for example in string theory, even if his construction of the physical details was not able to transcend the bounds of contemporary science.

The self-reflexive process of exponentiation as a symbol of natural evolution is construed by Schelling as being infinitely developing, a process of progression (*Höherentwicklung*) i.e. in Schelling's *Naturphilosophie* there is an infinite number of qualitative stages of exponentiation which evolve through self-reflexive processes. In his construction of space Schelling does not go beyond the third dimension, since, as he says literally, his phraseology reminding one very much of Hermann Grassmann's theory of extension, "space is a three-dimensionally extended quantity." (Schelling 1800, p. 31)

Justus Grassmann also characterises the dynamic process of "organic" natural evolution as a process of exponentiation. In his *Schulprogrammschrift* on the number concept (1827) we discover an interesting scheme which attributes the additive and subtractive mode of linking to the mechanical mode of linking, multiplication and division to the chemical mode of linking and exponentiation and its inverse operation to dynamic (= evolutionary) mode of linking (Grassmann 1827, p. 18). What Grassmann had in mind was obviously a mathematisation of natural history whose basic premise was that "evolution" is "exponentiation."

THE MATHEMATICAL CONSTRUCTION OF THE QUALITATIVE DIVERSITY OF MATTER AND THE SEARCH FOR A QUALITATIVE MORPHOGENETIC MATHEMATICS

Together with the notion of the unity of nature and reason, Kant's *Metaphysische Anfangsgründe* gave a boost to a further main aim of the *Naturphilosophen*: their striving to construct the qualitative complexity of matter theoretically, i.e. *a priori*.

In his *Kritik der reinen Vernunft*, Kant had denied emphatically the possibility of an *a priori* - i.e. a mathematical - construction of the qualitative diversity of matter. Kant had distinguished strictly between qualitative and quantitative knowledge. There is no other way to gain access to the qualitative diversity of nature than through the affectations of the senses. This applies not only to colours, smells and sounds, but also to material differences in matter. It was, among other things, for this reason that Kant considered a science of chemical processes to be virtually impossible. What we perceive qualitatively via our five senses is something given that cannot be constructed theoretically, since in Kant's view only quantities can be constructed, i.e. the mere formal side of nature. 'Constructing' means for Kant that a concept is depicted *a priori*, for which a **non-empirical** form of *Anschauung* is required, a form of *Anschauung* expressed not in terms of concrete phenomena but a medium from which objects originate. According to Kant's *Kritik der reinen Vernunft*, these media can only be space and time as the pure forms of *Anschauung* of the human mind. In these media *a priori*-synthetic - i.e. creative - functions are feasible which we do not empirically become aware of as mere factually existing shapes.

Thus as far as Kant is concerned, only mathematicians are able to produce objects in a cognitive-creative way, since it is the mathematicians who do their constructing in the non-empirical forms of space and time: the geometrician constructs his objects within the pure *Anschauung* of space, the arithmetician within the pure *Anschauung* of time.

Naturphilosophen like Schelling, Achim von Arnim and Kielmeyer, Steffens, Krause, Wagner or Reil were not happy with this separation of quantity from quality: their quest was for scientific-constructive methods for use in chemistry, biology and the evolution of nature, i.e. for the genesis of new qualitative units. These methods were supposed to make possible a theoretical- constructive creation of a qualitatively diversified type of matter that was formed from a single unstructured precursive medium. Schelling stated that it was, at the end of the day, a quest for the uniform construction of the "self-constructing activity of nature."

Kant's dynamism of the *Metaphysische Anfangsgründe* provided Schelling with a convenient escape-route. If matter really was the product of basic forces - it must be possible to construct the qualitative diversity of matter quantitatively as the product of these forces. Thus the media used by the forces would not merely be space on the one hand and time on the other but the unity of space, time and force.

We have already seen that Schelling short-circuits, as it were, space and the dynamic processes of expansion and contraction. Space itself as a "three-fold extended quantity" is the result of dynamic processes, or, to be more exact, the metaphysically perceived but not empirically manifested forms of magnetism, electricity and chemistry. Magnetism constitutes the line, electricity constitutes the surface and chemistry as an integration (and multiplication of) magnetism and electricity constitutes three-dimensionality.

Grassmann, too, is of the opinion that the first dimension can be depicted as magnetism, the second as electricity. In order to do justice to this fact, we will have to take into consideration the cultural context of this thinking. It was, in fact, this very theory of dimension of Schelling's that had brought his *Naturphilosophie* into disrepute in the eyes of many, in particular as a consequence of the thinking of Fries and the Fries school of philosophy. Grassmann writes:

> "At this stage the question arises whether or not the effect of the forces in one direction and in two directions could still become apparent even if no object is formed as a result, and we can hardly escape the notion that this really does happen in the guise of magnetism as the one force working in only one direction and in the guise of electricity, which only ever becomes real as a surface. This view of things, one must concede, requires rather more convincing evidence to back it up." (Grassmann 1829, p. 163)

Achim von Arnim attributes the role of the forces to dimensions differently. In one of the papers in his *Nachlaß* (no. 03/416, p. 2), he proposes this construction of the three-dimensionality of physical space: conduction is to be seen as the constructing (*das Construierende*) in one dimension, light as the constructing in two dimensions, and magnetism, like crystallization, is the constructing in three dimensions.[3] He published his critical remarks about Schelling's dimension theory in an essay to be found in the *Annalen der Physik* (see Arnim 1800). Grassmann obviously prefers to take up Schelling's line of argument.

The punchline is that on the basis of Kant's dynamism the qualitative diversities of nature became accessible through an *a priori* construction. The way this worked was that one equipped the *Ausdehnungen*[4] of space with a dynamic process, so that a qualitative notion of space and time

---

3 For sending me copies of some of the papers in Arnim's *Nachlaß*, I thank Michael Gerten (Bamberg) and Klaus Stein (Jena).

4 The German notion of 'Ausdehnung' has the double meaning of 'extension' and 'expansion.'

could be formed. The next step, namely towards a qualitative theory of *Ausdehnung* and morphogenetic mathematics that was independent of quantitative relationships, was thus influenced to a certain extent by speculative *Naturphilosophie* (Heuser-Keßler, 1994a).

CONCLUSION

What motivated me to take a closer look at this topic was not so much that I wanted to show that Grassmann had a number of "philosophical ancestors" or that the Schelling school of philosophy exerted a productive influence on the development of scientific activity. I would rather prefer to take the view that Grassmann's mathematical thinking - his "vision" - expressed certain philosophical concepts more concretely, refined them and carried them a stage further. But not only this: these mathematical ideas were to be translated back, so to speak, into dynamic *Naturphilosophie* in the form of concrete topics for discussion so as to broaden the canvas of *Naturphilosophie* and attempt to formulate a more profound understanding of nature theory. This, however, can only be undertaken in the course of a large-scale research project which is now underway. At this stage, my sole aim was to show what Justus Grassmann's "Combinatorial Theory" has to do with the speculative philosophy of the Schelling school of thought.

Institut für Philosophie, Universität Düsseldorf; Sonderforschungsbereich 230, Universität Stuttgart

REFERENCES

Achim von Arnim, *Versuch einer Theorie der elektrischen Erscheinungen*, (Halle: Johann Jacob Gebauer, 1799).

Achim von Arnim, "Uebersicht der magnetischen nicht-metallischen Stoffe," *Annalen der Physik*, Halle 1800, *5*,393-394.

Kenneth L. Caneva, *Conceptual and Generational Change in German Physics: The Case of Electricity, 1800-1846*. Dissertation (History), Princeton University, December 1974.

Michael J. Crowe, *A History of Vector Analysis. The Evolution of the Idea of a Vectorial System* (New York: Dover Publication, 1967).

Friedrich Engel, *Grassmanns Leben*, Hermann Grassmanns gesammelte mathematische und physikalische Werke, 3. Bd., 2. Teil (Leipzig: B.G. Teubner, 1911).

Justus Günther Grassmann, *Raumlehre für die untern Klassen der Gymnasien und für Volksschulen*, 2. Teil: *Ebene räumliche Größenlehre* (Berlin: G. Reimer, 1824).

Justus Günther Grassmann, *Über den Begriff und Umfang der reinen Zahlenlehre*. Beilage zu dem Programm für Michaelis 1827 bei dem Gymnasio zu Stettin (Stettin 1827).

Justus Günther Grassmann, *Zur physischen Krystallonomie und geometrischen Combinationslehre* (Stettin: Friedr. Heinr. Morin, 1829).

Marie-Luise Heuser-Keßler, *Die Produktivität der Natur. Schellings Naturphilosophie und das neue Paradigma der Selbstorganisation in den Naturwissenschaften* (Berlin: Duncker & Humblot, 1986).

Marie-Luise Heuser-Keßler, "Schelling's Concept of Self-Organization," *Evolution of Dynamical Structures in Complex Systems*, Eds. Rudolf Friedrich/ Arne Wunderlin (Berlin/ Heidelberg/ New York: Springer Verlag 1992), 395-415.

Marie-Luise Heuser-Keßler und Wilhelm G. Jacobs (Hrsg.), *Schelling und die Selbstorganisation. Neue Forschungsperspektiven* (Berlin: Duncker & Humblot, 1994).

Marie-Luise Heuser-Keßler, "Geschichtliche Betrachtungen zum Begriff "Topologie": Leibniz und Listing," *Topologie. Ein Ansatz zur Entwicklung alternativer Strukturen*, Hrsg. Sonderforschungsbereich 230 der Universität Stuttgart und Tübingen (Stuttgart: Sprint Druck Gmbh, 1994a), 1-13.

Marie-Luise Heuser-Keßler, "Spekulative Konstruktion und mathematische Physik, Philosophie und Wissenschaft," *Interagierende Perspektiven spekulativer und transzendentaler Philosophie und experimenteller Wissenschaften*, Hrsg. Jörg Sandkühler (Frankfurt a. M.: Lang, 1995).

Johann Christian Reil, "Von der Lebenskraft," *Archiv für die Physiologie* (Halle: Curtsche Buchhandlung), 1796, *1*, 8-162.

Friedrich Wilhelm Joseph Schelling, *Sämmtliche Werke*, Hrsg. K. F. A. Schelling (Stuttgart/Augsburg: Cotta, 1856-1861).

Friedrich Wilhelm Joseph Schelling, "Allgemeine Deduktion des dynamischen Processes oder der Kategorien der Physik" (1800), *Sämmtliche Werke*, Bd. IV, Hrsg. K. F. A. Schelling (Stuttgart/Augsburg: Cotta, 1859), 1-78.

Friedrich Wilhelm Joseph Schelling, "Miscellen" (1800a), *Sämmtliche Werke*, Bd. IV, Hrsg. K. F. A. Schelling (Stuttgart/Augsburg: Cotta, 1859), 525-565.

Friedrich Wilhelm Joseph Schelling, "Fernere Darstellungen aus dem System der Philosophie" (1802), *Sämmtliche Werke*, Bd. IV, Hrsg. K. F. A. Schelling (Stuttgart/Augsburg: Cotta, 1859), 331-566.

Friedrich Wilhelm Joseph Schelling, "Ideen einer Philosophie der Natur als Einleitung in das Studium dieser Wissenschaft" (1803), *Sämmtliche Werke*, Bd. II, Hrsg. K. F. A. Schelling (Stuttgart/Augsburg:Cotta, 1857), 1-343.

GERT SCHUBRING

THE COOPERATION BETWEEN HERMANN AND ROBERT GRASSMANN ON THE FOUNDATIONS OF MATHEMATICS

The innovations introduced and the conceptual development established by Hermann Grassmann in his *Ausdehnungslehre* transcended, as is well known, the conceptual level of contemporaneous mathematical thinking in several respects and in a remarkably profound manner. This great departure from the understanding of mathematical concepts in Grassmann's time is particularly striking in his introduction of an n-dimensional geometry, i.e. the break with the traditional notion of a three-dimensional space (see Lewis 1996). Even Gauß did not discuss or mention this innovation when responding to Grassmann's sending of the book to him.[1]

One might thus assume that Grassmann's isolated move forward must have been firmly based on convictions concerning the philosophical foundations of mathematics and that these must have been of a very special character. Since F. Engel's biography of H. Grassmann (Engel 1911) it has been usual to identify Daniel Schleiermacher's philosophy - and in particular his book on *Dialektik* - as having constituted this firm basis. The influence of Schleiermacher's philosophy has indeed been a major factor in the emergence of Grassmann's view and conception of mathematics. In fact, Albert Lewis has shown, in his seminal study (Lewis 1977), that there exist a number of correspondences between Schleiermacher's *Dialektik* of 1839 and Grassmann's *Ausdehnungslehre* of 1844. The most characteristic correspondences are the symmetrical division into polar opposites and their application as constructive principles of his mathematical theory (cf. Lewis 1977, p. 110).

SOME OBJECTIONS TO SCHLEIERMACHER'S INFLUENCE

There are several reasons, however, why Schleiermacher's *Dialektik* cannot have exerted the sole decisive influence.

First, the source on which Engel relies for his claim of

[1] Cf. Gauß's letter of 14. December 1844, partially published in: H. Grassmann 1896 [I.2], pp. 397-398. Curiously enough, Grassmann's letter which had accompanied the book is not extant in Gauß's *Nachlaß*.

G. Schubring (ed.), Hermann Günther Graßmann (1809–1877): Visionary Mathematician, Scientist and Neohumanist Scholar, 59–70.

Schleiermacher's decisive influence is not entirely reliable. Its validity is rather limited since it is prejudiced to some extent. The source is a *curriculum vitae* composed by Grassmann himself. According to Engel, Grassmann presented this text together with other examination papers to the Stettin consistory for the first of his two theological examinations in April 1834 (Engel 1911, p. 8). The second *curriculum vitae*, the "curriculum vitae et studiorum ratio," was written in 1831 for his application to the first teaching examination and likewise extensively used by Engel (ibid.).

I found copies of both documents - thusfar the only extant remains of the Grassmann Family *Nachlass*.[2] The purpose of the second document[3] is evident throughout its entire text and gives away its bias immediately - the first theological examination. Schleiermacher was unanimously regarded by then as the leading theological authority for the protestant religion. It was thus a natural and understandable strategy for a candidate, facing the probable predilections of his examination board within the Protestant Church, to not only enumerate Schleiermacher's lectures that he had followed but to emphasize a substantial formative effect of these studies.

Moreover, the text of this *curriculum vitae* as a whole shows that it does not correspond to the usual factually oriented structure of such reports as does the first document of 1831 but rather that it is intended to be a reflection on Grassmann's inner spiritual and moral development. Without intending to impute distortions to Grassmann it is evident, however, that such an introspective report is apt to stress points that correspond to the expected views of the audience.

In fact, the "Curriculum" composed two years earlier is characterized by a quite different tendency: this text, addressed to the "wissenschaftliche Prüfungskommission Berlin," the board for examining future *Gymnasium* teachers composed of professors and teachers, mentions the study of theology and its representatives Neander and Schleiermacher only marginally and even devalues their importance. Grassmann pointed out that he need not comment on these studies since they occurred in a period when he uncritically accepted his professors before he began to make his own way and changed from the study of

2 Engel published large parts of the two documents (Engel1911, p. 8 and passim). The two transcriptions (in typescript) from the originals, probably made for Justus Grassmann, Hermann's eldest son, are preserved in the Szczecin Public Library (Książnica Szczecińska Wojewodzka i Miejska Biblioteka Publiczna).

On the fate of the Grassmann *Nachlass* see the special report in this book.

3 According to handwritten notes on the transcription the original was written on March, 23rd, 1833 and, hence, for the application to the theological examination.

theology to that of philology (cf. Engel 1911, p. 24). And in this text, in contrast to his previous curriculum, he gave his intentions, ideas, and realizations for studying Greek, Latin and mathematics in utmost detail.[4]

Grassmann's two *curricula vitae* constitute, therefore, sources of limited reliability and cannot be regarded as objective presentations.

A second reason to search for additional influences affecting the emergence of his particular philosophy can be seen in the fact that, by 1831 - i.e. *after* his studies with Schleiermacher, Grassmann had not yet established any clear philosophical conception of his own. The teaching examination regulations then in force required Grassmann to pass, at the end of 1831, also an examination in philosophy. Those who passed it were entitled to teach "philosophical propedeutics" in the upper classes. The extensive certificate[5] shows that the commission refused him a teaching license in philosophy: it attested that he had a sufficient philosophical "Bildung" but lacked both a knowledge of the effects of philosophical enquiry and a capacity "to guide exercises for stimulating and developing scientific thinking" (quoted from Engel 1911, p. 41).

A third reason, finally, is suggested by Schleiermacher's *Dialektik* itself: there is no conception which is specific to mathematics. *Dialektik* meant for Schleiermacher "the principles of the art of philosophizing" (§ 17), its task is to investigate the process of knowledge, the formation of concepts and judgements as well as their relation to being. Except for some marginal remarks, Schleiermacher did not apply these investigations to mathematical concepts or discuss problems of philosophy of mathematics.

Consequently, even in his *curriculum vitae* for the theological examination, where he had emphasized the authority of Schleiermacher as particularly important for him, Grassmann had ascribed only a general methodologically orienting function:

> "I realized how one can learn from Schleiermacher, since he gave not so much positive instruction but rather made oneself able to begin each investigation at the right point and continue it independently and to find the positive result by oneself" (quoted from Engel 1911, p. 22).

---

4 A further contradiction to Grassmann's reports consists in the fact that he first matriculated in the Philosophical Faculty while he changed to the Theological Faculty only after 1 1/4 years, at the end of 1828 (Archiv der Humboldt University, AZ 1830, vol. LXVVII, fol. 157-161).

5 The short form in the annual report of the commission for the ministry just noted an unspecified "good" (GStA Berlin (ex-Merseburg), Rep. 76 VI, Sekt. 19, x, Nr. 1, vol. I, fol. 104)

ROBERT'S WAY TOWARDS PHILOSOPHY

Given this situation, I maintain that the essential impetus for the formation of Hermann Grassmann's particular philosophical conception of mathematics came from his interaction with his younger brother, Robert. It is known that the two brothers studied the *Dialektik* together in 1840 (ibid., p. 91) - but I am envisaging a much more intensive cooperation than just trying to understand a book by a joint reading of it.

Robert Grassmann (1815-1901), one of the twelve brothers and sisters of Hermann but intellectually the nearest one to him, received - in a manner remarkably analogous to the way his elder brother became engaged in research on the *Ausdehnungslehre* - a decisive impetus to becoming involved in the philosophy of mathematics by the tasks assigned to him in the teachers' examination. Moreover, the examinations both brothers were to pass became more and more parallel so that an interaction proved to be fruitful for both of them.

Robert's studies in school and university can be reconstructed by his teachers' examination file.[6] Contrary to his brother, Robert did not ask to be examined in Berlin but in Greifswald - quite near to Stettin - where the examination board for Pomerania was established in 1838 and where he was examined in 1840 as but their fifth candidate.

Robert had engaged himself, even since school days, in a more

---

[6] This file is preserved within the archives of the *Wissenschaftliche Prüfungs-kommission für die Provinz Pommern* where I uncovered it in a hitherto unknown and unordered collection (in the: Universitätsarchiv Greifswald; cf. Schubring 1991, pp. 24 and 222).

Robert has published autobiographical accounts in several of his books (see, e.g., the "Vorwort" in R. Grassmann 1890a and 1890b) but they had been used before only in the study: Birjukov/Birjukova 1982, in its sixth chapter (Robert Grassmann: scientist and editor, personage of public life and popularizer of science, pp. 56-61). Beyond the reproduction of these reports, there has been no research into his biography. It is not clear, hence, why he did not rise in his professional career as his father and his brother did: He began to teach at public schools early in 1841, at the *Friedrich-Wilhelms-Schule*, the important Stettin *Realschule*. In 1843, however, as part of a swap of teaching posts between several teachers which brought Hermann into his position, he changed to the *Höhere Töchterschule* and remained there until 1852 when he gave up his teaching career entirely - for unknown reasons - and turned to that of an *entrepreneur*: as editor of political journals and later as a publisher. He added the role of a local politician. He could permit himself to publish an enormous number of books of his own. While these were in a first phase school textbooks, covering almost all school disciplines, Robert undertook in a second phase to challenge academic knowledge and to publish an exhaustive encyclopedic series of his own propagating his "rigorous" new system.

Unfortunately, I did not succeed in tracing a portrait of this polyhistor. His own announcement to have inserted his portrait in the first volume of his definitive encyclopedia (R. Grassmann 1890a, p. XIX) proved not to have been realized in the copies consulted by me. Even the lengthy obituary published in the *Pommersche Zeitung*, one of his own newspapers, on August 16th 1901, did not have his portrait.

intensive and successful study of mathematics and the sciences than Hermann. His teachers certified him in the *Abitur*, the leaving exam, as meeting high standards in mathematics and in physics.[7] In philosophical propedeutics, his abilities in "logical reasonings" were judged unfavourably. In the autumn of 1834, Robert began to study at the Bonn university - then the Prussian center for studies in the sciences (cf. Schubring 1989). Officially, he studied theology and the sciences in a parallel manner, but in reality he devoted himself to mathematics and the sciences. Following the advice of the mathematics professor K. D. von Münchow, he studied textbooks by Euler and Biot. That he really intended to study the sciences is shown be the fact that he became, already in his first semester, a member of the *Seminar für die gesammten Naturwissenschaften* - the unique institution for an integrated study of the exact and natural sciences. Robert had been chosen by the professor for natural history to become his assistant in the mineralogical *Kabinett* in the second semester so that a career in the natural sciences might have seemed to be tenable for him. A heavy blow, the death of his sister Alwine made him, however, change his studies entirely: he turned to a study of theology, at first exclusively. In his third and fourth semesters he opened himself somewhat and followed some courses in philosophy too: history of philosophy, metaphysics, psychology, logic, and anthropology.

After these four semesters in Bonn, he changed to Berlin University where he studied for two more semesters and studied theology exclusively, mainly with Neander (Schleiermacher had died in 1834). He finished these studies by passing, in March 1838 in Berlin, the first theological examination for future ministers. Returning to Stettin, he became a "Hilfslehrer" in the *Gymnasium* teacher-training *Seminar*. There, imbued, as he explained in his "curriculum vitae," by "love of mathematics," he decided to leave a minister's career and to turn to the study of mathematics.

He began, like his brother, by reading his father's textbooks and then travelled to Greifswald in order to ask the advice of the mathematics professor J.A. Grunert concerning his autodidactic studies. Robert mentions having studied - following Grunert's advice - analysis (Lacroix, Ettingshausen, Cauchy), analytical geometry (Biot), and mechanics (Francoeur), but most influential was Lagrange's theory of analytical functions. In fact, Lagrange's books caused the first intensive

---

7 "Er lieferte überzeugende Beweise, daß sein mathematisches Wissen das für die Universitätsreife vorgeschriebene Maß überschreite." and in physics: "Mit den Erscheinungen der Körperwelt [...] und mit den zu ihrer Erklärung dienenden Naturgesetzen ist er auf erfreuliche Weise bekannt" (Abitur-Zeugnis, 22. 9. 1834; part of his teachers' examination file, Universitätsarchiv Greifswald).

cooperation between Robert and Hermann: as Robert depicts in his "curriculum vitae," he wished to study Lagrange's analytical mechanics too, due to his admiration of the author. Not having time to independently acquire this knowledge he asked Hermann - who, he recounts, was just then busy studying this work - to teach him these theories so that he succeeded in learning the main parts of Lagrange's mechanics.[8]

It was not only the work of the two brothers in elaborating their dissertations that occurred largely in parallel; just as Hermann got the impetus to develop his ideas on the extension theory by the mathematics examination, Robert became analogously attracted to the foundations of mathematics by his examination in philosophy. J.A. Grunert, then head of the Greifswald teachers' examination committee, presented Robert's application of 28.12.1839 to his colleagues and proposed that they examine him in mathematics, physics and philosophy. The member for philosophy proposed as subject for that dissertation: "to specify and to criticize the reasons for the assertion that mathematics is no philosophical science." This member was Ernst Stiedenroth (1794-1858), who was since 1825 professor for philosophy at the Greifswald university and was ascribed to the school of J.F. Herbart. The task of elaborating this dissertation became a true challenge for Robert. Two times he applied for a prolongation of the deadline. Eventually, he handed in a voluminous but unfinished paper, of at least 100 pages. Unfortunately, this dissertation is not extant but was presumably destroyed when most of the examination papers in the Greifswald teachers' examination files were cleared out.[9] One can obtain, however, certain ideas about its content by the commentary which Robert himself gave in a letter to Grunert (then *Rektor* of the Greifswald university) of 24 July 1840 accompaning his three examination papers.

In this commentary, Robert explained that he had changed the subject to the relation of mathematics and philosophy:

---

8 Since Robert's "curriculum vitae" was written at the end of 1839, this shows that Hermann did study Lagrange's mechanics only by the second half of 1839. This is confirmed by Robert's later report that Hermann studied Lagrange's and Laplace's mechanics in the Winter 1839/40 (Rob. Grassmann 1895, p. IV). One can conclude, hence, - contrary to Engel's opinion (Engel 1911, p. 84) - that Hermann's study of the theory of tides was instigated only by the examination task committed to him in March 1839.

9 At an unknown time, probably between the two World Wars, this destruction of great parts of the files was effected, maybe in order to save storage place. Whereas for some candidates one or two papers were preserved, in Robert's case all were destroyed - given their enormous volume. From the size of the gap in the binding of his file one can estimate their volume to have been at least 200 pages.

"The third dissertation, the philosophical one, has grown to a size that I find frightening and which was probably caused by a misunderstanding. As Your Magnificence[10] will see from the introduction, I have been led to understand my task as being to elaborate a historical exposition of the relations between mathematics and philosophy, to explain the positions of the individual philosophers and to adjoin my own exposition of that relation."

The historical exposition thus had the burden of exposing the doctrines of the philosophers on the basis of their own writings so that the length of his dissertation as well as the long duration in composing it is understandable. Stiedenroth judged this paper very positively and praised the "zealous and successful research on the history of philosophy." Robert was attested to be "eminently" able to teach philosophical propedeutics, particularly because of his "highly vivid and sensible original thinking." The mathematics examination, however, showed gaps, but only gaps, in his knowledge. Grunert, apparently on good personal terms with the Grassmann family, took the liberty of according Robert nevertheless the full teaching *facultas* in mathematics and physics.

OTHER PHILOSOPHICAL INFLUENCES

One can derive two main conclusions from the respective intellectual developments of the two brothers, both instigated by their examination tasks:

1. The leading part in the cooperation between the brothers Hermann and Robert, beginning in 1840, was taken by Robert with regard to philosophical and foundational questions of mathematics. It is therefore legitimate to assume that the embedding within a philosophical discourse which is characteristic for the A_1 resulted from Hermann's interaction with Robert in the philosophy of mathematics. In fact, the difference in style between the A_1 of 1844 and the *Theory of Tides* of 1840 is striking: the 1840 text is in general composed according to the traditional style of mathematical exposition.

2. Though we cannot reconstruct the precise content of Robert's philosophical dissertation of 1840, it is evident that in it he studied the relevant philosophical systems and reflected the opinions of their main representatives on the relation of philosophy to mathematics.

I am therefore lead to the reasonable hypothesis, that Robert reflected in particular the contributions of Jacob Friedrich *Fries* (1773-1843) to this subject (see Scubring 1981, pp. 119ff.). Fries was, namely, the only contemporaneous philosopher who studied mathematics and its philosophical analysis and interpretation in an intensive and competent

---

10 Traditional title of the *Rektor* at German universities.

manner. Fries had expressed, already in 1811, his concept of a "philosophy of mathematics" in his *System der Logik*, and then in 1822 in an extensive and remarkable form in his book *Die mathematische Naturphilosophie*. That both brothers knew Fries's philosophy of mathematics is all the more probable as their father Justus had discussed Fries in his important paper on number theory; this discussion is particularly revealing since Justus there *denied* Fries's assertion in that 1822 book of a non-limited number of dimensions ("Stufen") for the connection of operations (J. Grassmann 1827, p. 35; Fries 1822, p. 97).

There are several pivotal elements in Fries's philosophy of mathematics which are in structural agreement with Grassmann's conceptions. Fries stands, like so many German philosophers, in the tradition of Kant but he has independently analysed the process of concept formation in mathematics and the sciences. A preeminent task for him is to develop a philosophy of *pure* mathematics. This means primarily the study of the "laws of the connection forms" ["Verknüpfungsformen"]. And Fries emphasizes: "Mathematics is the complete system of the mathematical forms." Among all abstraction processes, pure mathematics finds its ultimate success in detecting "the pure forms of the composition of things" for the mind (Fries 1822, pp. 49-50). The correspondence between Fries's and Grassmann's conception of mathematics as "Formenlehre," hence, is striking.

Deeply influenced by the view of mathematics propagated by the combinatorial school, which had also impressed Grassmann's father, Fries elaborated a decisive innovation: he was the first to develop the notion of a self-standing fundamental branch of mathematics, underlying the other branches - arithmetic, algebra and geometry - as "independent." Fries called this foundational discipline 'Syntaktik' and ascribed to it the task of establishing "the doctrine of the pure laws of arrangement of given parts" (ibid., p. 65). This *Syntactic* constituted for him the genuine foundational discipline: "The syntactic includes the most general abstraction which one can achieve in mathematical knowledge" (ibid., p. 70).[11]

Fries was constrained, in his analysis of connection forms, to combinatorial forms of arrangement. The first discussion in mathematics of different connection forms in a general and abstract way was published in 1823 in the book *Grundlinien des typischen Kalküls*, by Eduard Collins (1791 - 1840), who originated from England and was a

---

[11] In his 1827 paper, Justus Grassmann had, in a somewhat analogous manner, argued for combinatorics as basic discipline underlying the other mathematical branches (J. Grassmann 1827, note on p. 7).

member of the Petersburg Academy.[12] One can conclude that Fries's conception of syntactic not only constituted the first form of a mathematical logic but that it can also be seen as a model for H. Grassmann's philosophical introductory part of the A_1.[13]

THE EVOLUTION OF COOPERATION

It is highly revealing that it was logic as a foundational discipline in which the further collaboration between the brothers Hermann and Robert proved to be the most intensive - in cooperation and in division of labor.

In fact, the cooperation between the two brothers concerned numerous domains. Their joint political work is well known: they began publishing a conservative newspaper in 1848 for fighting against revolutionary and liberal ideas. Their common production of school textbooks for German and Latin grammar has not yet been studied. But for us the most important domain has been their intensive collaboration, after 1847, in striving for a comprehensive exposition of mathematics based upon the new conception of the *Ausdehnungslehre*. Because of the division of labor agreed upon by the two brothers, we are led to a new explanation why the A_2 of 1862 was published without the philosophical introduction and without being embedded into a philosophical style characteristic of the A_1: its function had meanwhile been assigned to the logic part, and the composition and publication of this logic had been mandated to Robert!

Robert has several times described this close mathematical collaboration but these reports have been neglected by F. Engel. The most voluminous report is given in the preface to the third part of the 1890 Formenlehre:

"In the year 1847, the brothers Hermann and Robert Grassmann joined forces in order to rigorously deduce the *Ausdehnungslehre*, independently of the geometry, as a self-standing branch of pure mathematics by development of form ["Formentwicklung"] and to develop it to the limits of its domain of validity. The manuscript booklet of 132 pages

---

12 By 'typischer Kalkül,' Collins meant in fact the analysis of the *signs of operations* and the search for the general behavior of operations. He introduced abstract signs for operations, like |, \, /, and different levels of operations and studied general laws like commutativity, associativity, and transitivity for the operations. Collins's influence has not yet been analyzed, but he seems to have had a certain public in Germany (Schubring 1983, pp. 227 and 245f.). For the biography of Collins and a description of his mathematical work see Ožigova 1980, pp. 38-55.

13 The only philosopher who is explicitly mentioned in the A_1 is so as a negative case: "Hegel and his school" are criticized as model for "obscurity and arbitrariness," yielding philosophical debates of mathematical questions fruitless (A_1, p. 15).

which we then worked out is still in the possession of the author" (R. Grassmann 1895, p. IV).

Subsequently Robert described in detail their common results, in particular the different types of multiplication, and commented that

"...already in this year [1847], all branches of the *Ausdehnungslehre* were established in a rigorous form, [...] although the details of these branches were only partially established. Which share each of the two brothers had as a result of this joint work can no longer be determined since sometimes the one and sometimes the other took decisive action and mastered difficulties. Each of the two brothers felt that he alone would weary, [...], and that they were only able to overcome the problem by joint forces. After all, both have acted together and contributed their part" (ibid., p. V).

Robert underlined that he did not claim to share the invention of the *Ausdehnungslehre* and added:

"It falls to Robert Grassmann's share only to have worked towards the generality of the conception and towards the rigour of the form and to have thus contributed to the solution of the difficulties" (ibid., p. VI).

As Robert reports, they took up their joint work, after the years of revolution and their aftermath, in 1855 and 1856. They practised now a new form of collaboration: no longer in direct communication but in a Bourbaki-like style - while one of the brothers elaborated the exposition during one week and reported it afterward to the other brother who commented and criticized the results, while the next week their roles changed. They thus established the common basis and principles of the intended comprehensive mathematical treatise and decided to charge each of themselves, in division of labor, to compose the respective textbooks: Hermann the two mathematical branches - arithmetic and the *Ausdehnungslehre* - and Robert the two philosophical branches: logic and the combinatorial theory (ibid.).

Robert admitted that his brother was the first one to resolve his task, by publishing his two parts in 1860 and in 1862, while he succeeded only in 1872 in publishing his own two parts (logic and combinatorics), together with a rough summary of the entire undertaking. In 1890 Robert claimed to be entitled, after his brother's death in 1877, to publish his own, reworked version of the *Ausdehnungslehre* (ibid., p. VII).

It is highly regrettable that Engel barely considers this and other reports by Robert and that he did not even take notice of the 1847 joint manuscript.[14] It is evident from the treatise published by Robert in 1872 that it is clearly composed as complementary to the A_2. Only in his publications of the years 1890 and later did Robert undertake to expose

[14] See Engel 1911, pp. 131-132. It is possible, that Engel was deterred by Robert - as a particularly marked example of a "Grassmannianer" - a sort of narrow sectarian by which he and other co-workers in the edition were horrified.

also the subjects of the A_2 in his own writings.

One might wonder, on the other hand, why Hermann barely mentioned his cooperation with Robert - except in his arithmetic of 1860/61 - and why, in particular, he did not present his revision of the *Ausdehnungslehre* as a part of the common project. One can assume that the previously close relation between the two brothers had become colder, by the 1860s, probably since Hermann had withdrawn from the newspaper editorship (Engel 1911, p. 147).

Questions still remain open regarding the emergence of the A_2, and a need to further explore the respective contributions of the two brothers.

Institut für Didaktik der Mathematik, Universität Bielefeld

REFERENCES

B. V. Birjukov, L. G. Birjukova, ""Učenie o formach (veličinach)" German i Roberta Grasmanov kak predvoschiščenie konstruktivnogo napravlenija v matematike, I.," *Voprosy kibernitiki. Kibernetika i logičeskaja formalizacija. Istoričeskie i metodologi českie aspekty* (Moskva 1982), pp. 36-92; II: ibid. (Moskva 1984), pp. 45-111. ["Die Formenlehre (Grössenlehre)" der Bründer Hermann und Robert Grassmann als Vorwegnahme der konstruktiven Richtung in der Mathematik].

Eduard Collins, *Grundlinien des typischen Kalkuls* (St. Petersburg: Druckerei der Kaiserl. Akademie der Wissenschaften 1823).

Friedrich Engel, *Grassmanns Leben*, Hermann Grassmanns Gesammelte Mathematische und Physikalische Werke. Dritten Bandes zweiter Teil (Leipzig: B. G. Teubner 1911).

Jacob Friedrich Fries, *Die mathematische Naturphilosophie, nach philosophischer Medthode bearbeitet* (Heidelberg: C. F. Winter 1822). Reprint in: J. F. Fries, *Sämtliche Schriften*, Hrsg. G. König, L. Geldsetzer, Band 13 (Aalen: Scientia Verlag 1979).

Hermann G. Grassmann, *Die lineale Ausdehnungslehre, ein neuer Zweig der Mathematik* [A_1], id., Gesammelte Werke, Ersten Bandes erster Theil (Leipzig: B. G. Teubner 1894).

Hermann G. Grassmann, *Die Ausdehnungslehre, vollständig und in strenger Form bearbeitet* [A_2], id., Gesammelte Werke, Ersten Bandes zweiter Theil (Leipzig: B. G. Teubner 1896).

Robert Grassmann, *Das Gebäude des Wissens. Erster Band: Die Wissenslehre oder die Philosophie. Erster Teil: Das Verstandeswissen oder das formale Wissen* (Stettin: R. Grassmann 1890). [1890a]

Robert Grassmann, *Das Gebäude des Wissens. Erster Band: Die Wissenslehre oder die Philosophie. Zweiter Teil: Die Denklehre* (Stettin: R. Grassmann 1890). [1890b]

Robert Grassmann, *Die Ausdehnungslehre oder die Wissenschaft von den extensiven Größen. Der niedere Zweig der Synthesis.* Das Gebäude des Wissens. 23. Band: *Die Formenlehre oder Mathematik, in strenger Formelentwicklung. Dritter Zweig* (Stettin: R. Grassmann 1895)

Albert C. Lewis, "H. Grassmann's 1844 *Ausdehnungslehre* and Schleiermacher's

Dialektik," *Annals of Science*, 1977, *34*, 103-162.

Albert C. Lewis, "Grassmann's n-Dimensional Vector Concept," *Actes du Colloque: Nombre Complexe et Vecteur* (Paris 1996, forthcoming).

E. P. Ožigova, *Matematika v Petersburgskoj Akademii Nauk v konce XVIII - pervoj polovine XIX veka* [russ., Mathematics in the Petersburg Academy] (Leningrad: Nauka 1980).

Friedrich Daniel Schleiermacher, *Dialektik*. Aus Schleiermachers handschriftlichem Nachlasse herausgegeben von L. Jonas (Berlin: G. Reimer 1839).

Gert Schubring, "The Conception of Pure Mathematics as an Instrument in the Professionalization of Mathematics," *Social History of Nineteenth Century Mathematics*, ed. H.Mehrtens, H. Bos, I. Schneider. (Basel: Birkhäuser 1981), 111-134.

Gert Schubring, "Das mathematische Leben in Berlin. Zu einer entstehenden Profession an Hand von Briefen des aus Erlangen stammenden Martin Ohm an seinen Bruder Georg Simon," *Erlanger Bausteine zur Fränkischen Heimatforschung*, Jahrbuch 30 (1983), 221-249.

Gert Schubring, *Die Entstehung des Mathematiklehrerberufs im 19. Jahrhundert. Studien und Materialien zum Prozeß der Professionalisierung in Preußen (1810-1870)*. (Weinheim: Beltz 1983). Zweite, korrigierte und ergänzte Auflage (Weinheim: Deutscher Studien Verlag 1991).

R. STEVEN TURNER

THE ORIGINS OF COLORIMETRY: WHAT DID HELMHOLTZ AND MAXWELL LEARN FROM GRASSMANN?

The modern science of color vision dates essentially from the 1850s, when Clerk Maxwell and Hermann von Helmholtz introduced the modern techniques of colorimetry and popularized the first comprehensive and broadly successful theory of color vision -- Thomas Young's trichromatic theory. Both Helmholtz and Maxwell, however, acknowledged the powerful influence of Hermann Günther Grassmann and his seminal article, "Zur Theorie der Farbenmischung," which appeared in Poggendorff's *Annalen der Physik* in 1853. That article has powerfully shaped all subsequent study of color, yet it presents many historical problems. It was Grassmann's first and only real contribution to the field. It contained many assumptions and assertions that were immediately rejected by others. And in it Grassmann himself claimed to do nothing more than rigorously deduce certain universally-known and accepted principles of color-mixing which Newton had already laid down. Understanding the significance of Grassmann's contribution, therefore, requires placing it within the larger context of thought about color and colorimetry and beginning, not with Grassmann, but with Helmholtz.

HELMHOLTZ ON COLOR IN 1852

Helmholtz, already the *Wunderkind* of German natural science, published his first two papers on color in 1852. One of these papers attacked the well-known and strongly anti-Newtonian theory of color which had been advanced by the English scientist Sir David Brewster (Helmholtz 1852a). Helmholtz's second paper on color, more central to Grassmann's story, was closely related to the first. In that paper, Helmholtz analyzed the several possible senses of the term "primary colors" or *Grundfarben*, and sought to determine through experiment the minimum number of such colors that must be postulated (1852b). His experiments indicated that there exist no three colors in nature (or any four) from which all other colors without exception can be mixed. That

G. Schubring (ed.), Hermann Günther Graßmann (1809–1877): Visionary Mathematician, Scientist and Neohumanist Scholar, 71–86.

conclusion, as Helmholtz observed, not only flew in the face of centuries of experience in mixing pigment colors; it also eliminated theories like Brewster's, which postulated three primary colors and identified them with distinct forms of light; and theories like Thomas Young's, which postulated three primary colors and identified them with the specific sensation-energies of discrete classes of retinal receptors.

To reach these radical conclusions Helmholtz had experimentally generated series of overlapping spectra in which all possible binary combinations of spectral colors were displayed. On examining these binary combinations, Helmholtz found that yellow and indigo blue combined to produce true white, but that no other binary pair would yield more than a whitish impression tinged with a distinct hue. Thus he concluded that the solar spectrum possessed only one pair of "complementary" colors. The result that yellow and indigo-blue **light** combined to produce white clashed with the well-established fact that mixtures of yellow and blue **pigments** combine to yield green. This contradiction led Helmholtz to enunciate the fundamental distinction between additive and subtractive color mixtures and so to clarify the divergent results obtained from mixing pigments and mixing spectral lights.

In his paper Helmholtz also described the results of slightly more elaborate experiments, in which he had tried to make compounds which would match the various spectral colors by using spectral red, green, and violet as primaries. Helmholtz found that he could match the **hue** of any spectral color with appropriate mixtures of these spectral primaries, but that the compounds were always whitish and desaturated in comparison to their spectral, monochromatic equivalents. Helmholtz conceded that three **pigment** primaries can be chosen so that mixtures of them roughly approximate the flat colors of natural objects, but he observed that no pigment mixes can yield the spectral colors in their full richness. If one desires a set of primary colors from which all the spectral colors can be mixed at their full spectral saturation, then it is necessary in practice to adopt at least five monochromatic lights as primaries. Even their mixtures only approximate full spectral saturation (Helmholtz 1852b; cf. Kremer 1993, Sherman 1981).

NEWTON AND THE TRADITIONS OF COLOR-MIXING

Scientific papers are significant for what they do **not** say. As Grassmann was soon to point out, Helmholtz's article of 1852 featured one deafening silence, namely, the absence of any reference to the venerable Newtonian theory of color mixing. In the *Opticks* of 1704, and in subsequent editions, Newton had laid down a clear procedure for predicting the color of the compound which would result from any mixture of homogeneous lights, providing that the components of the mixture were known. In Book One, Part II, Prop. VI, Newton directs the reader to construct a circle (see Fig. 1), the center *O* of which will be taken to represent white and the circumference the spectral colors (Newton 1730, 154-61). The circumference is to be divided into seven intervals in proportions exactly specified by Newton; these proportions, he claims, correspond to the seven musical tones or intervals of the octave. Each part of the circumference will represent one of the seven

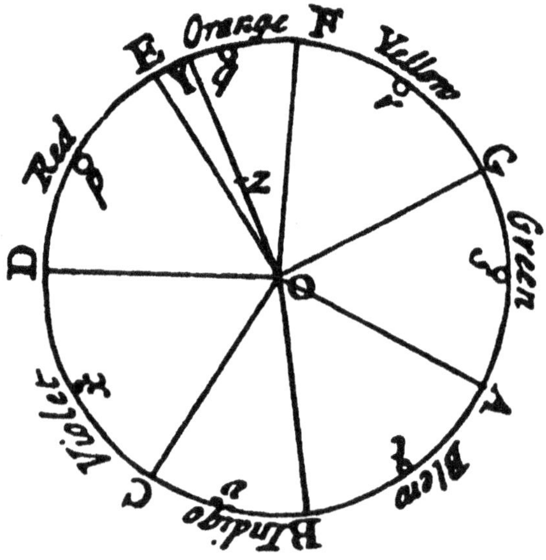

Fig. 1. Newton's illustration of his color circle (1704). *Source*: Turner 1994, 27.

"primary colors" which Newton declares to exist in the spectrum, namely red, orange, yellow, green, blue, indigo, and violet. Especially important is the fact that Newton further instructs the reader to locate the point on each primary-color arc which will represent the center of gravity of that arc; for the primary color red, for example, that point will be *p*.

In order to use Newton's color-circle construction to predict the outcome of any mixture of spectral colors we proceed as follows. We take the proportion of "red-making Rays in the Mixture" and represent that proportion as a weight suspended from the point *p*, the center of gravity of the red arc. We carry out the comparable procedure for any of the seven primary colors around the circumference of the color circle which are known to be present in the mixture. Then we locate the center of gravity of those seven weights. The location of that point in the color circle will, Newton assures us, represent the color of the mixture. If the center of gravity is the point *Z*, the hue of the mixture will be represented by the point at which the radius from *O* through *Z* strikes the circumference. The distance *OZ* will give the "fullness or intensity" of the mixture, i.e. in modern terminology its saturation. However, Newton offered this mix-procedure only as an empirical approximation. It was, he said, a "Rule I conceive accurate enough for practice, though not mathematically accurate" (Newton 1730, 158). He advanced no theoretical justification of the procedure, nor for the particular division of the circumference that he prescribed, except for its alleged correspondence to musical intervals.

The study of color and color-mixing after Newton unfolded as three overlapping traditions (Hargreave 1973, 187-201; Sherman 1981, 60-80; Kremer 1993, 220-27). One tradition studied subjective and pathological color phenomena; it is unimportant to the Grassmann story. A second tradition continued the discussion of the solar spectrum that took its cue from Newton. This tradition intensely debated the number of "primary colors" which can be said to exist within the spectrum. Newton's earlier writing had stressed the existence of an infinite number of colors, corresponding to the infinite number of degrees of refrangibility. His later writings, and his discussion of the mix procedure in particular, had played down this notion of continuous color-change and had emphasized a discrete spectrum characterized by the existence of seven "primary colors." The absence of a uniform color terminology further complicated the issue, but, as historian David Hargreave insists, most later Newtonian commentators adopted the notion of primary colors, even though they fiercely debated whether the true number of such colors was three, four, six, or seven. Intermediate gradations of color were explained as arising from the overlap of the primary colors in the spectrum.

The third European tradition of color and color mixing drew on the experience of craftsmen and dyers in mixing pigment colors. (Hargreave 1971, 117-30; Sherman 1983, 60-80.) It insisted on exactly three primary colors (red, yellow, blue) from which, as experience showed, all other hues could be mixed. Theorists in this tradition sometimes illustrated color-relationships in terms of "color triangles" that placed one of the primary colors at each vertex. The interior of the triangles was subdivided into small cells colored with tertiary combinations of the three primary colors. The relative proportions of each primary in each cell-mixture decreased with the distance from the respective vertices. As even this brief discussion suggests, this third European tradition of color study was representational and taxonomic in orientation, and it tacitly incorporated Newton's principle of barycentric mixing (Hargreave 1973, 117-30; Sherman 1981, 60-80).

HELMHOLTZ AND NEWTON

Against this larger background, Helmholtz's work of 1852 can clearly be seen as a campaign to defend Newton's claims about light and color. Not only did Helmholtz attack Brewster's anti-Newtonian theory, but he also adopted a seven-color characterization of the spectrum similar to Newton's. In the *Opticks* Newton had vigorously denied that complementary spectral colors exist and claimed that he had never been able to produce pure white experimentally from any binary mixture. Helmholtz, repeating these experiments, reported very similar results; he had been able to obtain pure white from only one of many possible binary pairs. Like Newton, he had been able to mix the "primitive colors" of the spectrum only from their immediate neighbors. Helmholtz's results conformed entirely to Newton's.

Why did Helmholtz, in a paper on color mixing that was devoted to a defense of Newton, make no reference to the famous Newtonian mix-procedure? One plausible hypothesis is that he considered his results to have shown up the mix procedure to be an even cruder and more limited empirical approximation than Newton himself had believed it to be. After all, Helmholtz had shown that pigment mixes (subtractive) and light mixes (additive) obey fundamentally different rules; therefore the Newtonian mix-procedure could no longer naively be assumed to apply to both correctly, as Newton and all who came after him had done. Despite its origins with Newton and studies of the spectrum, barycentric mixing was primarily associated with the tradition of pigment mixing, color triangles, and acceptance of red, yellow, and blue primaries. Helmholtz did not doubt that these concepts retained a rough validity where pigment mixing was concerned, but his results suggested that they could not be applied to the spectral lights. For all these reasons, one may speculate, Helmholtz tactfully declined to discuss the obvious limitations of the mix procedure; such a discussion would have been out of place in his larger programme to defend Newton and his theory of color.

If Helmholtz interpreted many of his results as undermining the universality and rigor of Newton's mix procedure, he almost certainly did not interpret his inability to produce more than one pair of spectral complementaries as doing so. Although some commentators tacitly assumed the existence of complementary colors (against Newton), they rarely or never justified their existence by appeal to the mix-procedure. On the contrary, authorities with which Helmholtz was certainly familiar, such as J.-B. Biot and James Forbes, agreed that Newton's original construction predicted no complementaries at all. In Newton's formulation, mix-procedures may be initiated only from the discrete center-of-gravity points of the color circle, not from arbitrary points along the circumference. But in the original construction, no two center-of-gravity points lie on a diameter, and so to the extent that the circle accurately describes mix-relationships, binary pairs can yield approximate but never true whites (Biot 1816, 3:445-55; Forbes 1849). Some commentators questioned the rationale behind the arc-lengths which Newton had assigned to the various primary colors around the circumference (Forbes 1849, 164). If even slightly different arc-lengths are assigned to the colors, then their center-of-gravity points shift their relative positions, so that they might or might not lie on common diameters. Under those circumstances, Newton's mix procedure would then make no firm predictions at all about the existence of spectral complementaries; the issue could be settled only by experiment, and that was what Helmholtz believed he had done.

GRASSMANN, HELMHOLTZ, AND NEWTON

Grassmann presented his first public pronouncement on light and color in a lecture to the Physikalische Gesellschaft of Stettin in October 1852. To my knowledge the contents of that lecture are known only through the brief notice reported in the society's *Mittheilungen über ... Thätigkeiten* and reprinted in Grassmann's *Gesammelte ... Werke* (3.1: 251-52). According to the anonymous author of that notice, Grassmann's lecture provided an historical overview of different opinions about color and color mixing, stressed the necessity of distinguishing subjective and objective color impressions, and reviewed various correspondences which others had drawn between the principal intervals of the octave and the supposedly harmonious color combinations of the spectrum. Grassmann also discussed Helmholtz's paper of 1852. He noted that Helmholtz had experimentally demonstrated two spectral complementaries, yellow and indigo blue. Grassmann went on, however, to argue from Helmholtz's own data that there must exist one other pair of complementary colors which Helmholtz for some reason had failed to detect, specifically a red and a green. The reporter suggests that Grassmann found this a satisfying result, as he urged that these four colors be regarded as the primary colors of the spectrum, and that others be treated as mixtures of those four.

Although little is known about Grassmann's October lecture, the presentation is historically significant for suggesting how very traditional was Grassmann's initial approach to color science; his concerns of 1852 with establishing tonal analogies and primary colors conformed entirely to established traditions of color research. It also hints that between

October, 1852, and the publication of his seminal article in 1853, Grassmann had come to conceive the problem of color in a very different and less traditional way. The article of 1853, for example, opens with a direct criticism of Helmholtz that Grassmann had either not made in his public lecture of 1852, or that the reporter had failed to communicate. Grassmann charged in 1853 that Helmholtz, far from confirming Newton's principal results, had in fact falsely attacked and rejected the Newtonian theory of color mixing which had been "universally accepted" since Newton's day (Grassmann 1853). After opening his 1853 article with these charges against Helmholtz, Grassmann went on to support his criticism with a short synopsis of Newton's color mixing procedure. In fact, Grassmann's description of that procedure departed radically from Newton's original. With neither comment nor apology, Grassmann jettisoned Newton's conception of a discrete spectrum divided into primary colors, in favor of one in which colors change continuously along the circumference of the color circle. Grassmann imposed this continuity-condition so rigorously as to insist that the most extreme violet of the spectrum is phenomenologically identical to the most extreme red, so that the color circle is definitively closed. Much more important, Grassmann allowed the barycentric mix-procedure to be initiated between any two points of the color circle. It is no longer to be initiated only from the center-of-gravity points of the discrete primary color arcs; in fact, Grassmann made no mention of these points. If Grassmann was conscious of misrepresenting Newton's original procedure, or embarrassed about doing so, he gave no sign in the 1853 treatise.

Grassmann's revised presentation of Newton's barycentric mix procedure makes it obvious that every spectral color must possess a spectral complementary. Indeed, there must be in principle as many complementary pairs as there are possible diameters of the circle. Helmholtz's failure to detect these many complementary pairs, Grassmann charged, could only be due to experimental error. He argued that the one pair of complementaries which Helmholtz had detected could, from their descriptions and relations to the Fraunhofer lines, plausibly lie along a diameter of the color circle in the form that Newton had originally constructed the circle. Grassmann argued, as he had in 1852, that Helmholtz's own results, properly interpreted, suggest that other complementary pairs surely existed. In 1853, however, he was convinced that Helmholtz's data pointed to more than one such pair. The several binary pairs from which Helmholtz had obtained gray or whitish mixtures indicated they contained actual complementary wavelengths, and Grassmann used Helmholtz's results to argue that these approximate pairs of complementaries lie roughly along diameters of Newton's original color circle (Grassmann 1853, 76-78).

Grassmann also included an important theoretical discussion on color which apparently had not been part of the 1852 lecture. He insisted that three variables suffice to completely define color sensations phenomenologically: *Intensität* (brightness), *Farbenton* (hue), and what he called *Intensität der beigemischten Weiss* (saturation). Hence a space of three dimensions suffices for a geometrical description of color relationships. In demonstrating those relationships, Grassmann did not simply assume the truth of the Newtonian mix-procedure under the assumptions of continuity. Instead, he rested its validity upon four postulates or "laws." These were simple, phenomenological relationships among colors and their mixtures that Grassmann adopted on the grounds of their intuitive plausibility alone. Space limitations prohibit any review or discussion of Grassmann's laws here. Their importance, however, lies in the fact that the four together guarantee that color mixtures will obey the rule of barycentricity. That in turn allowed colors to be represented as directed lengths on the color plane, and their mixtures to be represented as the resultants or "geometric sums," according the simple procedures Grassmann had outlined in his *Ausdehnungslehre* of 1844. Alternately, he showed, intensities could be expressed as weights suspended at the ends of directed lines of the same lengths, with the center of gravity expressing the hue and intensity of the mixture.

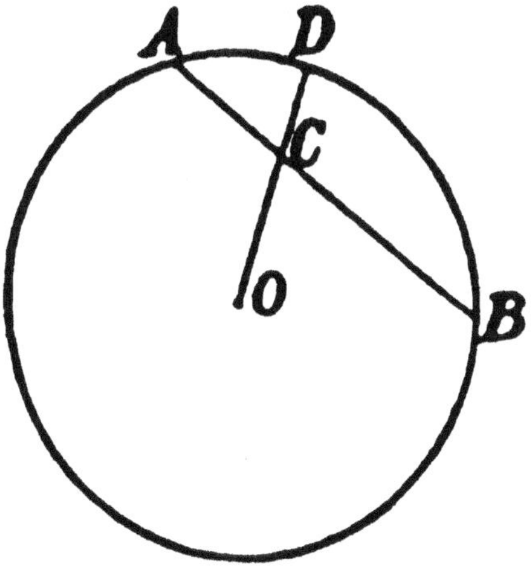

O = white point
a units of color A
b units of color B
C is the color of the mixture

(a+b)OC is the chromatic intensity of the mixed color
(a+b)CD is the "intensity of the intermixed white"
(a+b)OD (where *OD*=1) is the total intensity of the mixture

Fig. 2. Grassmann's geometrical representation of relationships on the color plain. *Source*: Grassmann 1853, 82.

Grassmann then drew the color circle so that homogeneous colors of unit intensity were ranged along the circumference. Under those circumstances, the center-of-gravity construction yielded the hue, the total intensity of the mixed color, and the proportions of the total intensity represented by the white and the chromatic proportions. In Grassmann's Fig. 2, for example, consider a particular mixture of *a* units of color *A* and *b* units of color *B*. The hue of the mixture is represented by the color-point *C*, their center of gravity. $(a+b)OC$ represents the chromatic intensity of the mixed color, $(a+b)CD$ the "intensity of the intermixed white," and $(a+b)OD$ (where $OD = 1$) the total intensity of the mixture. Grassmann concluded that from his four laws, each firmly grounded in experience, he had deduced results which agreed overall with Newton's "empirical rule," and that he had done so "in a purely mathematical way" (Grassmann 1853, 82-84).

GRASSMANN AND THE SUBSEQUENT TRADITION OF COLOR SCIENCE

Grassmann's analysis powerfully influenced all subsequent studies of color. That influence can be established most directly for Helmholtz. He apparently took no offense at Grassmann's criticism of his work, for he saw that Grassmann's approach vastly strengthened his own programme to defend Newton's theory of color. He immediately adopted Grassmann's continuity-interpretation of the mix procedure, and with the same disdain for niceties of historical accuracy that Grassmann had displayed, he happily labeled it "Newtonian" and tacitly ignored the elements of innovation it contained. In a lecture to the British Association for the Advancement of Science given within months of Grassmann's paper, and again in his next published paper of early 1855, Helmholtz acknowledged that Grassmann had "proved rigorously and in detail" that Helmholtz's former results were inconsistent with the Newtonian law of color mixing (Helmholtz 1855). Helmholtz introduced new experimental results in which, just as Grassmann had predicted, he had been able to observe numerous pairs of spectral complementaries and measure their wavelengths by comparing them to the Fraunhofer lines.

But Helmholtz rejected almost as many of Grassmann's conclusions as he accepted. He insisted that the greens have no spectral complementaries and that their complementaries (the purples) must be interpolated by fiat if the curve of homogeneous colors is to be closed. Much more important, Grassmann had defined a color to possess unit brightness if when mixed with its complementary it yields a unit white. Helmholtz proposed that "equal unit brightness" be defined in a different way, namely as "equally bright to the eye." His newest experiments, however, showed that under this convention **and** the barycentricity rule, the spectral colors cannot be considered as equally saturated and so cannot be located equidistant from the white point on the color plane. Newton's color circle is then no longer a true circle. Its form, however, can be deduced very roughly from the proportions of the components in the complementary mixes. Fig. 3 shows Helmholtz's rough sketch of the new, non-circular spectral locus, with the purples interpolated as a straight line.

Grassmann's fourth law postulated that the brightness of a light mixture is the sum of the brightnesses of its components. Helmholtz urged that this law be treated as a convention of measurement, a definition of total brightness, and he demonstrated the far-reaching implications of this convention. If we postulate positions on the plane of Newton's color circle for **any** three, non-collinear points (colors), and if we measure their respective brightnesses in any way we like, then that information, plus the convention of barycentric mixing, allows us to express the position of any color-point on the color plane in terms of the three initial points. That applies to colors which can actually be mixed from the three initial colors (and so lie inside the triangle they form) and those which cannot be so mixed (and hence lie outside the triangle).

Helmholtz did not introduce any techniques in 1855 by which this severe mathematical procedure might be realized in experimental practice; Clerk Maxwell, however, took that step in a revolutionary paper published almost simultaneously with Helmholtz's own. Since the early 1850s the young Maxwell had used the rotating color disk to additively combine three, arbitrarily-chosen primary colors in specified intensity ratios so that they matched some fourth color. In 1855 Maxwell showed that these matches could be expressed as "color equations," which in turn could be manipulated as if "negative intensities" were possible (Maxwell 1855b, cf. 1855a). Maxwell showed how, if his three pigment primaries were represented as any three points on the color plane, the color equations derived from them allowed any other color to be mathematically located on the plane in relationship to them. This also applied to the homogeneous or spectral colors, so that, as Helmholtz also had shown, the exact locus of these colors on the color plane became an empirical problem, susceptible in principle to experimental determination.

More striking still, Maxwell boldly reasserted the truth of Thomas Young's three-receptor theory as the physiological basis of human color vision, despite the fact that Helmholtz had considered and rejected Young's theory in 1852. He did so by simply abandoning Helmholtz's tacit assumption of 1852, that the color sensations corresponding to Young's primaries must be spectral colors, i.e. must lie on the spectral locus. Instead, Maxwell imagined the three color points corresponding to

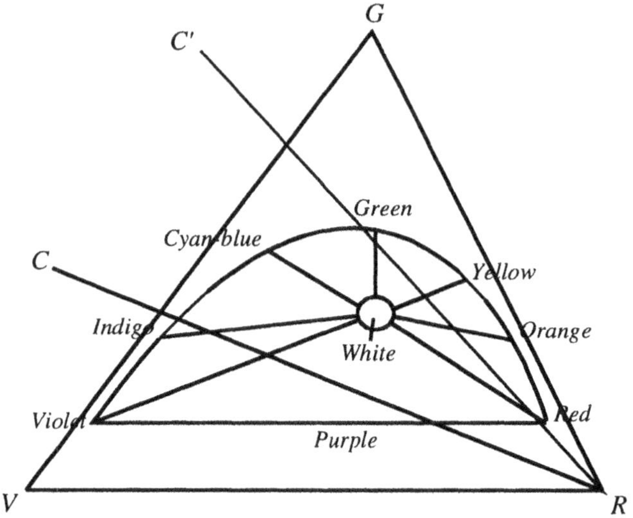

Fig. 3. Helmholtz's spectral locus of 1855, located inside a hypothetical triangle of fundamental sensations. *Source*: Turner 1994, p. 292.

Young's fundamental color sensations as defining a triangle on the color plane large enough to wholly enclose the curve of homogeneous, spectral colors. Fig. 3 represents a heuristic sketch of one of the infinite set of such hypothetical primaries (the points V, G, and R). Any such set allows all the spectral colors to be represented as desaturated, barycentric mixtures of Young's fundamental sensations. Phenomenologically, however, that assumption was a radical one; it implies that the fundamental sensations themselves are far purer, far more saturated colors than those evoked by spectral lights, or indeed any colors ever seen in nature (Maxwell 1855b, 136, 150-52). That radical assumption had apparently not occurred to Helmholtz in 1855, but he publicly adopted Young's theory in Maxwell's amended formulation in 1858.

Did Maxwell's work owe anything to Grassmann, or, for that matter, to Helmholtz? There is no evidence as to when Maxwell first wrote color equations or began to think about geometrical representations in color space, but that development probably came late and may have been influenced by his German colleagues. Maxwell read Grassmann's paper sometimes in 1854, after it was translated in the April number of the *Philosophical Magazine*. He subsequently credited Grassmann for the insight that color sensations are wholly determined by three variables, the insight which makes possible their geometrical representation in three-space. Maxwell apparently read Helmholtz's treatise of 1855 in the original, just prior to bringing out his own article; he acknowledged that paper in 1855, but expressed no debt to it (Maxwell 1855a, 124-25; 1855b, 141-42, 152-3).

Was Maxwell led to modify and adopt Young's theory on the basis of hints he found in the work of Grassmann (or Helmholtz)? Again, the timing suggests the possibility, for Maxwell apparently came to Young's theory late. In March, 1854 Maxwell still regarded red, yellow and violet as the primary colors of vision, a choice which suggests that he still had not yet adopted Young's theory, although he almost certainly had abandoned Brewster's, which he had held in 1849 (Maxwell 1990, 112, 264). Almost certainly Maxwell's readiness to identify Young's primaries with extra-spectral color sensations as Young's primaries was facilitated at least indirectly by the abstract, mathematical analyses of color relations in color space which he found in Grassmann and Helmholtz. Ultimately, however, the lines of mutual indebtedness are impossible to unravel.

Much larger and more definite claims can be made for Grassmann's long-term influence on color science. Grassmann's paper essentially ended the European obsession with the number and nature of the "primary" colors in nature. Under the barycentric mix-law with Grassmann's continuity-assumption, the spectrum no longer possessed "privileged" colors, i.e. colors which possessed special properties that made them alone suitable as primaries. To be sure, theories like Young's would continue to postulate "physiological primaries" identified with the action of particular retinal receptors, but these were no longer endowed with particular mathematical or physical significance for colorimetry. Grassmann himself had been deeply concerned with spectral primaries at the time of his October lecture; the change in his own thinking during the following months thus prefigured the transition through which the science of color itself was soon to pass.

Grassmann's paper signalled a turning point for color science in one other important respect. The transformation of color research which occurred in the 1850's and '60s resulted in large part from the cooption of that fragmented field by physicists, who successfully imposed their quantitative and mathematical methods upon many if not all aspects of the study of vision. That physicalist tradition, with its emphasis on mathematical modeling of visual capacities, persists deep into the twentieth century as perhaps the dominant theme in the study of color and color vision (Turner 1994, 237-39, 266). That tradition presupposed the tri-variability of vision and the rigorous barycentric relationships that were postulated first and most definitively by Grassmann in 1853. In this respect, the stamp of Grassmann's thought is probably more definitive even that of Helmholtz, which it influenced.

Not all subsequent pioneers of vision research were happy about what had sprung from Grassmann's work. Researchers who advocated more phenomenological and psychological approaches to vision research charged that the physicists' methodological need for simple, easily-modeled correlations between the properties of the physical stimulus and the sensory response they elicit had led to naive assumptions that perverted every real understanding of vision. These they traced squarely to Grassmann: that a definite wavelength always excites the same hue; that the brightness of a sensation is proportional to the physical intensity of the light which excites it; that saturation and hue do not vary with brightness. Ewald Hering, the leading figure in this tradition, denied in 1887 that Grassmann's laws as he had stated them suffice to deduce the barycentric property; the additive constancy postulated by the third law required further assumptions of multiplicative constancy (Hering 1887; cf. Grassmann 1893-1911, 3.1:251-55). Intense controversy over the third law and the limits of its validity arose in the later 1880s, and the issue was temporarily settled only when Johannes von Kries announced the duplicity theory of vision in 1894 (Turner 1994, 196-217).

GRASSMANN AS AN INNOVATOR

In conclusion, let us return to Grassmann in 1853 and ask how he was able to move away from his traditional concerns of 1852, to impose the essential continuity condition upon Newton's mix procedure, and so point others in the direction of modern colorimetry. The answer has everything to do with Grassmann's status as an outsider to the contemporary traditions of color research. Color theory per se was always a means to an end for Grassmann and never his central concern. The barycentricity of the Newtonian color plane provided a new avenue by which he could bring the elementary techniques of the *Ausdehnungslehre* before the public and save them from the oblivion into which they were threatening to fall. In a letter to Möbius of 22 May, 1853, he wrote:

"I have recently discovered an interesting application for the barycentric calculus in the field of optics and have submitted an article about it to Poggendorff's *Annalen*. In it I mathematically deduce from a few postulates a rule for color mixing which Newton introduced in his *Opticks* as an empirical rule adequately in agreement with experience.... In this way more and more points of connection between geometrical analysis and the laws of nature gradually present themselves, just as is to be expected, if this analysis conforms to nature" (Grassmann 1894-1911, 3.2:162).

In short, Grassmann came to the study of color with the search for "points of connection" between the laws of nature and the forms of geometrical analysis which he had developed foremost in his mind. That allowed him to read Helmholtz's paper of 1852, and Newton's of 1704, in ways that no one else had done, and to pose entirely new questions and new analytic approaches. But it also isolated him from the on-going tradition of color research. Much later, in 1877, when Grassmann made his only other foray into the field of color research, he was still primarily concerned with showing how the forms of geometrical analysis could be applied to relationships in color space (Grassmann 1891-1911, 2.1:213-21). His general understanding of color vision had scarcely changed since 1853, and by 1877 that rendered him a marginal figure at best in the on-going field of color research.

Department of History, University of New Brunswick, Fredericton

REFERENCES

Jean-Baptiste Biot, *Traité de physique expérimentale et mathématique*, 4 vols. (Paris: Deterville, 1816).

Heinrich Wilhelm Dove, *Darstellung der Farbenlehre und optische Studien* (Berlin: G. W. F. Müller, 1853).

Ernst Gottfried Fischer, *Lehrbuch der mechanischen Naturlehre*, 2 vols. (Berlin and Leipzig: Nauck's Buchhandlung, 1819).

James David Forbes, "Hints towards a Classification of Colours", *The London, Edinburgh and Dublin Philosophical Magazine*, 1849, 3rd ser. *34*:161-78.

Hermann G. Grassmann, "Zur Theorie der Farbenmischung", In Grassmann, 2.1:161-173. (Original: *[Poggendorffs] Annalen der Physik und Chemie*, 1853, *89*:69-84.)

Hermann G. Grassmann, 1894-1911. *Hermann Grassmanns gesammelte mathematische und physikalische Werke*, ed. Friedrich Engel. 3 vols. in 6. (Leipzig: B. G. Teubner, 1894-1911).

David Hargreave, *Thomas Young's Theory of Color Vision: Its Roots, Development, and Acceptance by the British Scientific Community.* Ph.D. Diss., University of Wisconsin 1973.

Hermann von Helmholtz, "Ueber Herrn D. Brewster's neue Analyse des Sonnenlichts." In Helmholtz 1892-95, 2:24-44. (Original: [*Poggendorffs*] *Annalen der Physik und Chemie*, 1852, *86*: 501-23.) [1852a].

Hermann von Helmholtz, "Ueber die Theorie der zusammengesetzten Farben." In Helmholtz 1892-95, 2:3-23. (Original: [*Poggendorffs*] *Annalen der Physik und Chemie*, 1852, *87*:45-66.) [1852b].

Hermann von Helmholtz, "Ueber die Zusammensetzung von Spectralfarben." In Helmholtz 1892-95, 2:45-70. (Original: [*Poggendorffs*] *Annalen der Physik und Chemie*, 1855, *94*:1-28.) [1855].

Hermann von Helmholtz, *Wissenschaftliche Abhandlungen von Hermann von Helmholtz*, 3 vols. (Leipzig: J. A. Barth,1892-95).

Ewald Hering, "Ueber Newton's Gesetz der Farbenmischung", *Lotos,* 1887, *7*:177-268.

Richard L. Kremer, "Innovation through Synthesis. Helmholtz and Color Research", *Hermann von Helmholtz and the Foundations of Nineteenth-Century Science*, ed. David Cahan (Berkeley: University of California Press 1993) 205-258.

James Clerk Maxwell, "On the Theory of Colours in Relation to Colour-Blindness", In Maxwell 1890, 1:119-25. (Original: *Transactions of the Royal Scottish Society of Arts*, 1856, *4.3*:420-28.) [1855a].

James Clerk Maxwell, "Experiments on Colour, as Perceived by the Eye." In Maxwell 1890, 1:126-54. (Original: *Transactions of the Royal Society of Edinburgh* 1855, *21*:275-98.) [1855b]

James Clerk Maxwell, *The Collected Papers of James Clerk Maxwell*, ed. W. D. Niven. 2 Vols. (Cambridge: Cambridge University Press, 1890). Reprint. New York: Dover, 1965.

James Clerk Maxwell, *The Scientific Letters and Papers of James Clerk Maxwell. Vol. 1: 1846-1862*, ed. P. M. Harman (Cambridge: Cambridge University Press, 1990).

Isaac Newton, *Opticks, or a Treatise on the Reflections, Refractions, Inflections & Colours of Light*, 4th ed. (London 1730). Reprint. New York: Dover, 1952.

Claude Servais Mathias Pouillet, *Elémens de physique expérimentale et de météorologie*, 2 vols. (Paris 1828/29).

Alan E. Shapiro, "The Evolving Structure of Newton's Theory of White Light and Color", *Isis* 1980, *71*:211-235.

Paul D. Sherman, *Colour Vision in the Nineteenth Century: The Young-Helmholtz-Maxwell Theory* (Bristol: Adam Hilger, 1981).

R. Steven Turner, *In the Eye's Mind. Vision and the Helmholtz-Hering Controversy* (Princeton: Princeton University Press, 1994).

QVOD DEVS OPTIMVS MAXIMVS FELIX FAVSTVMQVE ESSE IVBEAT

CLEMENTISSIME INDVLGENTE

AVGVSTISSIMO ET POTENTISSIMO

DOMINO

CAROLO

REGE WÜRTTEMBERGIÆ

RECTORE VNIVERSITATIS MAGNIFICO

OSCARO SCHÜPPEL

REGII ORDINIS CORONÆ WÜRTTEMBERGICÆ EQVITE SECVNDÆ CLASSIS

MEDICINÆ ET CHIRVRGIÆ DOCTORE PATHOLOGIÆ GENERALIS ET ANATOMIÆ PATHOLOGICÆ PROFESSORE PVBLICO ORDINARIO

VENIAM LARGIENTE EA QVA POLLET POTESTATE A REGE EI CONCESSA

CANCELLARIO VNIVERSITATIS MAGNIFICO

HENRICO GVSTAVO DE RÜMELIN

REGIORVM ORDINVM FRIDERICIANI COMMENDATORE PRIMÆ CLASSIS CORONÆ WÜRTTEMBERGICÆ COMMENDATORE

PHILOSOPHIÆ ET ŒCONOMIÆ PVBLICÆ DOCTORE REGI A CONSILIIS PVBLICIS COLLEGII STATISTICI PRÆSIDE HONORARIO

FACVLTATIS PHILOSOPHICÆ DECANO SPECTATISSIMO

VILHELMO SIGISMVNDO DE TEVFFEL

REGII ORDINIS CORONÆ WÜRTTEMBERGICÆ EQVITE PRIMÆ CLASSIS

PHILOSOPHIÆ DOCTORE LITTERARVM ANTIQVARVM PROFESSORE PVBLICO ORDINARIO

PROMOTORE AD HVNC ACTVM LEGITIME CONSTITVTO

RVDOLFO DE ROTH

ORDINVM CORONÆ WÜRTTEMBERGICÆ EQVITE PRIMÆ CLASSIS SANCTI STANISLAI PRIMÆ CLASSIS EQVITE

PHILOSOPHIÆ DOCTORE LITTERARVM ORIENTALIVM PROFESSORE PVBLICO ORDINARIO

AMPLISSIMVS PHILOSOPHORVM ORDO

IN REGIA VNIVERSITATE EBERHARDO-CAROLINA TVBINGENSI

VIRVM PRÆNOBILISSIMVM ATQVE DOCTISSIMVM

DOMINVM

HERMANNVM GRASSMANN

MARIANI GYMNASII STETTINENSIS PROFESSOREM

QVI RARO HAC ÆTATE EXEMPLO MATHEMATICÆ PERITIAM CONIVNXIT CVM SCIENTIA RERVM PHILOLOGICARVM

ET IN VTROQVE STVDIORVM GENERE SCRIPTOR EXTITIT CLARISSIMVS

MAXIME VERO ACVTISSIMA VEDICORVM CARMINVM INTERPRETATIONE NOMEN SVVM REDDIDIT ILLVSTRISSIMVM

HONORIS CAVSA

DOCTOREM PHILOSOPHIÆ ET ARTIVM LIBERALIVM MAGISTRVM

RITE CREAT

HOC IPSO DIPLOMATE SOLLEMNITER RENVNTIAT

EIQVE OMNIA IVRA ET PRIVILEGIA CVM DOCTORIS GRADV CONIVNCTA CONFERT

TVBINGÆ

SVB MAIORE FACVLTATIS PHILOSOPHICÆ SIGILLO

DIE XVIII MENSIS IVLII ANNI MDCCCLXXVI

Illustration 8: The *Ehrenpromotion* diploma of Tübingen University (1876)
The text of the Laudatio is repeated on page 351.

ERIKA HÜLTENSCHMIDT

HERMANN GRASSMANN'S CONTRIBUTION TO THE CONSTRUCTION OF A GERMAN "KULTURNATION". SCIENTIFIC SCHOOL GRAMMAR BETWEEN LATIN TRADITION AND FRENCH CONCEPTIONS

This paper[1] describes the content and context of Hermann Grassmann's *Grundriss der deutschen Sprachlehre*. It appeared in 1842 as a school program for the *Ottoschule* in Stettin where Grassmann had been a teacher since 1836. At a time when the structure of schools was still fluctuating and there were local variations among schools,[2] the Ottoschule held a position between that of a "Bürgerschule" und "höherer Bürgerschule":

"It sought to provide a somewhat better education for its students than the *Bürgerschule* but without fully attaining the goals of a *höherer Bürgerschule* as determined by the ministerial regulation of 8 March 1832. In order to give the boys something of an academic advantage over the girls, while there were, in addition to three primary or children's grades, two higher girls' grades, there were three higher boys' grades".[3]

Two of the three "higher boys' grades" offered *privatim* Latin and French, that is additional lessons in these subjects, for those desiring them. Girls could study French. H. Grassmann, as an *Ordinarius* or grade teacher, taught religion, German, mathematics, practical arithmetic, physics, and Latin.[4] In addition he was a private teacher (*Privatlehrer*) for the optional Latin and French.

It is not possible to determine from this information if the *Grundriss* was intended for the hands of the teachers or those of the pupils,[5] or even if it had any directly pedagogical goal. The publication of a school tract served as a way for such "literary" writings of a teacher to enhance

---

1 I am indebted to Albert C. Lewis, R. Steven Turner and Günter Seib for preparing the English version of my paper.

2 For an overview of this see Lundgreen 1980.

3 Engel 1911, S. 58.

4 Latin was evidently taught beginning with the first grade, but only to boys.

5 This is less likely since the text contains hardly any examples.

G. Schubring (ed.), Hermann Günther Graßmann (1809–1877): Visionary Mathematician, Scientist and Neohumanist Scholar, 87–113.

the reputation of the school and would demonstrate the teacher's own productivity independently of the classroom. What is certain is that H. Grassmann pursued the general goal of a scientifically-oriented method of teaching language and, as will be shown, he did this in a neohumanistic context. In any case the *Grundriss* is to be seen as providing an orientation. We can further say that it not only contains a systematic foundation for the textbooks Grassmann later on composed for his language teaching but also clear traces of it can be found again in his later writings on comparative philology. For example, it can be seen in his Indo-Germanic studies such as the "Grassmann'sches Aspiratengesetz" (1863) which even today is counted among the most important achievements in the field.

By his publication of the *Grundriss* in 1842, H. Grassmann was building upon his earlier work, *Lehre vom Satze* (1831),[6] which nevertheless remained unpublished. The *Grundriss* in turn was the starting point for a number of texts relating to the teaching of language: the *Leitfaden*, for the teaching of German, that appeared in three revised editions (done together with his brother Robert Grassmann), and the *Deutsches Lesebuch*, in eight editions (together with Wilhelm Langbein[7]). Furthermore, he revised a *Leitfaden* for instruction in beginning Latin (two editions). It is also noteworthy that in 1870 he produced the voluminous work, *Deutsche Pflanzennamen*, in which he

---

6 Hermann Grassmann's writings on language teaching are, according to (Engel 1911): 1. *Die Lehre vom Satze.* (1831). 2. *Grundriss der deutschen Sprachlehre.* Programm der Otto-Schule Stettin. (Stettin 1842. 54 pp.). 3. *Leitfaden für den ersten Unterricht in der lateinischen Sprache.* (Stettin 1843). (2nd ed. Stettin 1846. 52 pp.). 4. *Leitfaden für den ersten Unterricht in der deutschen Sprache.* (together with Robert Grassmann). (Stettin 1843). (2nd ed. Stettin 1848. 40 pp.). 5. *Leitfaden der deutschen Sprache, mit zahlreichen Übungen versehen.* (together with Robert Grassmann). (Stettin 1852. 78 pp.). 6. *Deutsches Lesebuch für Schüler von 8-12 Jahren.* (zusammen mit W. Langbein). (Berlin 1846, 394 pp.). (2nd ed. Stettin 1852. 422 pp. 3rd ed. 1855. 4th ed. Stettin 1857. 5th ed. Berlin 1861. 6th ed. Berlin 1868. 7th ed. Stettin 1877. 8th ed. Stettin 1879. 420 pp. Under the title: *Deutsches Lesebuch für Schüler von 8 bis 12 Jahren*). 7. *Deutsche Pflanzennamen.* (Stettin 1870. 288 pp.). In addition, Hermann Fritsche, Director of the Friedrich-Wilhelms-Schule, in his school history for 1840-1890, that appeared in the school Festschrift of 1890, mentions two other textbooks, done with W. Langbein: *Leitfaden für den deutschen Unterricht* und *Lateinisches Lesebuch.* In spite of the most intensive investigations, copies of these books are extremely rare: if any, there exist only some isolated copies, even for the reedited books. The least preserved are those on grammar. We could locate to date as the only items: 2 and 6 (1879 ed.), in the Szczecin municipal library, 3 in the Bibliothek für bildungsgeschichtliche Forschung Berlin, and 7 in the Greifswald University Library. Item 1 is probably an unpublished manuscript.

7 W. Langbein, 1815 - 1889, also a teacher at the Friedrich-Wilhelms-(Real-)Schule and important Realschule politician as well as editor of two journals on Realschule matters.

put forward a German-language alternative to the Linnaean nomenclature for those plants belonging to the vegetation in Germany - for the benefit of those not versed in Latin (both students and the general public). Measured on the basis of the number of editions they went through, his texts in the field of German-language teaching were Grassmann's most wide-spread writings.

There seems to be no explicit reason evident in Grassmann's writings for his composition of the *Grundriss einer deutschen Sprachlehre* while he was active at a *Bürgerschule*.[8] His pedagogical observations are confined to mathematics education: in the Foreword to his *Lehrbuch der Arithmetik* of 1861 he insists that the textbook and the methods it follows be "rigorously scientific"[9] (i.e. systematic and consistent, each step following from another) and that learners should be able to understand the formulaic development as based on the conceptual development. Their minds should be made capable "of an overview of a scientific system."[10] In order to achieve this goal, "an easily comprehensible and rigorous systematization" must be recognizable in teaching, "which grows organically out of the nature of things and is not tailored according to some extraneous stereotype."[11] Systematic and rigorous, understanding of connections as well as reference to the concrete object, these are the key points of this exposition: they signify a conception of education oriented to the sciences as they were understood at the time. In its fundamental ideas this textbook in arithmetic arose from joint work with Robert Grassmann.

Pedagogical discussions applying especially to German-language teaching occur in Robert Grassmann's writings with whom he had a lifelong close co-operation.[12] Robert explicitly addressed questions of German instruction, even if not until 1875 and 1890.[13] These writings

---

8 Around 1842 the teaching of German was still a relatively new phenomenon and it was not yet a professionalized teaching specialty (cf. C. Bartels 1989). There were also no generally recommended and widespread textbooks that transcended particular regions. It was usual for a teacher to put out his own writings and their distribution was very limited.

9 H. Grassmann 1861, p. 295.

10 loc. cit., p. 298.

11 loc. cit., p. 295.

12 On the co-operative work of the two brothers, see Schubring's contribution in this volume.

13 In the *Sprachlehre* (third book of the *Denklehre*, which latter is the first part of the *Wissenschaftslehre*), 1875b, pp. 75-88. This *Sprachlehre* contains only general reflections. A grammar was advertised as a separate monograph and in effect appeared in 1890 as part of *Gebäude des Wissens* (1890c). In this the Foreword, pp. iii-ix, can be cited. Reflections on German as a scientific language can be found in the fourth book of *Denklehre*, entitled *Das Formdenken*. (1875c, pp. 152-154).

imply that from the beginning both brothers acquired their fundamental conceptions on teaching German and on the construction of a grammar as a common undertaking.

During the composition of the *Grundriss* both brothers were active at Stettin schools. As a part of their teaching they studied together the leading German grammars of their time, namely those by K. F. Becker,[14] S.H.A. Herling[15] and Jacob Grimm.[16] "The fruit of this common work is written up by H. Grassmann in the program of the *Ottoschule* of 1842."[17] Together they produced the German-language textbook and also the short Latin grammar.[18] In 1890, after the death of his brother in 1877, R. Grassmann published a German grammar as part of his *Gebäude des Wissens* that was much more explicit and comprehensive than either H. Grassmann's small grammar of 1842 or the subsequent joint writings, though largely agreeing with these in its systematic construction.[19] The multi-volume, though still incomplete, *Gebäude des Wissens* was an encyclopedia of the sciences. Robert Grassmann's 1900 German grammar was a monograph edition of the same text as the grammar of

---

14 Appearing through 1841: Konrad Ferdinand Becker, *Die deutsche Wortbildung oder Die organische Entwicklung der deutschen Sprache in der Ableitung.* (Frankfurt a.M. 1824). *Organism der Sprache. Als Einleitung zur deutschen Grammatik. Deutsche Sprachlehre* Vol. 1. (Frankfurt a.M. 1827). *Deutsche Grammatik. Deutsche Sprachlehre.* Vol. 2. (Frankfurt a.M. 1829). *Schulgrammatik der deutschen Sprache.* (Frankfurt a. M.1831). *Leitfaden für den ersten Unterricht in der deutschen Sprachlehre.* (Frankfurt a.M. 1833). *Über die Methode des Unterrichts in der deutschen Sprache, als Einleitung zu dem Leitfaden für den ersten Unterricht in der deutschen Sprachlehre.* (Frankfurt a.M. 1833). *Ausführliche deutsche Grammatik als Kommentar der Schulgrammatik. Statt einer zweiten Auflage der deutschen Grammatik.* Abteilung 1 - 3. (Frankfurt a.M. 1836-39). *Organism der Sprache.* 2nd revised ed. (Frankfurt a.M. 1841). For a complete bibliography see Vesper, 1980, p. 190.

15 S.H.A. Herling, *Grundregeln des deutschen Stils oder der Periodenbau der deutschen Sprache. Ein Lehrbuch für den stilistischen Unterricht.* (Frankfurt a.M. 1823). *Die Syntax der deutschen Sprache.* Part 2. (Frankfurt a.M. 1827) (= 2nd ed. of the text of 1823). *Erster Cursus eines wissenschaftlichen Unterrichts in deutscher Sprache.* (Frankfurt a. M. 1828). *Die Syntax der deutschen Sprache.* 1st part: *Syntax des einfachen Satzes.* (Frankfurt a.M. 1830). *Die Syntax der deutschen Sprache.* 2nd part: (Frankfurt a.M. 1832) (= 3rd ed. of the text of 1823).

16 Jacob Grimm, *Deutsche Grammatik.* (Göttingen: Vol. 1: 1819, Vol. 2: 1826, Vol. 3: 1831, Vol. 4: 1837).

17 R. Grassmann, 1890, p. IX.

18 Cf. footnote 6.

19 Cf. the bibliography at the end of this paper.

1890. Robert also planned an historical German grammar but this was never realized.[20]

The close co-operation between the two brothers allows us to draw upon Robert Grassmann's 1875 work in *Wissenschaftslehre* and his 1890 work in *Gebäude des Wissens* in looking at the pedagogical foundation of Hermann Grassmann's *Grundriss*. To be sure these works were published nearly 30 to 50 years after the *Grundriss* work, but nevertheless, with a certain proviso, they can still be utilized since, as it seems, they have changed as little with respect to fundamentals as the common systematic foundation of the brothers' philological works. For both brothers the time around 1840 marked a period of development of their interests. Moreover, one has to take into account that Robert may well have been more radical in a number of his views than Hermann.[21]

THE VISION OF AN ENLIGHTENED PEOPLE

In 1875 Robert Grassmann espoused a vehement Prussian-German national ideology that combined elements of neohumanism, idealism, the Lutheran Reformation, and the anti-French mythology of 1813; it must be remembered that his work appeared shortly after the 1871 proclamation in Paris of the German Empire under Prussian leadership. Yet it is legitimate to maintain that ideas of this kind had also motivated the work of the Grassmann brothers during the decade of the '40s; all the elements of this ideology were present and complete by 1840. Both brothers were convinced proponents of the "German movement," i.e. were so-called "Deutschtümler."

When Hermann Grassmann's *Grundriss* appeared in 1842, German instruction in the schools was still relatively new as a social institution. German-language teaching was part of the national awakening that had been proceeding since the beginning of the century, just as the institutionalization of French instruction in France after 1789 had been part of the nation-building process there.[22] Instruction in the national language -- naturally in standardized and literary form -- was an essential part of the political and cultural modernization process of the nineteenth century. But it was never obvious that German instruction in a

---

20 R. Grassmann 1890b, p. 3. H. Grassmann likewise planned a never-realized Sanskrit grammar which was to have supplemented his "Wörterbuch des Rigveda." Engel 1911, p. 306.

21 Robert grew up with his uncle, the Stettin school superintendent Friedrich Heinrich Grassmann, who himself was the author of grammars.

22 Not all Germanists favored the institutionalization of German instruction. Especially well known is the opposition of Jacob Grimm, who identified language instruction with its normative tradition and never attempted to advance innovative ideas in this field.

Bürgerschule was to be "wissenschaftlich" or "systematic" in approach. Yet the *Grundriss* followed the same systematic conception of instruction that had been laid down in Hermann Grassmann's Introduction to his *Arithmetic*. The grammars that the brothers had studied during their preparation for the work, namely those by K. F. Becker and S.H.A. Herling, were grammars in which the language was often represented as a systematic whole, as an "organism." In the final analysis these texts approached the language they laid out for instruction as an object in its own right. They were not, as for example Diesterweg's *Praktischer Lehrgang* had been,[23] oriented to the social milieu of the people and to the developmental stage of the students. Diesterweg considered the works of Becker and Herling as suitable for instruction in elite, academic schools, but not for the "education of the people." The Grassmann brothers, who drew on systematic or "wissenschaftliche" grammars, shared no such dualistic view; they envisioned the *Grundriss* as suited to instruction in "Bürgerschulen," even though these did not qualify as an academic or "höhere" school.

In accordance with the scheme developed in Robert Grassmann's encyclopedic *Wissenschaftslehre* (1875), linguistics or *Sprachlehre* forms part of the "theory of thought" (*Denklehre)*:[24] language is the communal form of thought - it is the common medium of thought. As a form or sign it is an externality. Perception and imagery, on the other hand (the investigation of which in the *Denklehre* precedes that of *Sprachlehre)* constitute an internality. An act of spiritual freedom is required to transcend these internalities, limited as they are to the individual, and to transform them into the externalities of supra-individual, communal thought. To do so the individual must command the intellectual and spiritual freedom to intuit and appropriate the unique and particular thoughts of other individuals, that is, he must carry out the hermeneutic act. From the standpoint of linguistic philosophy, the capacity to carry out this act is what distinguishes humans from animals. These ideas derive from the theologian and philosopher Schleiermacher, with whom H. Grassmann studied in Berlin from 1827-30. In his linguistic theory, Schleiermacher accepts the proposition of Plato in the *Theaetetus* that thought is interior speech. It follows that speech is

---

23 F.A.W. Diesterweg 1828-1830. Cf. Erlinger/Knobloch 1991b, p. 62.

24 "The theory of thought treats those representations given through the senses and their articulation through speech and the theory of forms, i.e. mathematics." R. Grassmann 1875a, p.15.

externalized thought.[25] R. Grassmann made these ideas the foundation of both his linguistic theory and his conception of the ultimate aim of education in the schools. Linguistic exchanges of one people with another also fall into the category of inter-individual differences: the other language forms have to be learned and other thoughts have to be translated into one's own, so that the process of translation must, too, be understood as a hermeneutic act. A complete commonality of understanding (communication based upon identical language and ideas) had its existence just only in the distant realm of origins: the historical comparison of different languages ought therefore to lead to the discovery of common genealogical roots. On the basis of this philosophical conception, the theory of language (meaning the science of linguistics) can be divided into four branches: 1. the theory of sounds (phonetics), 2. the theory of syntax, 3. a derivative branch which Grassmann calls the "theory of ideas" or "Begriffslehre,"[26] and 4. an historical-comparative branch. Specifically the last refers to historical-comparative linguistics, in particular the study of Indogermanic. The first three branches are realized in the writings of the Grassmann brothers on the German language. The object of H. Grassmann's *Grundriss* and R. Grassmann's *Sprachlehre* is simultaneously syntax and "Begriffslehre," conceived as dual aspects of the platonic unity of discourse, existing first as a formal linguistic act and second as an act of thought. R. Grassmann's *Sprachlehre* also contains a substantial chapter on phonetics, and in his later studies of Indogermanic, Hermann Grassmann

---

25 On Schleiermacher cf. André Laks and Ada Neschke (Eds.), 1990. One can trace back this reference to Plato to the ideas of Herder and find them again in W. v. Humboldt.

In the philosophy of Schleiermacher, "Verstehen" (understanding) denotes the interpretation of the meaning of the subjective and the particular. On the other hand, "Erklären" (explanation) refers to general and objective relationships. The concepts stand to one another as freedom to necessity. Schleiermacher's philosophy legitimized the historical sciences as sciences of the Subject, as the subject produces meaning, and thereby as sciences of the individual. In this intellectual framework, philosophical linguistics is on the one hand a science of objective and collective speech, or more precisely, of collective discourse in the sense of discursive thought, by which the linguistic community produces speech. On the other hand, it is equally a science of speech in its subjective and particular sense, in which speech acts to produce individual discourse. In both cases it is Intellect or Spirit ("Geist") which is the creative agent; "spirit" functions as a genetic conception in this intellectual tradition. Cf. Schleiermacher 1835 (the printed text of his 1829 lecture on hermeneutics before the Berlin Academy) and Schleiermacher 1838.

26 The term of *Begriffslehre* is linked to the language of German philosophy, as created by Christian Wolff. Beginning at least with Descartes, the French equivalent is *idée*, not something like "concept," dating in this time from scholastic philosophy. The English translation has also to be *idea*, since Locke and the Locke-Tradition use this term.

developed the fourth element in this four-part typology. Nevertheless, R. Grassmann's project to develop an historical comprehensive grammar was never realized, and Hermann Grassmann restricted himself to the limited field of Indogermanic. During the first third of the nineteenth century, figures like F. Schleiermacher, W. von Humboldt, and August Boeckh had promulgated an idealist, neohumanist conception of science, in terms of which *Wissenschaft* would constitute an encyclopedia of the human spirit. Ultimately that conception proved unrealizable, and it was further undermined by the growing specialization of scientific practice.

R. Grassmann wrote further that in the school curriculum language instruction always took the central place, because language has been understood as the best and easiest means of communicating thoughts. And yet the goal of "authentic intellectual formation" ("wahrhafte Geistesbildung") has not been reached. The reason, according to Grassmann, lies in the difference between what is one's own ("the Proper," the "Eigenes") and what is foreign, alien to one's self. Following the interpretation of Plato, intellectual formation presupposes linguistic formation. It follows that so long as a "foreign language," i.e. Latin, remains the focus of instruction, then one's own thought cannot develop. The natural medium in which immediate, nascent, individual thought realizes itself is one's own language, one's "mother tongue." To that language, therefore, should be accorded the first place in education.[27]

The Grassmann brothers mobilized the hermeneutic relationship between the Proper and the Foreign, in the defense of the Proper against the Foreign. Their determination to do so can be traced to the contemporary historical context, and especially their enthusiastic participation in the "German Movement" of the 1840s. For Robert Grassmann, it was the Greeks who represented the most renowned model of a cultivated and original people, one fully individualized and wholly self-authentic.[28] The Greeks were "the most cultivated people on earth," the people who had achieved "the greatest things in the realm of the spirit." For Grassmann the Greek xenophobia that held other people for "barbarians" and could grasp the Foreign only in terms of the Proper became the marvellous precondition for national greatness:

---

27 "There is only one language that can truly form the spirit; that is the mother tongue. We live, talk, think in our mother tongue; the whole range of representations in our head is based on the mother tongue. The mother tongue must therefore be the starting point, the means, and the goal of all language instruction. Only if a child attains fluency in the mother tongue, can he also attain fluency in thinking. Talking in a foreign language does much more to corrupt thought than it helps and promotes." R. Grassmann 1875b, p. 77.

28 See also pp. 78, 81f and R. Grassmann 1890c, p. IIIf.

"The people of Greece ... learned no foreign language at all, but with the soundest instincts turned all their cultural power upon the formation of its own language and its own spirit."[29]

Grassmann's "people" is neither brutish, nor rustic, nor unrefined; on the contrary, its organic constituents are its philosophers and scholars, because these employ no "foreign language" (in Grassmann's time Latin or French), but rather use the language of the original community. According to Grassmann, even the technical terms of Greek philosophy, as for instance the terminology of Aristotle, was drawn from these common roots. His vision of "an enlightened people," (*gebildetes Volk*) toward the formation of which public education is "one of the most important means,"[30] rested on a romantic-idealistic, holistic or organic conception of the people, not on a sociological one. Grassmann never thought of the "people" as a unity distinct from its elite, but rather as the nation itself. As in the Greek paradigm, so in the German nation the double task falls to philosophers, scholars, and scientists, first to educate the people, and second to bring them to national greatness through outstanding cultural achievement based upon the indigenous national genius.[31] Corresponding to this high goal, the collective formation of the people is to be a scientific one.

The opposition of "the foreign" and "the proper," imitation and autonomy, was central for the modernization of culture and education. In the Germany of the nineteenth century, and especially in Prussian Germany, it had a very specific and uniquely anti-Latin emphasis. Cultural and political modernization represented itself through conceptual polarities that derived from antiquity as well as the modern period. If Germany followed the model of ancient Greece, according to "Deutschtümler" like the Grassmann brothers, then France followed the Roman model. The German nation as a cultural entity was to be formed in opposition to the Latin world. In Germany the modern conception of the nation that came to the fore with the French Revolution was interpreted (or more accurately, assimilated) through the concept of the people as an "organism." In the revolution of 1848 the Grassmann brothers came out with their own newspaper, in the pages of which they

---

29 1875b, p. 77.

30 ibid.

31 The discussion refers to the reception of the thought of Fichte, Humboldt, Schleiermacher during the first third of the nineteenth century. This catalogue of cultural-nationalistic concepts was available long before F. Meinecke's 1908 work, *Weltbürgertum and Nationalstaat* which popularized the self-concept of "Kulturstaat." What for the former individuals was a vision of the future, constituted for Meinecke a retrospective look at a received tradition.

opposed the contemporary revolutionary developments.[32] They abhorred the French Revolutions that had occurred since 1789, especially the revolutionary tendency to interpret national existence as culminating in a democratic or republican electoral system and a rational constitution that suited the framework of modernization. They stood by their king, from whom they expected a "modernization from above." Later Robert Grassmann, who outlived his brother, would become an enthusiastic supporter of Bismarck.

In his *Wissenschaftslehre* Robert Grassmann argued against the use of Latin and French as scientific languages, as well as against the preferential place accorded French in modern language instruction.[33] The national impulse proceeding from the German wars of liberation against Napoleon in 1813 had been (according to Robert G.) the cause of the transition to German as the language of science and scholarship. The events of 1813, therefore, appeared as a precondition for the possibility that the German people as a cultural entity might be formed on the Greek model. On the way to a new national greatness based upon the restoration of self-authenticity, the German people could cling to the memory of their one historical period of freedom from foreign domination, namely the German empire of the twelfth and thirteen century in the age of the Hohenstaufen. This age marked the "flowering of the German minnesingers and of German chivalry," when Germans "first trod the summit of political life among the European states."[34] For this reason instruction in German, as well as the academic study of German linguistics, is properly assigned a definite place in national history.

This radical vision of the organic culture-nation was one of those nineteenth-century utopias built upon the concept of progress. Real developments did not fully conform to it. Robert Grassmann complained in 1875 that the schools had in fact remained bastions of the Foreign, sites of predominantly foreign-language education (Latin and French).[35]

The place which Robert Grassmann accorded to Hermann Grassmann's 1842 *Grundriss* in the history of grammatical studies furnishes a striking example of how the hypothetical unity of Greece and Germany provided a focus for the national ideology and its adherents. In his *Wissenschaftslehre* Robert claimed to have created a new philosophical *Organon*. He wanted to divide his own encyclopedia into

---

[32] The paper bore the title, *Deutsche Wochenschrift für Staat, Kirche und Volksleben*. It was edited by H. and R. Grassmann. Cf. Engel 1911, p. 140.

[33] R. Grassmann 1875b, p.. 78ff

[34] See also p. 79f.

[35] See also p. 81.

four parts, on the model of Aristotle's.[36] All human knowledge was to be regarded as the combination of empirical observation and rational thought. To philosophy is assigned the problem of revealing the common ground of all knowledge and so uniting the many branches of knowledge into a single whole. Therefore philosophy or "Wissenschaftslehre" occupies the first place in the structure of Robert Grassmann's encyclopedia. Because language is the form of thought, then the theory of language (philosophical linguistics) similarly takes a central place in the *Wissenschaftslehre*.[37] With Aristotle, too, began, according to Robert Grassmann, all scientific grammar. The Indian grammarian Pâṇini, who published a grammar at about the same time as Aristotle,[38] also appeared to Grassmann as worthy of mention, but Pâṇini had reached no general or philosophical theory of language but rather had only described a single language, Sanskrit. Then R. Grassmann's historical survey vaulted obliviously over two thousand years to the German grammatical studies of his own time:

Since Aristotle and Pâṇini, "the scientific theory of language or philosophical grammar has remained almost unexplored. Karl Ferdinand Becker ... stands almost alone in this field, and even he based his work more on examples from a single language than on general laws of thought. After Karl Ferdinand Becker, the brothers Hermann and Robert Grassmann together attempted in winter 1841-42 to specify the laws of language, at least in so far as they appear in Greek, Latin, and German....."[39]

As these words show, the *Grundriss einer deutschen Sprachlehre* of 1842 was written to be a philosophical grammar, and to possess a significance that would place it in direct connection to the Greek philosopher.

The Grassmann's evaluation of their own achievement is staggering in its remove from historical reality. They were not aware of, or at least do

---

36 "Following the model of Aristotle, that old master of human knowledge, I divide the structure of human knowledge into four parts: 1) the theory of knowledge, or philosophy... 2) the science of nature, or physics... 3) the science of the state, or politics... (4) the science of the divine, or theology...." See also p. 14.

37 He also divided philosophy into four parts: 1. the *Denklehre* (a philosophy of perceptions and ideas, including language and mathematics), 2. the *Wissenslehre* (natural philosophy), 3. the *Erkenntnislehre* (moral philosophy), 4. the *Weisheitslehre* (theology). Notice that Grammar and mathematics are analogously placed in the *Denklehre*. Philosophy is simultaneously a science of (genetic) causes and also a meta-science.

38 A German edition of the Pâṇini text was published in 1839/40 by Otto Böhtlingk, under the title *Pâṇinis Acht Bücher grammatischer Regeln*. Pâṇini lived at the beginning of the fifth century B.C.; Aristotle was born in 384 B.C.

39 1890c, p. VIIIf. The *Grundriss* contains nothing on Latin or Greek grammar.

not mention, the most important developments in the history of grammatical studies during the seventeenth and eighteenth centuries, namely the general or philosophical grammars that were inspired by the "Port Royal grammar" of 1660, by Antoine Arnauld and Claude Lancelot.

The center of this development lay in France of course; German scholars who drew upon it doubtless appeared to R. Grassmann as slaves to foreign cultural domination. Moreover, the structure of R. Grassmann's *Wissenschaftslehre* of 1875 and his *Gebäude des Wissens* of 1890 expresses in general a philosophical or encyclopedic conception of science,[40] so that he must be considered a late, indeed an outdated, representative of a conception of knowledge characteristic of the first third of the nineteenth century. The two brothers tried to improve upon the work of K. F. Becker, whom Robert Grassmann considered "the modern Aristotle," by achieving greater scientific rigor. For the purpose of contextually embedding the *Grundriss* in the history of grammatical studies, it would doubtlessly be necessary to compare the *Grundriss* and its high claims to the work of K. F. Becker. However, for want of previous research on the Grassmann brothers as authors of German grammatical works, the following discussion can offer only a first step toward this goal: an attempt to locate the *Grundriss* within the more global context of early modern and modern German grammatical studies and their history. The discussion attempts this through an historical contextualizing of the syntax contained in the *Grundriss*. At the same time, the discussion examines how the claims to "rigor" are manifested in the structure of the *Grundriss*. By rigor the brothers understood an economy in the analysis, an economy such that each element of the contents occupies a unique and unequivocal position in an holistic structure that is built on a definite foundation. No element can be presented twice. A comparison with K. F. Becker could be undertaken only within the framework of a comprehensive study and a detailed consideration of the refinements introduced by Robert Grassmann.

THE PLACE OF HERMANN GRASSMANN'S *GRUNDRISS* WITHIN THE HISTORY OF GRAMMAR

Hermann Grassmann's *Grundriss der deutschen Sprachlehre* can be characterized generally as a systematic grammar structured according to a symmetry of oppositions, or, in his own words, to a system of

40 Hültenschmidt 1990.

"contradictions."[41] Besides, he links two perspectives in his systematic which are considered to be incompatible by today's historians of 19th-century German philology: a functional-syntactical and a historico-etymological one (i.e. one pertaining to the history of words). The former is easily linked to his having read K.F. Becker's and S. H. Herling's works, in the latter he takes up the ideas of J. Grimm. Their common basis resides in the concept of *Sprachgesetz* (law of language): syntax is just as systematic or lawful as the historical derivation of the words. According to general opinion, syntax, especially in the form K.F. Becker gave to it, has become the basis of German school grammar until far into the 20th century. The historical grammar has become the scientific grammar of "academic" German philology.[42] Yet, in 1842, when Hermann Grassmann published his contribution, both types of grammar were still held to be "scientific," because they were systematic - at least for the reviewers of the above works in the authoritative Berlin *Jahrbuch für wisssenschaftliche Kritik*.[43] These two, syntax and historical (historico-comparative) grammar were the most significant innovations of research into the German language at the time the *Grundriss* was published.

The *Grundriss* is organized into two main parts placed in symmetrical opposition, one titled "simple forms and their meaning," and the other "the simple ideas of language and their representation in language." In its totality, the *Sprachlehre* thus contains a grammar of forms (*Formenlehre*) and a theory of ideas (*Begriffslehre*).[44] The relationship between these two parts is based in an explicit semiotic presented at the beginning of the *Begriffslehre* within which the basic function of language, according to Hermann Grassmann, is to be an image or representation (*Abbild*). The relation which realizes this function is a ternary one: the linguistic sign is the representation of a mental unit (thought, idea), and this again is the representation of a unit in the external world (fact, what is given). He names the two partial relations "contradictions" (*Widersprüche*). (Contradiction between sign and mental unit, and contradiction between mental unit and unit in the external world.) The mental unit forms the "internal," the linguistic sign

---

41 A comparison with Schleiermacher's *Dialektik* whose lectures Hermann Grassmann had listened to during his studies at Berlin, might make philosophico-logical stimulations from this source visible for his grammar. Cf. Engel 1911: 20.

42 W. Vesper 1980: 91, and G. Schmidt, 1985: 257.

43 e.g. R. H. Hiecke 1831.

44 The *Formenlehre* contains a section on orthography. As to terminology, "Formenlehre" traditionally means "theory of grammatical forms," i.e. parts of speech and inflexions. As to the expression "Begriffslehre" see note 26.

and the unit in the external world form the "external." The two "external" terms of the image relation differ with regard to the activity or passivity of the speaking subject: the units of the external world are only (passively) given to him while the subject (actively) produces the linguistic signs. Language is conceived here as a totality of form and idea: at the same time, it is an image of what is given to the subject and a construction executed by the subject. The ordering of Hermann Grassmann's theory follows the path from "outside" to "inside," that is from form to idea. The "first course," the grammar of forms, is an "external grammar" of the German language; the "second course," the grammar of ideas, is an "internal grammar." Schleiermacher's conception of language which is based on his interpretation of Plato is not only found in Robert Grassmann's legitimization of teaching the mother language, but also in the internal structure of the *Grundriss*.

The modernity of Grassmann's grammar can be made visible by comparing it to two of the most widely disseminated major works of the previous generation, to Johann Christoph Adelung's 1782 two-volume *Umständliches Lehrgebäude der Deutschen Sprache* which had been the most important grammar reference book for the teaching of German until the twenties of the 19th century, and to Johann Severin Vater's *Versuch einer allgemeinen Sprachlehre* (1801). These two grammars had been written in the context of late early modern education which in current periodization is designated as late Enlightenment (*Spätaufklärung*).[45] Early modern grammar instruction was both a propaedeutic of rhetorics and a propaedeutic of logic, the former of the two functions being quantitatively much more widespread. As propaedeutics of rhetoric, grammar instruction, which was essentially teaching Latin, was normative: the art of correct speech came before the art of good (beautiful) speech (or writing). Grammatically impeccable speech required a thorough knowledge of the grammar of forms, and generally of language usage in certain accepted models. The theory of forms was tied to the word as its basic unit (forms are accidents of the word). The *ratio* of this kind of grammar adhered to the classification of words into parts of speech (substantive noun, adjective, etc.).[46] As a propaedeutic of logic (and together with it, of philosophy and theology), that is on a higher level within the early modern hierarchy of disciplines, grammar was an applied logic and served a reflected use of language. Its basic unit was a fundamental idea from logic, that is the idea of judgement (which

---

[45] As types, however, these kinds of grammars have their origin in Humanism and early Enlightenment.

[46] This tradition created the practical distinction between a) grammars which have to do with parts of speech, and b) dictionaries which contain individual words.

contained that of "idea"). The *ratio* of this kind of grammars adhered to the classification of the terms of judgement (subject and predicate). The parts of speech were assigned to these terms: substantive noun - subject; adjective - predicate; etc. (*Est* was the operator which shaped the relationship between subject and predicate into a statement of existence which could be assigned values of truth.) The logical structure of the judgement was deemed to be an absolute, universally valid, that is necessary structure of the operations of the human mind. It defined the frame of what was considered empirically possible in the various individual languages and realized differently within these according to their own degree of perfection. If an individual language was examined in the context of general grammar, this was a special grammar. - The two types of grammar - formal grammar and general or philosophical grammar - were separated from one another by the secular historical distinction between empirics and philosophy. A special grammar did not mediate between empirics and philosophy, but was rather applied philosophy.

Adelung's *Umständliches Lehrgebäude* corresponds to the type of formal, empirical word-grammar. J.S. Vater's *Versuch* is a general grammar. Grammar in the teaching of the German language during the first twenty years of the 19th century was predominantly aligned to the empirical, formal type. It thus adhered to the model of the Latin grammar which prepared for "eloquence." Against that, Vater's grammar refers largely to French models and so continues the tradition of the Port-Royal grammar.[47]

Hermann Grassmann's *Grundriss*, against that, has its place in a fundamentally changed context which can be easily identified as a modern one. The finality of teaching the German language for which this

---

[47] Johann Christoph Adelung (1732-1806) successively was a *Gymnasialprofessor* in Erfurt (1759-61), a private scholar (until 1787) and a chief librarian in Dresden. Johann Severin Vater (1771-1826) was a professor of theology in Halle, Jena and Königsberg. His theological professorship included the teaching of the biblical oriental languages. The difference of formal and philosophical grammar as represented by the two scholars was not a disciplinary one in the modern sense: scientific disciplines, as we know them, did not emerge before the 19th century. The difference rests upon the general fact that during the early modern period every object could be treated in a double way, namely empirically and philosophically, philosophy being regarded as higher in status than empirical knowledge. During the first third of the nineteenth century, scientific modernity emerges from an "epistemological revolution," changing completely the relations of philosophical and empirical knowledge. Scientific disciplines, as we know them today, are a consequence of this profound revolution. The shared epistemological orientation of both scholars is salient from the fact that, without any problem or major shift, Vater continued the famous *Mithridates* after the death of Adelung in 1806. The *Mithridates* is a universal collection of examples from all known languages of the world, a kind of geographically organized *herbarium* of the world's languages.

grammar was written is not one of propaedeutics. Rather, it adheres to an educational goal of its own which has been characterized above, referring to his brother's writings, as that of the "*Kulturnation*."

With regard to epistemology, the opposition between formal grammar and grammar of ideas forms a totality. The two components are not separated from one another by the distance typical for a hierarchy. Rather, empirics and theory are unified by means of a common concept of science. Hermann Grassmann, just like the authors of the then new grammars for the German language who inspired him, did not privilege the absolute, but rather unifies theory and empirics in his search for systematical, regular connections in what is "given," form and idea being equally given as the external and the internal of language. The absolute or metaphysical opposition between formal and logical grammar no longer determines the grammar's *ratio*.

Grassmann's dialectic of form and idea permits him - doubtlessly inspired by K.F. Becker - to treat the formal grammar no longer in the frame of the theory of parts of speech, but rather to insert it, together with the theory of the parts of speech, into a functional syntax. The Latin grammars which had served as models for the German grammars of the formal type contained (almost) no syntax, that is "theory of the connection of words" (into the sentence). While Adelung, who had become aware of this deficit, had inserted a large chapter on syntax into his *Umständliches Lehrgebäude*, he had not changed his own grammar's *ratio* on that basis, and this is why his syntax also adheres to the structure of the theory of parts of speech: the sentence is an addition of words. While Vater's *Versuch*, being based on the structure of judgement, does contain a syntax, this is a logical one, and not a syntax of forms. Against that, Hermann Grassmann makes the sentence's functional structure the *ratio* of his entire grammar, that is for formal grammar as well: his *Sprachlehre* begins with formal grammar, and his formal grammar begins with an analysis of the relations within the sentence, not within the judgement.

Grassmann's penchant towards analyzing a totality as a system of symmetrical oppositions is very salient in his analysis of the sentence. He designates the individual relations of his syntax as *Verhältnisse*. The parts of speech are classified according to these relations. He determines four syntactical relations. Two parts of speech are assigned to each relation, which yields eight syntactically functional parts of speech in his theory. While this number has a long tradition, as Donatus's grammar, one of the bases of teaching Latin in the Middle Ages, has just this number as the one canonically handed down, the basis of classification in these two cases is totally different, a fact which results in only partial congruence of the parts of speech involved.[48] This is where Grassmann ought to be compared to K.F. Becker: Becker's grammar contains a relational formal grammar, too, but he anayzes only three such relations.[49] It is manifest here what the Grassmann brothers understood by continuing Becker's work in greater rigor.

Hermann Grassmann distinguishes the four syntactical relations into two "sentence relations" (*Satzverhältnisse*) and two "formal relations" (*Formverhältnisse*). The sentence relations are those which constitute the simple sentence: 1. predication (*Aussageverhältnis*) and 2. determination (*Bestimmungsverhältnis*). The formal relations are modifications which do not change the basic relations and their terms as given by the two sentence relations. Grassmann designates these terms as syntactical places (*Satzstellen*). The formal relations are: 3. the "hinting relation" (*Deuteverhältnis*). This is a pragmatic function which is based on the difference of what is known in communication, and what is new information,[50] and also englobes syntactic parataxis. 4. The extension of the simple sentence is based on the "joining relation" (*Fügeverhältnis*). It englobes both relations marked by prepositions (connections between things) and the relationship between main clause and subordinate clause (connections between facts). The parts of speech associated with the "relationships" are: 1. Noun and verb. 2. Adjective and adverb. 3. Pronoun and "conjunctive word" (*Bindewort*) 4. Prepositions and "sentence joiners" (*Satzfüger*). With the new parts of speech "conjunctive word" and "sentence joiner," the traditional part of speech which is the conjunction is subdivided into two new parts which are functional for this syntax. Grassmann's four relations, besides being realizable by special parts of speech, can also be realized by forms in the narrower sense (by so-called "flexions"). One group of such forms are the case-forms, for instance. The idea of the syntactical place articulates the two levels of sentence relationships and form relationships. By

---

48 Donatus, *Ars grammatica*. Donatus was a late Roman grammarian whose booklet was used for teaching beginners in the Middle Ages. It is based on the most eminent Roman grammarian Priscian. Grassmann, for instance, omits Donat's eighth part of speech, interjection, as the latter is no syntactically functional part of speech but stands outside syntax. (The *Ars grammatica* is accessible in Heinrich Keil's *Grammatici latini,*, Vol. IV, Leipzig 1864, repr. Hildesheim 1961).

49 These are subject, predicate, and object.

50 "Das, was dem Redenden und dem Hörenden bekannt ist, braucht nur noch angedeutet zu werden ; z.B.: Karl versprach, daß er kommen wollte. - Das Wort, welches etwas andeutet, ohne es zu benennen, heißt Deutewort." ("What is known to the speaker and the hearer needs only be hinted at; e.g. Charles promised that he would come. - The word which hints at something without naming it is called hinting word"). Grassmann 1842: 4.

means of this concept, all syntactical structures, however complex they may be, can be reduced to the four terms of the two sentence relations functioning as syntactical places.

It will be sufficient here to explain Grassmann's sentence relations in comparison to early modern grammars. Because of the abstract and scientific character of Grassmanns's theory of language, this comparison can be confined to the preceding general grammars for which that of Vater shall serve as a case in point. The first function which constitutes the simple sentence is a necessary one for Grassmann. This is (almost) the same relation as that which constituted the (logical) judgement for Vater: the relation between subject and predicate. There is, however, a small but essential distinction. Where Vater analyzes the sentence into three parts, as in the entire tradition of general grammar, Grassmann analyzes it into two parts. The reason for that is that the operator *est/is* which constitutes the assertion has been omitted. Vater's example is: "Der Tisch ist braun" ("The table is brown"). Grassmann's example, however, is "Schüler arbeiten" ("Pupils work"). Grassmann, for whom not a judgement, but a thought is expressed, is thus able to analyze the subject-predicate relation **within** the empirical German language, that is by conceptual systematization of what is empirically found. Vater, against that, maintains a very special distance to the empirical fact. In order to analyze a sentence like that given as an example by Grassmann, he must transform it into another which, while maintaining the logical meaning, deviates from German actual use: "Der Schüler ist arbeitend." ("The pupil is working.") He has to replace the one unit which is the verb by the two units which are the assertive operator (the so-called "Verbum substantivum") and the participle.[51]

From this difference in the starting point for analyzing the simple sentence ensues a specific difference with regard to the function of determination. Since Antoine Arnauld and Pierre Nicole published their "Logic of Port Royal" in 1662, grammarians analyzing the judgement were able to recur to the "law of Port-Royal" which determines the relationship between the logical ideas' intension and extension. Each logical idea "contains" a more or less large number of features resp. attributes. The smaller the number of features, the larger the extension of the ideas, and vice versa. For Vater, this means that the attributes of the idea of subject are expressed by adjectives in empirical language.

---

51 In general grammar, the elementary form of a judgement is expressed by the parts of speech which are the substantive noun and the attributive adjective. Both are linked by the "verbum substantivum"). Vater's example is: "Der Tisch ist braun" ("The table is brown"). In Grassmann, the verb *sein* (to be) is an auxilary verb having a hinting function so that it is treated in connection with the pronouns.

51 Hiecke 1831.

Attributes of the idea of predicate (i.e. of the adjective) are expressed by adverbs. For a logic whose field of application was not syllogistic (formal) reasoning, but rather exactitude of the "contents" of discourse, this novel analysis of ideas was fundamental.

If the idea of subject is assigned an idea of predicate, this is called "determination," (or *Bestimmung*).[52] Determination is a relation between the idea of subject and the idea of predicate, independent of whether an act of assertion is present, or not. As examples for a determination, Vater both gives the syntagma "brauner Tisch" ("brown table") and the sentence "Der Tisch ist braun" ("the table is brown"). Determination thus does not constitute a sentence resp. judgement itself, but refers to the basis of the judgement which lies in the logic of ideas. As attributes which can be added to the adjective are expressed by adverbs, the relationship between adjective and adverb is a determination as well: "Der Tisch ist sehr braun" ("the table is very brown") (or: "Der sehr braune Tisch" - "the very brown table"). An important consequence of using this logic of intensions in general grammar resides in the difference between "logical idea" (*Begriff*) and "grammatical meaning" (*Bedeutung*). On the logical level (level of ideas), there are, in the case of a sentence containing more than two meaningful words, too, only two ideas, that is the idea of subject and that of predicate. While they will be expressed on the level of language, in the simplest case, one by one word (part of speech) each (substantive noun and adjective), they can also be expressed by connecting several terms, like substantive noun + adjective, or adjective + adverb. From this view, the empirical language is marked by the drawback of consisting of arbitrary and conventional signs which makes that the signs fixated by language usage do not always correspond, in the intension of their meaning, to the intension of the logical ideas. Logical ideas and linguistic meanings relate to one another like natural law and customary law.

Hermann Grassmann, whose grammar is a totality of formal grammar and grammar of ideas, continues this analysis of general grammar in his formal grammar, albeit in another way. As his first syntactical distinction is a dual one, he does not distinguish between assertion and determination with regard to the relation of predication. He speaks of determination only with regard to the relations which again subdivide the first two members of a sentence into two parts each, that is with regard to the internal articulation of the sentence's subject and predicate. Here again, we find Grassmann's penchant for analyzing totalities in polar

---

[52] On "determination" cf. also Du Marsais' contributions *Détermination* and *Construction*. The two are published in Diderot's and d'Alembert's *Encyclopédie* of 1754. They were one of Vater's sources.

oppositions: he designates the relation of predication which constitutes the sentence as "sentence relation of first order," and the internal articulation of its terms which he calls determination (*Bestimmen*), as "sentence relation of second order." While the relation of first order is necessary for constituting the sentence, the relation of second order is only a possible one. The school teacher Grassmann's example for a simple sentence formed with determinations is: "Assiduous pupils like to work" ("fleißige Schüler arbeiten gern"): "Fleißig" determines "Schüler" and "gern" determines "arbeiten" ("Schüler" and "arbeiten" being "principal elements" (*Hauptstücke*), "fleißige" und "gern" "determining elements" (*Bestimmungsstücke*)). The four terms of his relational analysis of the simple sentence (two terms of "first order," and two terms of "second order"[53]) are the syntactical places (*Satzstellen*) which he occasionally names "syntactical parts" (*Satzteile*). There was no analogous expression in general grammar: the latter's analysis of functional relations was confined to the logical level, that is to the relation between ideas.

Against that, the basic syntactical relations, in Becker as in Grassmann, are not analyzed on the level of (logical) judgment, but rather on that of the (linguistic) sentence. In his designation of the individual terms, Grassmann first recurs to the logical tradition: he names the syntactical places of the sentence relation of first order "subject", and "predicate." The names for the syntactical places of second order are neologisms created by himself. He names them *Adjekt* (determination in the subject) and *Addikat* (determination in the predicate). As he integrates the parts of speech into his own syntactic formal grammar, he uses for their ensemble the new designation "sentence words" (*Satzwörter*). Where a traditional terminology is available, however, he keeps the old designations intact. The sentence words are substantive noun, verb, adjective, and adverb. In this Hermann Grassmann was very much less radical than Robert who consistently "germanized" the entire terminology of grammar (providing every term with a Greek equivalent). Where the relations between sentence relations and form relations are concerned, all the form relations take on the syntactical places which execute the function of determination, that is they replace *Adjekt* and *Addikat*. If the simple sentence, for instance, contains only one thing resp. substantive noun, however, the extended sentence can contain two substantive nouns which are linked to one another as "main thing" (*Hauptding* - sentence subject) and "counter

---

[53] The expression "simple sentence for both these syntactical orders is not to be found in Hermann Grassmann's very brief and condensed *Sprachlehre*, but in his brother Robert's more explicit *Sprachlehre* (R. Grassmann 1890c)

thing" (*Gegending* - sentence object). In this the sentence object takes on the syntactical place of the *Addikat*. While the syntactical places are not understood to be abstract structures which are replaced by variables, (no attempt to formalization is recognizable), but rather as general places in which words and sentences (or clauses in today's terminology) of different kind may have their "seat." Something like "proper" and "improper" parts of speech and sentences (clauses) are distinguished in this: the adverb is in this case the simple, proper holder of the place, while the substantive noun, whose proper place is the sentence subject, is the extending holder. Besides, the *Addikat*'s place can be extended by a subordinate clause. All improper place holders are marked by certain forms like preposition, case, conjunction, or word order.

The analysis of the syntactico-functional relations at the "external" level of form distinguishes Grassmann's grammar fundamentally from the two grammars developed by Adelung and Vater which were typical for the early modern period. If, however, functional syntax represents, as in Grassmann, the *ratio* of formal grammar, and is not longer treated in the frame of the theory of (logical) judgment, how precisely does he analyze the relationship between form and idea? And what is then the task alotted to a grammar of ideas? On the basis of his own dialectic conception of a totality of form and idea, Grassmann is able to abandon the quasi-ontological separation of logical idea from grammatical meaning which was so typical for the early modern period. The language empirically found is no longer seen as arbitrary and conventional, that is as a probabilistic collection without a rational organization of its own. (In general grammar, as represented by Vater, the rationality of language did not have its cause in language itself, but rather in the logic of judgment which was localized on a higher, philosophical level.) In Grassmann, language, as a form, itself forms a systematically (functionally) structured whole which is linked to the likewise systematically structured level of thoughts and ideas on the basis of the dialectic relation which he considers the semiotic function of imaging-representation to be. The new, grammatico-relational concept of syntactic place and the concomitant shift of the formerly logical ideas of subject and predicate from the logic of judgement to the syntax of discourse permit to dehierarchize the relationship of sentence and thought, of form and idea. (This also devalidates the universalist claim of the syntax of logical judgement which was the basis of general grammar.) For the factual grammatical analysis, this new view is interpreted in such a way that sentence and thought, word and idea are placed in a 1:1 relationship to one another, i.e. that a grammatico-"ideological" parallelism is assumed. An important conclusion which resulted from this kind of modernizing grammar was not drawn by the

Grassmann brothers themselves, but by the reviewers of Becker's and Herling's grammars in the *Jahrbücher für wissenschaftliche Kritik*: The interior analysis of languages must separate from general logic and develop a logic of its own.[54] A more complete study of Grassmann's grammar should have the difference between the logic in early modern grammars like Vater's and a "Grassmannian logic" as one of its objects.

The loss of the universal-absolute of the logical structure is compensated by anchoring the sentence's basic structure in the unity of first origin, which may be both origin in the system (e.g. relationship between simple and extended sentence) and origin in time. The central concept of such a systematic is the concept of derivation (*Ableitung*): derivation is a regular, lawful change taking place both "ideologically"-genetically and historically-morphologically, the two levels entertaining a 1:1 relationship here as well. As an example for a derivation, we should like to use Grassmann's example of the derivation of the substantive noun "Geruch" ("smell"). According to Grassmann, all substantive nouns are originally designations of concrete objects. Smell, against that, is an abstract noun as it expresses an active property (e.g. a flower's smell). The original part of speech used to designate active properties resp. activities of a thing is the verb. Consequently, "Geruch" is an abstract noun derived from a verb by changing the base "riech" by the ablaut u to "ruch" and adding the prefix (*Sproßsilbe*) "ge-". Semantic genesis, as seen by Grassmann, is completely visible in word formation. In this kind of analysis, Hermann Grassmann can be recognized as a reader of Jacob Grimm. In his later scientific research devoted to Sanskrit and generally to Indogermanic, Hermann Grassmann systematically and rigorously formulated in a successful way (as measured against continued recognition) an historical derivation in the shape of a so-called *Lautgesetz*: Grassmann's *Aspiratengesetz* which was published in 1863 in the well-read *Zeitschrift für vergleichende Sprachforschung* (*Kuhn's Zeitschrift*). According to Winfried P. Lehmann, the entire comparative grammar resp. Indo-european linguistics was decisively modified by the formulation of that law.[55] In his vicissitudinous scientific career, Hermann Grassmann received the largest recognition by far for this achievement. In the small school program of 1842 which can have been disseminated only locally because of its type of publication, two features can be recognized which announce the epistemological orientation and the methodological character of Hermann Grassmann's subsequent research: The conception of language as of a systematical totality in the sense described above,

---

54 Hiecke 1831.

55 Winfried P. Lehmann 1967:109sq.

and the methodological rigor of research (as well as his teaching). One may surmise whether Grassmann's rigor has a "Pythagorean" trait, whether it is connected to a mysticism of Number: the key to the brothers order of systems is the number four. The assymmetrical is excluded from this "theory of harmony."

APPENDIX: NOTE ON THE "ASPIRATENGESETZ"

The aspirates Grassmann's Law is about are puffs of breath articulated after occlusive consonants like p, b; t, d; k, g. They are normally represented by h - as in Greek trékho or Sanskrit budhna. (In today's English language there are in certain positions slightly aspirated consonants, too). H. Grassmann was the first to treat some of the "exceptions" left by Grimm's Law (1822), which is about certain regular phonetic relations between Sanscrit, Greek, Latin and Gothic. Research into H. Grassmann as an Indo-European linguist should examine whether his work contributed to motivating the movement of the "Junggrammatiker" who postulated, that all sound-laws should be regular, i.e. without exceptions. Grassmann's question was whether there were originally roots in Indo-European with aspirates initially and finally. In pursuing this question, he investigated a problem which had remained unsolved in the narrower context of classical philology: the explanation of the shift of aspiration in Greek, e.g. in trékhō, thréksomai. Classical philologists had explained this phenomenon by assuming hypothetical roots with initial and final aspirates. Since comparative grammar had shown, that for diachronic reasons roots of this kind were impossible in Greek, explanations of this phenomenon in the context of Greek language alone had also been regarded as impossible. Finding, too, no necessity or regularity linking the phenomena to Sanskrit, they were regarded as a monstrosity and the question was abandoned. They resisted to the fundamental espistemological conviction, that the world as viewed by scientific research had to be well-ordered. Since, the conviction, that there were no roots of this kind in Greek, became a kind of axiom.

H. Grassmann, questioning this subject again, maintained this basic epistemological conviction, but changed the idea predominant in Indo-European linguistics, which held that "facts" are only those transmitted by written documents, i.e. in historical language. In a certain way, he inverted the relations between "laws" and "facts," claiming that "facts" not occurring in documents ought to be reconstructed on the basis of a necessity or regularity being functional for explaining historical facts. Positing thus, that there had been roots with aspirates initially and finally in the prehistorical times of the assumed original Indo-European language, he provided a regular explanation for phenomena treated before as "monstrosities" or "exceptions." For Indo-European linguists, his explanation was a convincing one. Even scientists working in modern formal linguistics accept Grassmann's Law as a regularity worth to be formalized.[56] In present research on the history of linguistics, it should be curious to look upon Grassmann's Law (and its success) as a construction creating "facts" in the context of beliefs and myths about the structure of the world. (A perspective motivated by Fleck's observations.[57]). One should not be astonished to rediscover the idea of a symmetrical, harmonic structure, now represented by the "pure," original roots of Proto-Indo-European.

A comprehensive and detailed account of Grassmann's Law is given in Lehmann (1967) and Collinge (1985). A comprehensive, interesting and actual account of the role of Indo-European linguistics as related to Indian studies is given in Filliozat (1974). I will conclude by quoting from Lehmann, first on the role of Grassmann's Law in the history of Indo-European linguistics, and second on the style of Grassmann's work in this field. First: "Grassmann's (law) is one of the celebrated articles of linguistics. Rightly, because it largely led to the conviction that reconstructed languages must be set up for any language family. Before Grassmann, Sanskrit had served as the measure against which forms in the other languages were compared. [...] By demonstrating that Germanic actually was "older" in one phonological pattern than was Sanskrit, Grassmann undermined the position of Sanskrit as the language which was the earliest attainable in Indo-European linguistics. - By this demonstration Grassmann also undermined the notion that language developed from an analytic to a synthetic structure through an agglutinative. With it he did away with the close relationship that had been observed previously between genealogical and typological classification. [...] The appealing notion of a straightforward development of language had been abandoned."[58] And the second quote: "These contributions to general linguistic theory were achieved by explanation of one phonological problem [...]. His article is admirably composed. First he examines previous attempts at explanation, than the data. His examination of both is complete. When he presents his conclusions there is no question of their validity. - His convincing explanation led linguists to deal with entire forms, not merely with single

---

56 Cf. P. C. Stanley 1985.

57 Fleck 1935 / 1980.

58 Lehmann 1967: 109.

segments. [...] All "exceptions" have been dealt with. One could not ask for more rigourous methods."[59]

Unfortunately, there is a curious mistake in Lehmann's account of Grassmann's biography: "Hermann Grassmann (1809-1877) was a banker who was compelled to retire because of tuberculosis. In his leisure he occupied himself with mathematics and linguistics." Lehmann gives no justification for this entirely erroneous information - and it has already begun to be reproduced in recent secondary literature![60]

Institut für Wissenschafts- und Technikforschung, Universität Bielefeld

REFERENCES

Johann Christoph Adelung, *Umständliches Lehrgebäude der Deutschen Sprache. Zur Erläuterung der Deutschen Sprachlehre für Schulen* (Leipzig 1782. Reprogr. Nachdruck i.d. Serie Documenta Linguistica; Hildesheim/New York: Georg Olms, 1971).

Sylvain Auroux, "La querelle des lois phonétiques," *Linguisticae Investigationes* III:1. Amsterdam 1990, 1-27.

Claudia Bartels, "Deutschunterricht ohne Deutschlehrer," Erlinger et al. 1989, 21-38.

Herbert E. Brekle, Brigitte Asbach-Schnitker, Hans-Jürgen Höller, "Der Jansenismus und das Kloster Port-Royal," *Grundriß der Geschichte der Philosophie. Die Philosophie des 17. Jahrhunderts. Band 2: Frankreich und Niederlande*, ed. Jean-Pierre Schobinger (Basel: Schwabe, 1993), 475-528.

N. E. Collinge, *The Laws of Indo-European* (Amsterdam/Philadelphia: John Benjamins, 1985)

Berthold Delbrück, (Nekrolog auf H. Grassmann). *Beilage zur Allgemeinen Zeitung*, Augsburg 1877, Nr. 291 vom 18. Oktober.

F. A. W. Diesterweg, *Praktischer Lehrgang für den Unterricht in der deutschen Sprache. Ein Leitfaden für Lehrer.* 3 Teile, 1. Aufl. (Crefeld, 1828-1830).

Friedrich Engel, *Hermann Grassmanns gesammelte mathematische und physikalische Werke.* Dritten Bandes zweiter Teil: Grassmanns Leben (Leipzig: B.G. Teubner, 1911).

Hans Dieter Erlinger, Clemens Knobloch, Hartmut Meyer (Hrsg.), *Satzlehre - Denkschulung - Nationalsprache. Deutsche Schulgrammatik zwischen 1800 und 1850* (Münster: Nodus, 1989).

Hans Dieter Erlinger / Clemens Knobloch (Hrsg.), *Muttersprachlicher Unterricht im 19. Jahrhundert. Untersuchungen zu seiner Genese und Institutionalisierung* (Tübingen: Niemeyer, 1991a).

Hans Dieter Erlinger, Clemens Knobloch, ""'...Die von eigentlichen Forschern zutage geförderten Goldbarren in gangbare Münzen auszuprägen" - Diesterweg und der

59 Ibid.: 110.

60 As for example in N. E. Collinge 1985: 47.

muttersprachliche Unterricht," Erlinger / Knobloch 1991a, 61-78. (quoted as Erlinger / Knobloch 1991b).

Jean Filliozat, "La naissance et l'essor de l'indianisme," *Laghu-prabandha. Choix d'articles d'indologie.* Leiden 1974: 1-32

Ludwik Fleck, *Entstehung und Entwicklung einer wissenschaftlichen Tatsache. Einführung in die Lehre vom Denkstil und Denkkollektiv.* Mit einer Einleitung herausgegeben von Lothar Schäfer und Thomas Schnelle (Frankfurt a.M.: Suhrkamp, 1980) (1. Aufl. 1935)

Hermann G. Grassmann, *Grundriß der deutschen Sprachlehre.* Progamm der Ottoschule Stettin (Stettin: H.G. Effenbart's Erbin, 1842).

Hermann G. Grassmann, "Stücke aus dem Lehrbuche der Arithmetik" (1861), *Hermann Grassmanns gesammelte mathematische und physikalische Werke.* Zweiten Bandes erster Theil: Die Abhandlungen zur Geometrie und Analysis, eds. E. Study, G. Scheffers und F. Engel (Leipzig: Teubner, 1904), 295-438.

Hermann G. Grassmann, "Über die Aspiraten und ihr gleichzeitiges Vorhandensein im An- und Auslaute der Wurzeln," *Zeitschrift für vergleichende Sprachforschung auf dem Gebiete des Deutschen, Griechishen und Lateinischen*, 1863, *12.2*: 81-138.

Hermann G. Grassmann, "Concerning the Aspirates and Their Simultaneous Presence in the Initial and Final of Roots," *A Reader in Nineteenth Century Historical Indo-European Linguistics*, ed. Winfried P. Lehmann (Bloomington and London: Indiana University Press, 1967), 109-131.

Robert Grassmann, *Die Denklehre.* Erster Teil von *Die Wissenschaftslehre oder Philosophie* (Stettin: Druck und Verlag von R. Grassmann, 1875a).

Robert Grassmann, *Die Sprachlehre.* Drittes Buch der Denklehre (Stettin: R. Grassmann, 1875b).

Robert Grassmann, *Das Formdenken.* Viertes Buch der Denklehre (Stettin: R. Grassmann, 1875c).

Robert Grassmann, *Die Philosophie.* Erster Teil von *Das Gebäude des Wissens* (Stettin: R. Grassmann, 1890a).

Robert Grassmann, *Die Geschichte der Philosophie und erste Einleitung in die Philosophie* (= Einleitung in *Das Gebäude des Wissens*), (Stettin: R. Grassmann, 1890b).

Robert Grassmann, *Die Sprachlehre, d.h. die Lehre von den Arten der Laut-, Wort- und Satzbildungen, welche dem Menschen möglich sind, von ihren Formen und Gesetzen. Erstes Buch der Wissenslehre oder Philosophie.* Erster Band des *Gebäude des Wissens* (Stettin: R. Grassmann, 1890c. 2. Aufl. 1900).

Robert Heinrich Hiecke, Review of: S.H.A. Herling, Die Syntax der Deutschen Sprache..., *Jahrbücher für wissenschaftliche Kritik*, Berlin 1831/2: 589-612.

Erika Hültenschmidt, "Enzyklopädien, Wissensdifferenzierung und Sprachwissenschaft um 1800 (Frankreich), "*Einsamkeit und Freiheit" neu besichtigt. Universitätsreformen und Disziplinenbildung in Preussen als Modell für Wissenschaftspolitik im Europa des 19. Jahrhunderts*, ed. Gert Schubring (Stuttgart: Franz Steiner, 1991), 57-91.

André Laks, Ada Neschke (éds.), *La naissance du paradigme herméneutique. Schleiermacher et les sciences historiques* (Lille: Presses Universitaires de Lille, 1990).

Winfried P. Lehmann 1967: see Hermann Grassmann 1967.

Peter Lundgreen, *Sozialgeschichte der deutschen Schule im Überblick. Teil I: 1770-1918* (Göttingen: Vandenhoek & Ruprecht, 1980).

Friedrich Schleiermacher, "Über den Begriff der Hermeneutik mit Bezug auf F. A. Wolfs Andeutungen und Asts Lehrbuch." Gelesen in der Plenarsitzung (der Berliner Akademie) am 13. August 1829. F. Schleiermachers *Sämmtliche Werke*, III/3, ed. L. Jonas. (Berlin, 1835), 344-386.

Friedrich Schleiermacher, *Hermeueutik und Kritik mit besonderer Beziehung auf das Neue Testament.* Aus Schleiermachers handschriftlichem Nachlasse und nachgeschriebenen Vorlesungen herausgegeben von Dr. Friedrich Lücke (Berlin: Reimer, 1838).

Gunther Schmidt, "Sprachkonzeptionen und Schulpraxis. Bemerkungen zu sprachtheoretischen Gundlagen des Deutschunterrichts," *Sprachwissenschaftliche Germanistik. Ihre Herausbildung und Begründung,* eds. Werner Bahner und Werner Neumann (Berlin: Akademie-Verlag, 1985), 249-281.

Patricia C. Stanley, "Grassmann's Law, Verner's Law, and Proto-Indo-European Root Structure." *Indogermanische Forschungen.* Berlin/New York 1985, *90*: 39-54

Johann Severin Vater, *Versuch einer allgemeinen Sprachlehre* (Halle, 1801). Faksimile-Neudruck mit einer Einleitung und einem Kommentar von Herbert E. Brekle. Reihe Grammatica universalis, Bd. 3 (Stuttgart-Bad Cannstatt: Frommann / Holzboog, 1970).

Wilhelm Vesper, *Deutsche Schulgrammatik im 19. Jahrhundert. Zur Begründung einer historisch-kritischen Sprachdidaktik* (Tübingen: Niemeyer, 1980).

PART III

HISTORICAL INFLUENCES OF GRASSMANN'S WORK

RENATE TOBIES

THE RECEPTION OF H. GRASSMANN'S MATHEMATICAL ACHIEVEMENTS BY A. CLEBSCH AND HIS SCHOOL

The dissemination of Grassmann's ideas to the larger mathematical public in Germany intensified with the interest in this scholar's achievements shown by Alfred Clebsch (1833-1872) in the early 1870s. The premature death of Clebsch prevented him from deepening his adaptation, but the friends and disciples in the Clebsch school continued the reception of Grassmann's work. I intend to show the important role of Clebsch's school, and in particular that of Felix Klein (1849-1925) in making Grassmann's work accessible.

It was Clebsch's scientific biography of Julius Plücker (1801-1868)[1] which prompted an enhanced reception of Grassmann's mathematical works. Clebsch supported Viktor Schlegel's efforts to propagate Grassmann's results and wrote to him on 21 November, 1871:

"I hope that science will benefit from the methods of Mr. G: and your presentation of the same, and I am most interested in both...I would consider *every more precise presentation of any parts of G's methods and results* most desirable ..." (Grassmann 1911, p. 400).

In 1896 when Schlegel presented a history of the reception of Grassmann's *Ausdehnungslehre*, he listed 179 titles connected with this oeuvre in some way (Schlegel 1896). The propagation of Grassmann's ideas was further promoted by the *Gesammelte Werke* which were being published since 1894 (in connection with the 50th anniversary of the *Ausdehnunglehre*). In the meantime, related publications had become so numerous that Friedrich Engel (1861-1941), in the last volume of the *Gesammelte Werke* published in 1911, decided - contrary to the original

---

1 Clebsch commented in this obituary, first published in 1872, on the relation between synthetic analytic methods in geometry and underlined the key achievements of Grassmann:."For me these [the analytic methods] include the concept of general curves and surfaces which was only explained much later by purely synthetic means in Grassmann's profound works." In a note, Clebsch remarked: "Regrettably, the excellent works of this very important geometrician are still too little known. This must mainly be attributed to the fact that in Grassmann's presentation, these geometric results occur as corollaries to much more general and very abstract investigations which, in their unusual form, present the reader with considerable difficulties." (Clebsch 1895, p. XV).

G. Schubring (ed.), Hermann Günther Graßmann (1809–1877): Visionary Mathematician, Scientist and Neohumanist Scholar, 117–130.

plan - against including an index of publications initiated by Grassmann's work (see Grassmann 1911, p. VII).

1. THE SPECIAL INTEREST OF THE CLEBSCH SCHOOL IN PROPAGATING GRASSMANN'S IDEAS

Despite the short period between Clebsch becoming acquainted with the *Ausdehnungslehre* and his death in November 1872, he still managed to set an important starting point for the enhanced reception of Grassmann's results. He helped Grassmann obtain an important honor by having him accepted as a corresponding member of the *Gesellschaft der Wissenschaften* in Göttingen (2 December 1871) and, somewhat as his legacy, his own disciples became active in promoting Grassmann's work. It is quite revealing that Clebsch's friends emphasized, in their scientific biography of Grassmann, the latter's "extraordinary merits" in establishing the "process of general geometric development" (Clebsch 1873, p. 12).

The following mathematicians belonged to the first generation of Clebsch students who were particularly active in promoting and disseminating Grassmann's ideas: Paul Gordan (1837-1912), Olaus Henrici (1840-1918), Alexander Brill (1842-1935), Max Noether (1844-1921), Jakob Lüroth (1844-1910), Aurel Voß (1845-1931), Felix Klein and Ferdinand Lindemann (1852-1939).[2] Eduard Study (1862-1930), who used Clebsch-Aronhold symbolic in the invariant theory, also said that Clebsch had been his scientific mentor (Weiss 1930, p.54).

The majority of Clebsch's students were familiar with Grassmann's ideas and propagated them in various ways: by means of their own publications which were linked to Grassmann's works; by disseminating his theories in their own lectures; by actively participating in publishing his *Gesammelte Werke*; by encouraging doctoral theses on related topics; by describing Grassmann's mathematical methods in the great project of an encyclopedia of mathematics and their applications, and by presenting them in publications intended to popularize science.

Grassmann's system had model character for Clebsch, and for some of his disciples in their search for a systematic of their own. It prompted further a research program characterized by the search for connections between geometry, algebra and function theory, focusing, according to the Clebsch school, on "new" geometry (algebraic geometry) and "new" algebra (invariant theory).

---

2 For the relations among Alfred Clebsch's students see Section 2 in (Tobies 1992).

Uniting synthetic and analytic methods[3]

The basic tendency of the Grassmann method of applying calculus to geometry corresponded to the method of algebraic geometry substantiated and developed by Clebsch and his school. The usual separation between analytic and synthetic methods was abandoned; instead, algebraic geometry examined the linkage between projective geometry and invariant theory on the one hand, and the investigation of n-dimensional spaces and general complex numbers on the other (see Klein 1926, pp. 131-191).

Essential works on algebraic geometry had already been available when Clebsch and his students took notice of Grassmann's *Ausdehnungslehre*. This attention was due to several fortunate circumstances. Hermann Grassmann had sent his eldest son Justus to Göttingen at Easter 1869 to study mathematics, and had commissioned him to give a copy of his *Ausdehnungslehre* both to Alfred Clebsch and Moritz Abraham Stern (1807-1894). It was Clebsch who first gave Grassmann's ideas closer scrutiny after Stern had told Felix Klein about them, who in turn had called Clebsch's attention to them (Grassmann 1911, p. 311).

Theory of n*-dimensional space*

The concept of multi-dimensional space, established for the first time by Grassmann in his *Ausdehnungslehre* of 1844, was welcomed by Clebsch and his disciples in 1870, because they had been prepared for it by Julius Plücker's[4] and Bernhard Riemann's (1826-1866) works. The space of n dimensions became the general property of a "future-oriented younger generation" of mathematicians in the 1870s (see Klein 1926, p. 171).

Clebsch treated the problem of the invariant theory for n variables analogously to Grassmann's degrees ("Stufen"). In 1872, he integrated the Grassmann degrees explicitly into the invariant theory (see Klein 1927, p. 5). In his paper *Über eine Fundamentalaufgabe der Invariantentheorie,* Clebsch adhered to the concept of an $(n\text{-}1)$-dimensional manifold, examining so-called "basic elements" ("Grundgebilde"). In his

---

3 In 1845, Grassmann explained the essence of his *Ausdehnungslehre* in Grunert's *Archiv der Mathematik und Physik* stating: "The tendency of this method of calculation for geometry is to unite the synthetic and the analytical method i.e. to plant the advantages of the one into the soil of the other by placing a simple analytical operation at the side of every construction, and vice versa" (Grassmann 1894, p. 300).

4 In his book *System der Geometrie des Raumes* (1846), Plücker developed the concept of a 4-dimensional projective geometry.

comment, he particularly emphasized that such concepts can already be found in Grassmann's 1844 *Ausdehnungslehre* (Clebsch 1872a, p.9).

2. FELIX KLEIN'S EFFORTS IN PROMOTING GRASSMANN'S IDEAS

Klein expressed his full agreement with the general approach to mathematics upon which the *Ausdehnungslehre* is based: he emphasized that "... continuous magnitude - i.e. extension, space - is as much an original concept of human intellect as number..." using this concept to establish ***the raison d'être of the geometric way of thinking*** against the prevailing arithmetic tendency. Klein pointed out:

"The idea that this way inherited from Euclid is actually a detour to the construction of geometry, that it actually does not even lead to the goal -- to comprehension and mastery of the continuum -- this idea is an ever recurring tendency that runs directly counter to the arithmetization of mathematics prevalent today" (Klein 1979, p.165).

Subsequent to this, Klein particularly emphasized that Grassmann, on the contrary, had rejected the idea that geometry is only an application of arithmetic and that Grassmann had claimed the status of an independent discipline for his own *Ausdehnungslehre* (Klein 1926, p. 178). Klein referred in considerable detail to Grassmann. In particular, Klein connected Grassmann's *Ausdehnungslehre* to Plücker's line geometry and used it for his own work (see Blaschke 1926, p. 121).[5]

In his *Erlanger Programm*, Klein referred a number of times to Grassmann, in particular underlining in his § 10 on arbitrarily extended manifolds the essential identity: "...the theory of extended manifolds, or (in accordance with Grassmann) in short the theory called *Lineale Ausdehnungslehre*" (Klein 1921, p. 486). The *Erlanger Programm* was published in October 1872. Already before, in his articles on the foundations of geometry, Klein had dealt with manifolds of n dimensions and in particular, in his second note, completed on 8 June 1872, *Ueber die sogenannte Nicht-Euklidische Geometrie*, he referred to the general definition of manifolds in Grassmann's book of 1844. Klein emphasized the correspondence between his own concepts and Grassmann's method:

"His method is at least in many ways the projective method referred to here; what is here called the element of manifold is called extensive magnitude in [Grassmann]." (Klein 1921, p. 320)

In many of his lectures, Klein presented Grassmann's system, or his ideas and methods, in a more or less detailed manner. In the 1890s,

---

5 Here Blaschke implicitly referred to an essential statement already formulated by Klein in his *Erlanger Programm*, Note IV in the appendix: "Both theories - Grassmann's and Plücker's - have their own unique advantages; one uses both, changing back and forth, to one's advantage."(Klein 1921, p. 492).

starting with his lecture on Non-Euclidean Geometry II (summer term 1890), he presented in great detail Grassmann's *n*-dimensional geometry, and Grassmann's degrees [Stufen] (Klein 1890, pp. 9-14 and 1892/93, pp. 232-240).[6]

Another part of Grassmann's mathematical work which was only noticed and developed further after having been discovered and disseminated by Klein were his achievements on the so-called *Pfaff problem* in the A_2 (see Engel in Grassmann 1911, p. 233).

Application of Grassmann's "Determinantenprinzip" by Klein

Klein examined to what extent Grassmann's concept and use of determinants was suitable as a principle of classification for the case of an arbitrary group of linear transformations. In his *Erlanger Programm,* Klein mentioned transformations which had already been examined by Grassmann in his A_2 (1896, p. 278; Klein 1921, p. 478). Grassmann's idea of "change" ("Änderung", meaning parallel displacement) essentially corresponded to the idea of the transformation of a multi-extended space. The prerequisites which Grassmann set for his changes amounted to examining the characteristics of certain transformation groups. The algorithms taken by Grassmann belong to certain groups of transformations, to the general projective group, to the group of affine transformations (in the A_1) as well as to the group of rotations about a fixed point and to the more comprehensive group of Euclidean motions (in the *Geometrische Analyse* of 1845). Grassmann's ideas were a starting point for the classification of various geometric directions in Klein's *Erlanger Programm*[7] and in subsequent research. His classification system first did not contain, for example, *affine* geometry which one would have expected from his knowledge of the A_1. Klein

6 See also Klein 1928c, p. 33 and Blaschke pp.117-121, 156. Klein first expounded Grassmann's theories in his lectures in the summer term 1890 within the framework of the cycle of his geometry lectures at this time. He himself pointed out that he had not mentioned Grassmann's name in the preceding term but that he was one of the "most outstanding geometricians of this century..."(Klein 1890, p. 9). Klein's more intensive treatment of Grassmann's work led later to the publication of Grassmann's *Collected Works.*

7 See Engel's comment in (Grassmann 1894, pp. 405f). In a letter of 21 January 1911 to Engel, Klein judged Grassmann as "an affine geometrician and not a projectivist. I realized this in late autumn of 1871 and it helped me (along with the study of Moebius and Hamilton and the processing of all the impressions gained in Paris) develop my concept for the later *Erlanger Programm.* I differentiated (depending on the adjoint group) between different 'methods' of geometry" (Grassmann 1911, p. 312).

first "inserted" affine geometry between projective geometry and the usual metric geometry in his lectures on number theory in 1895-96.[8]

In his famous lectures *Elementarmathematik vom höheren Standpunkte aus*, Klein introduced the geometry lecture by expounding, in the first three chapters, what he called "das Grassmannsche Determinantenprinzip": using an algebraic concept, the determinants as a means to classify geometrical objects, in the plane as well as in space. Klein hence developed his approach of classifying geometrical transformations, even for rectangular coordinate transformations, as an application of Grassmann's methods for determinants (Klein 1925, pp. 21-73, 161ff). The detailed presentation is a remarkable, thorough, systematic classification in accordance with Grassmann's system. In his likewise famous lectures on the development of mathematics in the 19th century, Klein resumed this exposition (part I 1926, pp. 173-181; part II 1927, pp. 5-12).

Since he also used the vector calculus based on Hamilton, he explained which concepts came from Grassmann (line element, plane element (plane segments), "planar quantity" (Plangröße), inner and outer products) and which from Hamilton (scalar, vector, scalar and vector products), and explained in the end his view on how the theories of both these men are related:

"Whereas Grassmann pursues the invariant theory in his 'lineale Ausdehnungslehre' which is part of the group of affine transformations leaving invariant the co-ordinate origin, the later Grassmann in his 'complete linear extension theory' [A2], and Hamilton in his 'Quaternion calculus', base their observations on the group of rotations..." (Klein 1925, p. 57f).

Foundations of Arithmetic

Klein was active in spreading the view that Grassmann had been one of the first to investigate the foundational characteristics of arithmetic. In these investigations, Grassmann had attained the introduction of hypercomplex numbers. In numerous lectures, Klein presented Grassmann's introduction of alternating numbers and their close connection to the theory of determinants. His lecture *Funktionentheorie in geometrischer*

---

8 When his *Gesammelte Mathematische Abhandlungen* (Collected Papers) were published, Klein emphasized that he had not considered 'affine' geometry in his *Erlanger Programm*, due to a 'one-sided tradition': "I first did this in detail in 1895-96 in my [...] lectures about number theory and then later on always included it in the beginning. Only in this way can one really do justice to the work of *Moebius* and *Grassmann* and first here does the algebraic invariant theory finds its full geometrical interpretation (... if one confines oneself to entire homogeneous linear changes of variables)" (Klein 1921, p. 320f).

Behandlungsweise (winter term 1880/1881) contained in its introduction a section on *Allgemeine Theorie der komplexen Zahlen* with the explanation of the "two most important systems of numbers,... Hamilton's Quaternions and Grassmann's numbers..."(Klein 1880/81, 20-21). He also briefly treated, one year later, the extension of the number system in his lecture *Projectivische Geometrie, Teil I* (Klein 1881/82, p. 204).

With respect to Grassmann, there were several papers in the 1870s and 1880s on the axiomatic foundation of the number concept (E. Schröder, G. Peano).

For Klein, Grassmann's exposition of foundations of arithmetic in his *Lehrbuch der Arithmetik* (1861) was of particular interest because it included conceptual elements along with logical ones. In 1908, in his comments about *Die logischen Grundlagen des Rechnens mit ganzen Zahlen,* Klein differentiated between four interpretations and preferred that one which retained a minimum of intuition (Klein 1924, p. 11-18). In his lectures on the development of mathematics in the 19th century, Klein characterized Grassmann as one of the first - along with Martin Ohm - who had reflected about the foundations of arithmetic (Klein 1926, p. 178-180).

3. RECEPTION OF GRASSMANN'S METHOD OF GENERATING ALGEBRAIC CURVES

Grassmann's method of generating algebraic curves and surfaces, especially curves of the third order, was received more rapidly, earlier and more often than his other results. This statement is true in general and also for the Clebsch school. In Clebsch's correspondence with Schlegel (November 1871), it was at first primarily Grassmann's method of generating curves of the third order which Clebsch considered to be most remarkable (Grassmann 1911, p. 400). Clebsch went into detail on this in his lectures, thus imparting this information to his students.[9] In an article *Über zwei Erzeugnisarten der ebenen Curven dritter Ordnung* sent on 4 March 1872 to the journal *Mathematische Annalen* for publication, Clebsch compared the types of generation - in accordance with Otto Hesse (1811-1874) and Heinrich Schröter (1829-1892) - to those of Grassmann. Clebsch described Grassmann's method of generation as "mechanical", pointing out in particular "that *every* cubic curve can be generated in this way" (Clebsch 1872 b, pp. 424-426). This

---

[9] Some evidence how present Grassmann was in the mathematical life around Clebsch is given by the gift offered by Klein to Clebsch on his birthday on 19 January 1872: a model of the lineal generation of curves of the third order, following Grassmann's methods (Grassmann 1911, p. 312).

subject was also disseminated through Clebsch's lectures published by Lindemann (Clebsch/Lindemann 1876, pp. 536-541). Lindemann, who prepared the edition of these lectures under Klein's supervision in 1873 in Erlangen, twice read papers in Klein's mathematical seminar on "Grassmann's Generation of Plane Curves", in June 1873 and in March 1875.[10] For Clebsch, Grassmann's generation of planes by projective bundles of planes was - as it was for the Italian mathematician Luigi Cremona (1830-1903) - the basis for investigations into mappings of third-order surfaces onto the plane. Thus Grassmann's work influenced the beginning of the mapping theory, the geometry on cubic surfaces (see Clebsch 1873, p. 33).

Comments about the group of all point transformations in Klein's *Erlanger Programm* reveal the influence that Grassmann's "lineal" construction of algebraic elements in general had on Klein's thought processes. He stated:

"Projective geometry can likewise be obtained from the geometry of all point transformations by adjunction of manifolds of the planes, just as elementary geometry can be obtained from the projective geometry through adjunction of the infinite spherical circle. In particular, for example, from the perspective of all point transformations, we have to conceive of the designation of a surface as an algebraic one belonging to a certain order as an invariant relation to the manifold of planes. This becomes very clear if one links, as Grassmann did, the generation of algebraic quantities to their lineal construction." (Klein 1921, p. 483).

In his lectures Klein repeatedly described Grassmann's method of "lineal construction," or the generation of algebraic quantities (literal numbers), through "planimetric products", praising the comprehensibility of this method. Klein usually confined his comments to the plane and, in particular, to curves of second and third order. Thus, the lecture he held in Göttingen on *Riemannsche Flächen* contained in its second part (*Beziehung von Riemanns Theorie zur Lehre von den algebraischen Kurven*) a historical survey about the foundation of the theory in which he characterized Grassmann's "lineal" mechanism as a third way of expressing the concept of algebraic curves in a purely geometric way (Klein 1891/92, pp. 121-123). In his lecture *Anwendung der Differential- und Integralrechnung auf Geometrie (Eine Revision der Prinzipien)* in the summer term 1901, Klein explained the "lineal mechanism" in a particularly detailed manner and expressed his regret that Grassmann's concept was still not to be found in textbooks (Klein 1928, p. 132, and 1926, pp. 180-181).

---

[10] Summaries of the lectures are contained in Vol. I of the files of Klein's mathematical seminars (Mathematisches Institut Göttingen).

Doctoral theses on Grassmann inspired by Klein and Brill

Felix Klein and Alexander Brill encouraged students to undertake research into Grassmann's methods of the generation of higher order curves in their dissertations. One student was Friedrich Dingeldey (1859-1939) who devoted his attention in a special way to Grassmann's method of generation. In 1885, inspired by Brill and Klein, he defended his thesis with Klein on the generation of curves of the fourth order through motion mechanisms. In this thesis, he investigated different possibilities of generating corresponding curves and examined the question "...how Grassmann's motion mechanism can be modified to obtain curves with one, two or three double points..." (Dingeldey 1885, p. VI). Grassmann had only demonstrated that with his mechanism one can obtain general fourth-order curves. Dingeldey also answered the reverse question, how the elements of Grassmann's mechanism have to lie in order to generate a fourth-order curve (in particular for the rational curve). After completing his dissertation, Dingeldey continued research related to Grassmann's work.

Alexander Brill, who, in 1884, transferred from the Technical University in Munich to the University of Tübingen, also promoted additional geometrical research linked to Grassmann. In Tübingen, he entered a context where Paul du Bois-Reymond (1831-1889) had started to attract the interest of students to Grassmann. There, for example, Rudolf Mehmke (1857-1944) had defended, in 1880, his thesis on the topic *Anwendung der Grassmannschen Ausdehnungslehre auf die Geometrie der Kreise und der Ebene*. In 1886, Friedrich Kölmel was awarded the doctor degree for a thesis on Grassmann's methods of generating plane curves of the third order, in Tübingen.[11] Karl Kommerell (1871-1962), also one of Brill's students[12] likewise treated the geometric applications of Grassmann's *Ausdehnungslehre*.

Franz Meyer (1856-1934), who was influenced by Klein and Brill while in Munich and became acquainted with the methods of the Clebsch School, defended a thesis on applications of topology on the shape of algebraic curves in 1878. He had participated for three semesters (winter term 1876/77 to winter term 1877/78) in Klein's seminars and had delivered six papers. The last paper *Topologisches über algebraische Curven* already hinted at the direction his dissertation would take. Klein included Meyer in a list of his doctoral students which he prepared in 1913 (Klein *Nachlaß*, 22L:5, Bl. 1; Arndt 1935).

---

[11] Kölmel who worked as a teacher at several *Gymnasien* - like so many of Brill's students - also later published research on generation of plane curves of the third order.

[12] Kommerell was awarded his doctorate in 1897 in Tübingen with a thesis on curvature of two-dimensional structures in plane space of four dimensions.

4. THE EDITION OF GRASSMANN'S COLLECTED PAPERS

It is thanks to the efforts of a former student of Clebsch that Grassmann's *Gesammelte Werke* - the first volume of which appeared in 1894 at the occasion of the 50th anniversary of the original *Ausdehnungslehre* - were published; other mathematicians of the Clebsch School supported the project. Felix Klein's scientific and organizational commitment to making Grassmann's work widely known must be particularly stressed. As a successor of Clebsch and Carl Neumann (1832-1925), Klein - together with Adolph Mayer (1839-1908) - took over the managing editor position of the journal *Mathematische Annalen* in 1876 and was particularly interested in publishing Grassmann's articles as quickly as possible. (Tobies/Rowe 1990, p. 86f). The publication of Grassmann's *Gesammelte Werke* was the direct result of Klein's initiative. Klein obtained the permission from the Grassmann family to publish material from the *Nachlaß*, and the cooperation of family members; he succeeded in winning Friedrich Engel as the chief editor and the *Sächsische Gesellschaft der Wissenschaften zu Leipzig* as the sponsoring organization.

In addition to the editor-in-chief Friedrich Engel, Eduard Study and Jacob Lüroth, both Clebsch students, participated directly in the edition. While Study cooperated mainly in the edition of the A_1 and of the *Geometrische Analyse*, Lüroth was responsible for the edition of Grassmann's works on mechanics.

5. GRASSMANN WITHIN THE FRAMEWORK OF GREAT LITERARY UNDERTAKINGS

Grassmann's achievements received their due recognition in the ***Encyklopädie der mathematischen Wissenschaften.*** The major undertaking of the encyclopedia was the result of a joint initiative on the part of Felix Klein, Franz Meyer and Heinrich Weber (1842-1913) and was particularly shaped by Klein's efforts who assumed a certain final editorial responsibility (see Tobies 1994). The participation of many of the authors was mainly due to his personal intercession.

The index of names for Volume III (Geometry) shows that Grassmann was mentioned on 240 pages. The references to Grassmann are more extensive in the geometry volume than, for example, the references to Gauß. In the introductory remarks to Volume III, the two editors Franz Meyer and Hans Mohrmann (1881-1941) praised Grassmann as one of the "...authoritative German scholars..., who have restored German mathematics in the field of geometry to international renown." (Encyklopädie, Vol. III, p. VI).

In many different chapters, reference was made - depending, of course, on the topic - to Grassmann's individual achievements. All of Grassmann's essential results as well as their further development by other scientists were taken into consideration.[13] His entire geometric system was discussed in detail in section II, *Systeme geometrischer Analyse* (Encyklopädie, Vol. III, 1.2). From other sources, it becomes clear that Klein attached particular importance to this presentation. The original plan was that Heinrich Burkhardt (1861-1914), whom Klein had won as one of the most important editors for the project, should prepare this section.[14] Because of his premature death, the entire chapter *Systeme geometrischer Analyse* was handed over to the Viennese mathematician Hermann Rothe (1882-1923) who had studied with Klein for some years. The actual elaboration of the section on Grassmann's *Ausdehnungslehre* (in accordance with Rothe's concept) was left to Alfred Lotze (born 1882)[15].

Grassmann as a mathematician was also mentioned in works which were designed for a general public. This reference mainly pertains to Klein's work on the development of mathematics, which he wrote for the University Exhibition in 1893 in Chicago and which was also translated into English. In this work, he referred to Grassmann as an important representative of 19th century mathematics in Germany "...who compiled the results of more recent geometrical speculations into a system in his *Ausdehnungslehre* ..." (Klein 1893, p. 13). Klein's correspondence with American mathematicians during this period document how important Klein considered the propagation of Grassmann's results for which he had initiated the publication of the *Gesammelte Werke*.[16]

Here, one must also mention that another Clebsch student, Aurel Voß, also referred to Grassmann's *Ausdehnungslehre* as the beginning of *n*-dimensional geometry in his comments *Über die mathematische Erkenntnis* which he had written for the great encyclopedia *Kultur der Gegenwart* (Voß 1914, p. 109,110).

---

[13] It is not feasible to include a detailed list of references to Grassmann here. A number of the participating authors can, in a certain sense, be regarded as representatives of the Clebsch School of the second generation, in particular Julius Sommer (1871-1943) and Gino Fano (1871-1952).

[14] See Encyklopädie, Vol. III 2.1. (1906), p. 357.

[15] Both Rothe and Lotze, in their own research, were working on problems related to the *Ausdehnungslehre* (see Radon 1926, p. 172 and Lotze 1922).

[16] On 22 September 1892, Klein wrote, for example, to Simon Newcomb, that he was preparing to publish Grassmann's works. I wish to thank David E. Rowe for having found this letter.

FINAL REMARK

The special efforts of the Clebsch school members with regard to Grassmann's results is, in the final analysis, the product of Clebsch's openness to an extensive field of mathematical investigations, his openness to knowledge won in other schools of thought or by other individuals. Klein, in his introduction to Volume I of the Clebsch lectures edited by Lindemann, described Clebsch's approach as follows: "...from him we learned to take the work of others fully into consideration and to weave these results into our own..." (Clebsch/Lindemann 1876, p. IV). This approach was a component of the "...great program of unification (of persons as well)..." inaugurated by Clebsch (Klein 1893).

Fachbereich Mathematik, Universität Kaiserslautern

REFERENCES

Unpublished Sources

Niedersächsische Staats- und Universitätsbibliothek Göttingen, Handschriften-abteilung: Cod. Ms. Felix Klein: Felix Klein, *25 Jahre moderne Mathematik*. Manuskript einer Vorlesung vom 20. 2.1893.

Mathematisches Institut, Universität Göttingen: Protokollbände der mathematischen Seminare Felix Kleins. 29 volumes (Manuscript).

Bibliography

B. Arndt, "Wilhelm Franz Meyer zum Gedächtnis," *Jahresbericht der Deutschen Mathematiker-Vereinigung*, 1935, *45*, 99-113.

Wilhelm Blaschke (Ed.), Klein, Felix, *Vorlesungen über höhere Geometrie*. 3. Auflage, (Berlin: Julius Springer, 1926).

Alfred Clebsch, "Zum Gedächtnis an Julius Plücker," Julius Plückers *Gesammelte Wissenschaftliche Abhandlungen*. Bd. 1 *Mathematische Abhandlungen*, ed. Arthur Schönflies (Leipzig: B.G. Teubner, 1895) IX-XXXV.

Alfred Clebsch, "Über eine Fundamentalaufgabe der Invariantentheorie." *Abhandlungen der Kgl. Gesellschaft der Wissenschaften*. 2. März 1872, 3-62. [1872a]

Alfred Clebsch, "Über zwei Erzeugnisweisen der ebenen Curven dritter Ordnung," *Mathematische Annalen*, 1872, *5*, 422-426. [1872b]

(Clebsch 1873) "Rudolf Friedrich Alfred Clebsch. Versuch einer Darlegung und Würdigung seiner wissenschaftlichen Leistungen von einigen seiner Freunde." *Mathematische Annalen*, 1873, *7*, 1-55.

(Clebsch/Lindemann 1876), *Vorlesungen über Geometrie von Alfred Clebsch.* Bearbeitet und herausgegeben von Ferdinand Lindemann. Mit einem Vorworte von Felix Klein. Bd. 1: Geometrie der Ebene. (Leipzig: B.G. Teubner, 1876).

Friedrich Dingeldey, *Über die Erzeugung von Curven vierter Ordnung durch Bewegungsmechanismen*, Inauguraldissertation zur Erlangung des Doktorgrades der Philosophischen Facultät der Universität Leipzig (Leipzig: B.G. Teubner, 1885).

Encyklopädie der mathematischen Wissenschaften mit Einschluß ihrer Anwendungen. Bd. III: Geometrie. Drei Teile (Leipzig und Berlin: B.G. Teubner, 1903 - 1934).

Hermann G. Grassmann, *Gesammelte mathematische und physikalische Werke.* Friedrich Engel (Hsrg.), und unter Mitwirkung von Jacob Lüroth, Eduard Study, Justus Grassmann, Hermann Grassmann d. J., Georg Scheffers, 6 Bände (Leipzig: B.G. Teubner, 1894 - 1911). Reprint New York 1972. Band I, 1: 1894, Band I, 2: 1896, Band III, 2: 1911.

Felix Klein, *Projectivische Geometrie*, Teil I, Vorlesung, gehalten im Wintersemester 1881/82 (mimeographed copy, distributed by Teubner, Leipzig).

Felix Klein, *Nicht-Euklidische Geometrie I*, Vorlesung, gehalten im Wintersemester 1889/90. Ausgearbeitet von Fr. Schilling (Göttingen 1892). (mimeographed copy, distributed by Teubner, Leipzig)

Felix Klein, *Nicht-Euklidische Geometrie II*. Vorlesung, gehalten im Sommersemster 1890. Göttingen 1890 (mimeographed copy, distributed by Teubner, Leipzig).

Felix Klein, *Einleitung in die höhere Geometrie I*, Vorlesung, gehalten im Wintersemester 1892-93. Ausgearbeitet von Friedrich Schilling. Göttingen 1893 (mimeographed copy, distributed by Teubner, Leipzig).

Felix Klein, "Mathematik," *Die Deutschen Universitäten*. ed. Wilhelm Lexis. Band 2. (Berlin: Verlag von A. Asher & Co., 1893), 6-15. English translation in: id.:, *The Evanston Colloquium Lectures on Mathematics*, Reported by Alexander Ziwet. New York and London: Macmillan and Co., 1894, p.99-109.

Felix Klein, *Gesammelte Mathematische Abhandlungen*. Band 1 (Berlin: Julius Springer, 1921).

Felix Klein, *Elementarmathematik vom höheren Standpunkte aus*, Vol. I: *Arithmetik, Algebra, Analysis*. Ausgearbeitet von Ernst Hellinger. Für den Druck fertig gemacht und mit Zusätzen versehen von Fr. Seyfarth. 3. Auflage (Berlin: Julius Springer, 1924).

Felix Klein, *Elementarmathematik vom höheren Standpunkte aus*, Vol. II: *Geometrie*. Ausgearbeitet von Ernst Hellinger. Für den Druck fertig gemacht und mit Zusätzen versehen von Fr. Seyfarth. 3. Auflage (Berlin: Julius Springer, 1925).

Felix Klein, *Vorlesungen über die Entwicklung der Mathematik im 19. Jahrhundert* (Berlin: Julius Springer, Band I 1926, Band II 1927). English Translation: *Development of Mathematics in the 19th Century*. Part I. Translated by M. Ackermann (Brookline, Mass.: Math. Sci. Press, 1979).

Felix Klein, *Elementarmathematik vom höheren Standpunkte aus*, Vol. III: *Präzisions- und Approximationsmathematik*. Ausgearbeitet von Conrad H. Müller. Für den Druck fertig gemacht und mit Zusätzen versehen von Fr. Seyfarth. 3. Auflage (Berlin: Julius Springer, 1928).

Felix Klein, *Vorlesungen über Nicht-Euklidische Geometrie*, Für den Druck neu bearbeitet von W. Rosemann. (Wesentlich geänderte Fassung der autographierten

Vorlesung "Nicht-Euklidische Geometrie I" (Wintersemester 1889/90) (Berlin: Julius Springer, 1928). [1928c]

Felix Klein, *Riemannsche Flächen*. Vorlesungen, gehalten in Göttingen 1891/92. Hrsg. und kommentiert von Günther Eisenreich und Walter Purkert (Leipzig: B.G. Teubner, 1986).

Felix Klein, *Funktionentheorie in geometrischer Behandlungsweise*. Vorlesung, gehalten in Leipzig 1880/81, hrsg. von Fritz König. (Leipzig: B.G. Teubner, 1987).

Alfred Lotze, *Die Grundgleichungen der Mechanik, insbesondere starrer Körper. Neu entwickelt mit Grassmanns Punktrechnung*. (Leipzig und Berlin: B.G. Teubner, 1922)

Alfred Lotze, "Die Grassmannsche Ausdehnungslehre," *Encyklopädie der mathematischen Wissenschaften mit Einschluß ihrer Anwendungen*, Bd. III, 1.2. (Leipzig: B.G. Teubner, 1923) 1425-1550.

Jacob Lüroth, "Das Imaginäre in der Geometrie und das Rechnen mit Würfeln. Darstellung und Erweiterung der v. Staudt'schen Theorie," *Mathematische Annalen*, 1874, *8*, 145-214.

Johann Radon, "Hermann Rothe," *Jahresbericht der Deutschen Mathematiker-Vereinigung*, 1926, *35*, Abt. 1, 172-175.

Viktor Schlegel, "Die Grassmann'sche Ausdehnungslehre. Ein Beitrag zur Geschichte der Mathematik in den letzten fünfzig Jahren," *Historisch-literarische Abtheilung der Zeitschrift für Mathematik und Physik*, 1896, *41*, 1-21 and 41-59.

Richard Sturm, Ernst Schröder und L. Sohncke, "Hermann Grassmann. Sein Leben und seine mathematisch-physikalischen Arbeiten," *Mathematische Annalen*, 1879, *14*, 1-45.

Renate Tobies, "Felix Klein in Erlangen und München: ein Beitrag zur Biographie," *Amphora. Festschrift für Hans Wussing zu seinem 65. Geburtstag*, eds. S.S. Demidov, M. Folkerts, D.E. Rowe und Ch. J. Scriba. (Basel, Boston, Berlin: Birkhäuser Verlag, 1992), 751-772.

Renate Tobies, "Mathematik als Bestandteil der Kultur - Zur Geschichte des Unternehmens 'Encyklopädie der mathematischen Wissenschaften mit Einschluß ihrer Anwendungen'," *Mitteilungen der Österreichischen Gesellschaft für Wissenschaftsgeschichte*, 1994, *14*, 1-90.

Renate Tobies and David E. Rowe (Eds.), *Korrespondenz Felix Klein - Adolph Mayer. Auswahl aus den Jahren 1871-1907* (Leipzig: B.G. Teubner, 1990).

Aurel Voß, *Über die mathematische Erkenntnis* Abt. I, Teil III von *Die Kultur der Gegenwart*, hrsg. von Paul Hinneberg (Leipzig und Berlin: B.G. Teubner, 1914).

Eduard A. Weiss, "Ein Nachruf auf Eduard Study," *Sitzungsberichte der Berliner Mathematischen Gesellschaft*, 1930, 29, 52-77.

DAVID E. ROWE

ON THE RECEPTION OF GRASSMANN'S WORK IN GERMANY DURING THE 1870'S

INTRODUCTION

It has often been remarked that Grassmann's mathematics was not widely appreciated during his lifetime. Although awareness of the dimensions of his achievements began to spread in the early 1870's, even in Germany relatively few mathematicians appear to have been well acquainted with either the original 1844 edition of Grassmann's *Ausdehnungslehre* or the mathematically more accessible edition of 1862. The main reasons for this weak and rather delayed reception have been described often enough---Grassmann's isolated working environment and his nearly impenetrable language---but there are a number of related aspects that still deserve closer consideration.[1]

It ought to be noted, for example, that sheer bad luck played a role as well. As Grassmann himself recorded in 1877, in the introduction to the second edition of his original *Ausdehnungslehre*, his work might have become better known had not two of its distinguished proponents passed on unexpectedly.[2] Hermann Hankel (1839-1873) and Alfred Clebsch (1833-1872) had both drawn attention to the importance of Grassmannian ideas, but their efforts in this direction ended abruptly in the early 1870's when both died prematurely.[3] Afterward, no one of comparable stature arose to champion Grassmann's cause, and those who did so tended to promote a rather narrow, sectarian approach to his mathematics. Grassmann's own efforts to highlight the significance of his *Ausdehnungslehre* during the last years of his life only served to

---

The author wishes to thank Gert Schubring for numerous helpful suggestions in preparing this article.

1 For an account of Grassmann's work and its influence, see Crowe 1967, pp. 54-95.

2 Grassmann Werke I.1, pp. 18-19; quoted in English translation in Crowe 1967, p. 91.

3 Hankel presented Grassmann's theory in his (Hankel 1867); Clebsch commented on the theory in his (Clebsch 1872).

G. Schubring (ed.), Hermann Günther Graßmann (1809–1877): Visionary Mathematician, Scientist and Neohumanist Scholar, 131–145.

reinforce this tendency. The resulting tensions between a zealous band of purists, on the one hand, and those mainstream mathematicians, on the other, who sought to adapt Grassmann's concepts to other established branches of mathematics---affine and projective geometry, hypercomplex number systems, invariant theory, etc.---have often been mentioned, but never studied in any detail. The present essay takes up this theme during the earliest phase of the conflict in the 1870's. Before turning to this issue directly, however, a few general observations with regard to Grassmann's career and the mathematical world of his day should be made.

MATHEMATICS AT UNIVERSITY AND GYMNASIUM

The belated recognition of Hermann Grassmann's importance and stature led some of his closer followers to view him as a kind of martyr figure, a man who had been forced to toil away his life as a school teacher in Stettin and whose brilliant genius only began to be appreciated after his death in 1877.[4] Of course, no one today would dispute that Grassmann was a man of extraordinary gifts and impressive accomplishments. But one should also bear in mind the times and culture in which he lived. Grassmann may not have been particularly enamored with his teaching duties at the various schools in which he taught (he succeeded his father, Justus Günther Grassmann (1779-1852) at the Stettin Gymnasium upon the latter's death in 1852), but neither should he be compared with a figure like Heinrich Mann's Professor Unrat. In fact, Grassmann was the product of a culture that generally honored, and often even revered its school teachers, and furthermore attached extraordinarily high value to scholarly productivity. Indeed, the pursuit of higher learning lay at the heart of Germany's neohumanist tradition.[5] In Prussian secondary schools during the mid-nineteenth century this research ethos was cultivated by and transmitted through numerous scholars with considerable élan. These *Oberlehrer* were generally treated with considerable respect and deference, if not by their pupils then at least by their peers, and they often consorted with members of the upper echelon of local society.[6]

During Grassmann's day the gulf separating those who taught at the universities and their counterparts in the *Gymnasien* was not nearly so wide as it later became at the end of the century. Academically and

---

4 See, for example, the remarks in Klein 1926, pp. 181-182; see also Friedrich Engel: *Grassmann's Leben*, Grassmann III.2, pp. 311-334.

5 On this "research imperative" see Turner 1971.

6 For further details, see Schubring 1983.

socially, an *Oberlehrer* was only one remove from a university professor, and those who held the position were referred to by their proper title: *Herr Professor*. Gymnasium professors like Grassmann played a major part in Germany's swift ascent in the world of mathematics, which until around 1830 was dominated by France. Ernst Eduard Kummer (1810-1893) and Karl Weierstraß (1815-1897), the two most influential mathematics teachers in Germany during the 1860's and 1870's, had both begun their careers teaching at secondary schools. Other distinguished mathematicians like Hermann Schubert (1848-1911), inventor of the "Schubert calculus," spent their entire professional careers teaching their young charges elementary mathematics. Some highly significant, non-mathematical findings took place in these settings. Thus, Leopold Kronecker (1823-1891) was discovered by Kummer when he taught at the *Gymnasium* in Liegnitz, whereas Schubert befriended young Adolf Hurwitz (1859-1919), his most gifted pupil, in Hildesheim. These particularly striking examples attest to a pattern of truly impressive quality. Indeed, it is safe to say that had it not been for the high standards and intense dedication demonstrated by scores of now forgotten *Oberlehrer*, Germany's sudden emergence as the leading nation in mathematics during the second half of the nineteenth century would have been unthinkable.

As these remarks imply, the scholarly ideals that animated professors at the Prussian *Gymnasien* did not differ markedly from the research ethos at insitutions like Berlin University, where Grassmann studied theology, philosophy, and philology in the late 1820's. This background could, of course, hardly be considered ideal preparation for someone who aspired to a career in mathematics at the university level.[7] But, as a student, such a thought evidently never entered Grassmann's mind, and, in fact, only very few mathematicians in Germany seriously contemplated the possibility of pursuing a career within the German system of higher education, for two obvious reasons. First, there were simply too few positions available, and second, the long, arduous road that had to be travelled before one received a call ("Ruf") to a university professorship (it was considered bad form to apply for an open position) posed considerable hardships. Little wonder that even some of the most promising talents, like David Hilbert (1862-1943), opted to take the *Staatsexamen*, which qualified a person to teach in the secondary schools, rather than risk having nothing to fall back on later. Most of those who *did* choose to go on, by habilitating at a German university, came from families with fairly substantial means. Habilitation gave one

---

7 Although both Gauss and Jacobi pursued serious philological interests before ultimately turning to mathematics.

the right to teach as a *Privatdozent* and to collect fees from the students, but nothing more. It was strictly an unsalaried position, and many were discouraged by the fact that it often took many years before one could hope to receive a call to a genuine professorship. The fact that Grassmann never attained this more exalted position, although he was given serious consideration for a position at Greifswald in 1847, was certainly unfortunate, but in view of the novelty of his work and, even more, its obscurity, this outcome should hardly seem surprising.

PATTERNS OF RECEPTION

With regard to the slow diffusion of Grassmann's ideas within wider mathematical circles, it should be noted that this type of transmission pattern was closer to the norm rather than the exception throughout most of the nineteenth century. Even when the ideas issued from highly esteemed individuals, like Gauss or Riemann, it often took many decades before important new breakthroughs could be absorbed and understood. As a general rule, the more novel or "revolutionary" the ideas (e.g., Galois's theory of algebraic equations), the longer it took to assimilate them. But within Germany's decentralized institutional structures, even relatively routine transmission of information was not easy as there were few publication outlets and even fewer possibilities for meeting face to face. Before the founding of *Die Mathematische Annalen* in 1869, Crelle's *Journal für die reine und angewandte Mathematik* represented the only specialized mathematical journal in Germany, and not until 1890 did German mathematicians establish a national organization, the *Deutsche Mathematiker-Vereinigung*, which provided a regular forum for exchanging information, both formally and informally. Thus, prior to 1890 most university mathematicians conducted their research in rather closed, and isolated settings. Among the leading geometers of this period, Julius Plücker (1801-1868) worked in quiet solitude in Bonn, August Ferdinand Möbius (1790-1868) was employed as an astronomer in Leipzig, and K.G.C. von Staudt (1798-1867) taught for many years in the solitude of Erlangen. Much of Möbius's mathematical work only became more widely accessible in the 1880's, and it took several decades before Staudt's fundamental contributions to the foundations of projective geometry became widely known.[8]

No doubt Grassmann faced even graver difficulties than either Möbius and von Staudt, but these were largely due to the esoteric philosophical

---

[8] Klein, for example, failed to recognize the extent to which the ideas in his *Erlanger Programm* had already been anticipated by Möbius until the mid-1880's when he coedited the latter's collected works (see Klein 1921-1923, vol. 1, p. 497). On von Staudt's career, see Rowe 1996.

notions that motivated his work, ideas closely akin to Schleiermacher's thought (see Lewis 1977). Even those who were motivated enough to penetrate into the misty realm of ideas set forth in Grassmann's 1844 edition of the *Ausdehnungslehre* usually came away from it feeling like they had entered into a strange new world. One of the few foreigners who undertook the journey was William Rowan Hamilton (1805-1865), whose own work on quaternions had been partly inspired by an attempt to conceptualize a Kantian-inspired notion of pure time.[9] Hamilton no doubt possessed the proper philosophical temperment (and linguistic skills) needed to tackle Grassmann's work, but unfortunately his principal concern appears to have been to reassure himself that he alone had discovered quaternions. Still, he expressed considerable admiration for Grassmann's originality, an opinion that was also shared in Italy by Luigi Cremona (1830-1903) and Giusto Bellavitis (1803-1880) (see Crowe 1967, p. 88).

In Germany it seems fair to say that Grassmann's work failed to attract serious and sustained interest among leading contemporary mathematicians not so much for lack of sympathy with his goals as for the manner in which he approached them. Many expressed bewilderment when faced with the convoluted formulations found in the *Ausdehnungslehre*. Particularly telling in this regard was Möbius's reaction, since his *Baryzentrische Calcül* of 1827 revealed strong affinities with Grassmann's fundamental conceptions. Grassmann even visited Möbius after completing work on his *Ausdehnungslehre*, but when he asked the latter to write a review of the book Möbius answered:

> "I reply that I was sincerely pleased to have come to meet in you a kindred spirit, but our kinship relates only to mathematics, not to philosophy. As I remember telling you in person, I am a stranger to the area of philosophic speculation. The philosophic element in your excellent work, which lies at the basis of the mathematical element, I am not prepared to appreciate in the correct manner or even to understand properly. Of this I have become sufficiently aware in the course of numerous attempts to study your work without interruption; in each case however I have been stopped by the great philosophic generality."[10]

Despite this mild rebuff, Möbius's sympathy was apparent. Moreover, he clearly understood and appreciated Grassmann's work, or at least enough about it that he could offer him some valuable advice, namely that he should compete for a prize announced the previous year by the Jablonowski Society in Leipzig. This scientific society wished to receive a work which would develop the "geometrical calculus" Leibniz had sketched in 1679 in a letter to Huygens, first published in 1833 (Crowe,

---

9 Hamilton began reading Grassmann's book in 1852. See Hankins 1980.

10 A. F. Möbius to H. G. Grassmann, 2 February 1845, as quoted in translation in Crowe 1967, pp. 78-79.

pp. 3-5). Taking up Möbius's advice, Grassmann submitted his *Geometrische Analyse, geknüpft an die von Leibniz erfundene geometrische Charakteristik*. This essay won the prize (as the only entry submitted) in 1846 and was published the following year along with a commentary by Möbius.[11] Like the *Ausdehnungslehre*, however, it failed to elicit sustained interest, despite favorable comments from some leading European mathematicians.

Somewhat better known were a number of articles Grassmann published in *Crelle's Journal* between 1842 and 1856. During these years, this highly regarded journal carried fourteen papers by Grassmann, most of them concerned with methods for generating algebraic curves and surfaces. Although these results were both elegant and easy to comprehend, few geometers appear to have followed up on them. Nevertheless, many of these papers seem to have been read and appreciated.[12]

After 1856, Grassmann's mathematical productivity slackened, but certainly not his zeal for propagating his ideas. He presented his *Ausdehnungslehre* again in 1862, but in strictly Euclidean fashion and without reference to its philosophical roots. Yet even in this form, the unfamiliarity of Grassmann's terminology and the algorithms he employed posed major obstacles for his readers, most of whom gave up before they had gotten very far into the book. Like the 1844 edition, this account, too, failed to awaken substantial interest, and soon after its appearance Grassmann turned to other endeavors, most notably philological studies, discouraged by the indifference of professional mathematicians.

In the meantime, however, he and his brother, Robert, had turned to a different audience, hoping for a more enthusiastic response from mathematics teachers at the *Gymnasien*. Thus, in 1860 they published the first of a projected three volumes on elementary mathematics which were to serve as modern textbooks for the secondary schools (Grassmann 1860). Yet the response of the mathematics teachers proved just as discouraging, as that of the research mathematicians. Whatever they thought about the mathematical merits of this new, "fully rigorous" approach to arithmetic, the immediate consensus of opinion indicated that the presentation of it in this book was altogether inappropriate for teaching purposes (see Engel 1911, pp. 225-228). Despite this setback, Grassmann was convinced that his ideas could be and one day would be incorporated into the standard mathematical curriculum for *Gymnasien*; and others eventually came to share that conviction.

---

11 See *Grassmanns Werke*, I.1, pp. 321-398; Möbius's essay appears in I.1, pp. 613-633.

12 See, for example, Klein's comment, cited below.

The most important new convert was Victor Schlegel (1843-1905), who taught alongside Grassmann at the Stettin Gymnasium from 1866-68. Interestingly enough, Schlegel only began intensive study of Grassmann's *Ausdehnungslehre* in 1869, after he left Stettin to accept a position as *Oberlehrer* at the Gymnasium in Waren, a small town in Mecklenburg. Around this time he began work on a two-part textbook entitled *System der Raumlehre*, which he dedicated to Grassmann (Schlegel 1872). Part I of this work, published in 1872, dealt with elementary plane geometry, whereas Part II appeared in 1875, and presented a Grassmannian version of those portions of algebraic geometry closely connected with invariant theory (Schlegel 1875). By virtue of this two-part study, but especially the biography of Grassmann that he published in 1878, Schlegel emerged as the leading spokesman for the Grassmannian cause after the master's death.[13] Although Friedrich Engel, author of the definitive Grassmann biography published in volume 3 of the collected works, was rather dismissive with regard to Schlegel's *System der Raumlehre*---asserting that "[he] was not the man to put the old Grassmannian wine into new vessels"[14] ---he went out of his way to praise Schlegel's biography of Grassmann and his numerous efforts to promote interest in the man and his work.[15]

SCHLEGEL'S SYSTEM DER RAUMLEHRE

Like Grassmann, Schlegel hoped to show that the *Ausdehnungslehre* offered more than just another useful tool for the research mathematician. Indeed, he saw Grassmann's methods as applicable to mathematics at nearly every level, and consequently as holding the key to a much needed reform of the mathematics curriculum in the secondary schools. Thus, following the master's lead, Schlegel's *System der Raumlehre* had two principal objectives: to show the importance of Grassmannian ideas for pedagogical purposes, and to indicate the superiority of vector methods over techniques based on coordinate systems in presenting the main results of geometry, particularly projective geometry. He pursued the first of these goals in Part 1, which dealt with the basic theorems of school geometry, whereas the second objective was relegated to Part II, which only appeared three years later.

---

13 Schlegel's biography gives the most personal portrait of Grassmann available (Schlegel 1878).

14 Engel, *Grassmanns Leben, Grassmanns Werke*, III.2, p. 324.

15 Schlegel published over 25 works dealing with Grassmannian ideas over the course of his lifetime. Perhaps the most valuable for the historian is Schlegel 1896.

In the introduction to Part I, however, Schlegel made the two-fold purpose of his study more than clear, attacking both research mathematicians and pedagogues for ignoring Grassmann's work. Adopting a strident tone, he suggested that Grassmann's *Ausdehnungslehre* of 1844 had simply been too radical an advance forward for a "generation that viewed imaginary quantities as impossible and that shook its head in dismay at non-Euclidean geometry" (Schlegel 1872, p. vi). But this was not the only reason Schlegel gave to explain the indifferent reaction to Grassmann's theory: he also intimated that the playing field for mathematicians in Germany was far from fair, and that Grassmann's ideas had largely been neglected because he had not held a university chair, a circumstance that had prevented him from establishing a school for the promulgation of his ideas. Whereas younger mathematicians, as Schlegel observed,

"had been preoccupied with the elaboration of the theories set forth by Jacobi, Dirichlet, Steiner, Möbius, Plücker, Hesse, and others, all of whom gathered a circle of active pupils around them, fate failed to grant the author of the *Ausdehnungslehre* a similar influence" (ibid.).

Schlegel seemed to imply further that Grassmann could claim priority with regard to various methods usually ascribed to others. For, having ignored Grassmann's work, other mathematicians (he referred to Otto Hesse explicitly) had in the meantime developed alternative techniques very close to Grassmann's, but without fully realizing the advantages of the latter's coordinate-free, n-dimensional mode of representation. But priority claims were the least of Schlegel's concerns. Far more pressing, from his point of view, was the need to clarify the foundations for the treatment of extensive magnitudes in mathematics.[16] Thus, he decried the state of arithmetic and algebra, calling it a "conglomeration of loosely connected rules for calculation" supplemented by "a collection of arbitrary foundational principles, more or less vague explanations, and geometric artifices" (Schlegel 1872, p. xi). Rather than developing a sound basis for their analytic methods, he claimed that contemporary mathematicians tended to invent new symbolisms on an *ad hoc* basis, creating a Babel-like cacaphony of unintelligible languages.[17]

Schlegel went on to argue that this weakness had exerted an adverse effect not only on higher mathematics but on school mathematics as well. Indeed, he contended that such chaos inevitably reinforced the

---

16 Schlegel mentioned Martin Ohm's, *Versuch eines vollkommen consequenten Systems der Mathematik* as about the only noteworthy recent effort to clarify the foundations of mathematics. On Ohm, see Bekemeier 1987; and Schubring 1981.

17 This type of criticism had long been hurled at analytic geometers, leading many to conclude that the only legitimate approach to geometry was the purely synthetic one.

widespread view of mathematics as a kind of "black magic" accessible only to those with a predisposition for its abstruse formalisms. Schlegel further asserted that the neglect of foundations had led to the widely acknowledged lack of interest in mathematics in the schools. Echoing the views expressed in the preface of Hermann and Robert Grassmann's textbook, he proclaimed that the standard mathematics curriculum contained nothing whatsoever that could serve as a basis for a truly "scientific method." Thus his goal in presenting Part I of *System der Raumlehre* was to demonstrate how Grassmann's ideas could provide just such a foundation.

While Schlegel's claims were, no doubt, exaggerated and his remedy far-fetched, his book did offer some fresh perspectives that might well have borne fruit. Certainly it appeared at a propitious point in time, as by now the sorry state of mathematics instruction in the *Gymnasien* had become a matter of considerable concern. In fact, Schlegel could even call attention to an editorial statement in the newly founded *Zeitschrift für mathematischen und naturwissenschaftlichen Unterricht* indicating that the journal had come out strongly in favor of a rational reform of the teaching methods in geometry (Schlegel 1872, p. xii). But Schlegel also realized that any significant reform was bound to face strong resistance from those who had a vested interest in the *status quo*. Rather than mincing words, he confronted the conservative opposition head-on, stating that he was well aware that his book would not please "those who regard the present state of elementary mathematics as satisfactory" (*ibid.*, p. xiii). Drawing battle lines in advance, he appealed not to these readers, but to all those capable of "an *unbiased* [*vorurtheilsfreie*] consideration of new viewpoints" (*ibid.*, Schlegel's italics).

KLEIN'S CRITIQUE

As indicated already above by Engel's comments, Schlegel's book appears not to have engendered the kind of response its author had hoped for; nor did it garner accolades from the German mathematical community. Three years after its publication, however, it was reviewed at considerable length by Felix Klein (1849-1925) in the *Jahrbuch über die Fortschritte der Mathematik* (Klein 1875). Klein, then a fast-rising star in German mathematics, had in 1872 been called to Erlangen as an *Ordinarius* at just twenty-three years of age. In fact, a favorable review of his famous *Erlanger Programm* appeared in the *Jahrbuch* alongside the one he wrote on Schlegel's *System der Raumlehre*.[18] Although written for different audiences and with divergent goals in mind, these

[18] Klein 1872. The review was written by his friend, Otto Stolz.; *ibid.*, pp. 229-231.

two programmatic works offer suggestive insights into the dynamics surrounding the reception of Grassmannian ideas in Germany during the early 1870's. Indeed, a comparison of their overall orientations reveals a good deal about the clashing agendas of research mathematicians at the universities, on the one hand, and Grassmann's followers, most of whom were Gymnasium teachers, on the other.

Klein had been exposed to Grassmann's *Ausdehnungslehre* as a member of Clebsch's school. He had studied both the 1844 and 1862 editions (the latter was richer mathematically, the former more suggestive philosophically) and he clearly held Grassmann's ideas in high regard, referring to both works prominently in his *Erlanger Programm.*[19] Like Schlegel, he also favored a reform of the mathematics curriculum, and indeed the whole of philosophy behind mathematics instruction in the Gymnasien.[20] His reaction to Schlegel's book, however, was decidedly negative, though not so much due to its substance as its tone and tendency. Thus, while he found the proofs of familiar results, like the Pythagorean Theorem, by means of Grassmannian methods interesting, he also noted that these proofs often required extensive calculations. This, Klein wryly observed, stood in flat contradiction to the book's stated objective. He could have added that such excessive formalism was precisely what Schegel had criticized in the work of contemporary analytic geometers, a criticism that Klein and other members of the Clebsch school had heard often enough.

Aside from this particular point, however, Klein largely ignored the pedagogical issues that Schlegel's book sought to address, preferring instead to comment on matters that he personally found more interesting. In fact, his account suggests that he was far less interested in the book under review than in the convenient opportunity it offered for him to expound on the merits and shortcomings of Grassmann's mathematics as a whole, particularly from the perspective of his own *Erlanger Programm.* Thus, he pointed out the limitations of Grassmannian conceptions for projective geometry, noting that neither Grassmann nor Schlegel introduced imaginary elements in their treatments of projective constructions. He also called attention to their general neglect of key aspects of the theory, such as cross ratios, polar curves, etc. After remarking on these drawbacks, Klein went on to say:

"Perhaps instead of presenting Grassmann's conceptions as such, and only in elementary form, it would have been more important to show how these connect with and compare to similar directions that research has taken afterward, independent of Grassmann. So far as the present book is concerned, the principal accomplishments of Grassmann are

---

19 See Klein, *Erlanger Programm*, pp. 478, 480, 483, 489, 492.

20 For a detailed account of Klein's views at this time, see Rowe 1985.

essentially three. First, it is due to him that formal algebra gained an unsuspected depth, as he showed how to grasp the essence of the operations of addition and multplication in a much more general way than had been done before him. In this regard, Grassmann stands alongside the English investigators, such as Hamilton. Furthermore, he was the first to develop the theory of higher-dimensional manifolds, which contains as special cases the theory of space and, especially, the theory of linear (projective) manifolds. Finally, he opened a new, wide-reaching field of research through his methods for generating all algebraic structures by means of lineal mechanisms."[21]

This praise for Grassmann's achievements was followed by two critical observations. First, with regard to invariant theory, Klein found that Grassmann's methods---with the exception of the mixed product, which had not been fully incorporated into algebra---offered nothing essentially new.

"The two forms of representation [Grassmann's and those employed by invariant theorists] are, at bottom, barely distinguishable and even the formulas that express them are often identical."[22]

Second, with respect to Grassmann's notion of *n*-dimensional manifolds, Klein noted that, from the present-day standpoint, this marked only a beginning. Grassmannian manifolds were

"only the direct generalization to higher dimensions of ordinary space with its positional and metrical properties, whereas Riemann's investigations opened a much more general line of inquiry that has since been extended significantly in various directions" (ibid.).

In his *Erlanger Programm* Klein had sought to integrate a wide spectrum of geometrical research by placing the concepts of manifold, transformation group, and the associated group invariants at the center stage. This offered a bold new conceptual framework, but its design was clearly dominated by the particular field of research Klein knew best, namely projective geometry and the related theory of algebraic invariants of the projective group. Remarkably enough, the theory of affine spaces was not even mentioned in the *Erlanger Programm*, a clear indication that Grassmannian ideas played a rather peripheral role in Klein's overall conception of geometry at this time.[23] Klein's background and his strong predilection for projective methods must, therefore, be taken into account when reading these reflections on the significance of Grassmann's mathematics for geometrical research in the year 1872.

---

21 Klein 1875, pp. 233-234.

22 *Ibid.*, p. 235. This opinion was later repeated at considerable length by Eduard Study in his commentary on Grassmann's last mathematical works; see *Grassmanns Werke*, II.1, pp. 431-433.

23 Klein later openly admitted that this omission stemmed from the dominant role of projective geometry in his background. See Klein 1921-1923, I, pp. 320-321.

After these comparative remarks, Klein summed up his opinion of Schlegel's book in a single sentence, offering this decidedly negative verdict:

"If, like the author, one presents Grassmann's ideas divorced from such comparisons, it is this reviewer's opinion that the reader will tend to be repelled rather than attracted by them; he will be forced to accept that Grassmann's methods are absolutely superior, and such a claim always contains something improbable about it" (Klein 1875, p. 235).

Schlegel must have been dismayed that one of Clebsch's closest associates could have written such an unsympathetic, self-serving review of Part I of his book. Not only had Klein largely ignored the core issues at stake, namely the efficacy of Grassmannian methods for the presentation of elementary geometry, but the whole thrust of his final remarks seemed to say: why didn't the author write the kind of book I would have written instead of this one? Clearly, Klein had a very different mathematical agenda than the one Schlegel wished to pursue. The whole aim and tendency of his *Erlanger Programm* was to unite disparate strands of geometrical research within a broad conceptual framework. As noted above, in setting forth this new scheme Klein drew his main inspiration from projective geometry (with Galois theory serving as a loose structural analogy). Moreover, several passages in the *Erlanger Programm* criticized tendencies toward disciplinary fragmentation and methodological purism, trends Klein spent his whole life fighting.[24] Yet it was seldom, if ever, that Klein feuded with any of Grassmann's followers. He did not need to because this faction never attained any significant standing within the German mathematical community.

The same year when Klein's review appeared, Schlegel took the opportunity to respond in the introduction to Part II of his *System der Raumlehre*. He began by addressing the criticism voiced in Klein's closing judicium in which Schlegel was taken to task for having failed to draw comparisons between Grassmann's results and those obtained by others. After noting where this had been done in Part II, Schlegel curtly pointed out that with respect to Part I this criticism was vacuous, "since this part only comprises theories of elementary geometry" (Schlegel 1875, p. viii). As for the assertion that his book forced the reader to accept Grassmann's methods as absolutely superior, Schlegel replied straightforwardly that, in his opinion, Grassmann's methods offered "the shortest and easiest approach to the results of ancient and modern geometry and algebra" (ibid.). But he also emphasized that the reader stood under no compulsion to accept this view; all he asked of the mathematical public was that they study the material presented in the

---

24 See *Erlanger Programm*, pp. 461, 490-491.

book before drawing their own conclusions. Schlegel further added that the issue as to whether Grassmann's *Ausdehnungslehre* was merely of historical interest, perhaps

> "with the added regret, after a comparison with other methods, that it had not exerted its influence earlier ... [or whether it was] also at present worthy of further development and capable of being usefully employed in scientific advancements, that is a question that cannot be answered in three lines; nor can it be answered only on the basis of my merely introductory writings, but rather alone by a thorough study of Grassmann's original works..." (ibid.).

Klein apparently did not reply to these comments, at least not in print, nor did Part II of Schlegel's *System der Raumlehre* ever receive a review in *Jahrbuch über die Fortschritte der Mathematik*, despite the fact that its mathematical content went far beyond that contained in Part I. Needless to say, neither party in this dispute succeeded in persuading the other side to alter their views, and in any case it seems unlikely that any sort of compromise could have been reached. Klein later played a major role in engineering the appointment of Friedrich Engel as editor of Grassmann's *Gesammelte Mathematische und Physikalische Werke*, a disappointing turn of events for Schlegel.[25] In fact, Engel later adopted a position with respect to "fanatical Grassmannians" very similar to Klein's own.[26] There appears to be no extant evidence indicating how Grassmann himself responded to Klein's review of 1875, but if his last mathematical works serve as any indication, then he apparently felt very strongly that his methods and ideas, as presented by Schlegel, had, once again, failed to receive a fair hearing. Grassmann tried to recoup some of these loses in his final works, but his efforts probably only resulted in a hardening of opinion.[27] Since he considered Clebsch a strong and sympathetic supporter, Grassmann clearly had reason to be deeply disappointed by his premature death in November 1872.[28] Given these circumstances, it seems only plausible to assume that he felt stung by the critical remarks published by Clebsch's leading disciple three years later. If so, it may well be that these events marked a turning point in the

---

25 Grassmann's leading disciple made this displeasure known by noting that his own participation in the Grassmann edition had been restricted to the preparation of a bibliography (see Schlegel 1896, p. 3).

26 See, for example, Engel,1910, pp. 12-13.

27 In 1904 Engel wrote in regard to Grassmann's last works that these reveal "a striking imbalance between what is actually accomplished and the claims Grassmann made in them. This, however, has not restrained Grassmann's unconditional admirers from praising even these works far above their merits" (Friedrich Engel, *Grassmanns Werke*, II.1, p. vi). See also Engel 1910, pp. 11-12, and Study's critical remarks in *Grassmann's Werke*, II.1, 431-435.

28 See Grassmann's tribute to Clebsch in Grassmann 1872.

relations between Grassmann's growing contingent of followers and influential members within Germany's still fragmented and often divisive mathematical community.

Fachbereich Mathematik, Universität Mainz

REFERENCES

Bernd Bekemeier, *Martin Ohm (1792-1872): Universitätsmathematik und Schulmathematik in der neuhumanistischen Bildungsreform* (Göttingen: Vandenhoeck & Ruprecht, 1987).

Alfred Clebsch, "Zum Gedächtnis an Julius Plücker," *Abhandlungen der Königlichen Gesellschaft der Wissenschaften zu Göttingen*, 15 (1872); reprinted in Julius Plückers *Gesammelte Mathematische Abhandlungen*, ed. Arthur Schoenflies (Leipzig: Teubner, 1895), pp. ix-xxxv.

Michael J. Crowe, *A History of Vector Analysis* (Notre Dame: University of Notre Dame Press, 1967).

Friedrich Engel, "Hermann Grassmann," *Jahresbericht der Deutschen Mathematiker-Vereinigung*, 1910, *19*: 1-13.

Hermann G. Grassmann, *Lehrbuch der Arithmetik für höhere Lehranstalten* (Berlin: Enslin, 1860).

Hermann G. Grassmann, "Über zusammengehörige Pole und ihre Darstellung durch Produkte," *Göttinger Nachrichten*, 1872 (Nr. 28): 562-576 (reprinted in Grassmann's *Gesammelte Mathematische und Physikalische Werke*, II.1, 250-255).

Hermann G. Grassmann, *Hermann Grassmanns Gesammelte Mathematische und Physikalische Werke*, ed. Friedrich Engel, 3 vols. in 6 pts. (Leipzig: Teubner, 1894-1911).

Hermann Hankel, *Theorie der complexen Zahlensysteme* (Leipzig: L. Voss, 1867).

Thomas Hankins, *Sir William Rowan Hamilton* (Baltimore: Johns Hopkins University Press, 1980).

Felix Klein, *Vergleichende Betrachtungen über neuere geometrische Forschungen*, [Erlanger Programm] (Erlangen: Deichert, 1872). Reprinted in Felix Klein, *Gesammelte Mathematische Abhandlungen*, I, 1921: 460-497.

Felix Klein, "Review of Victor Schlegel, System der Raumlehre, Erster Teil," *Jahrbuch über die Fortschritte der Mathematik*, Jahrgang 1872, (Berlin: Georg Reimer, 1875), pp. 231-235.

Felix Klein, *Gesammelte Mathematische Abhandlungen*. 3 Bände (Berlin: Springer, 1921-1923).

Felix Klein, *Vorlesungen über die Entwicklung der Mathematik im 19. Jahrhundert*, Band 1 (Berlin: Springer, 1926).

Albert C. Lewis, "H. Grassmann's 1844 Ausdehnungslehre and Schleiermacher's Dialektik," *Annals of Science*, 1977, *34*: 103-162.

David E. Rowe, "Felix Klein's 'Erlanger Antrittsrede': A Transcription with English Translation and Commentary," *Historia Mathematica*, 1985, *12*: 123--141.

David E. Rowe, "In Search of Steiner's Ghosts: Imaginary Elements in 19th-Century Geometry," to appear (1996)

Victor Schlegel, *System der Raumlehre, nach den Prinzipien der Grassmannschen Ausdehnungslehre und als Einleitung in dieselbe dargestellt, Erster Teil: Geometrie* (Leipzig: Teubner, 1872).

Victor Schlegel, *System der Raumlehre, nach den Prinzipien der Grassmannschen Ausdehnungslehre und als Einleitung in dieselbe dargestellt, Zweiter Teil: Die Elemente der modernen Geometrie und Algebra* (Leipzig: Teubner, 1875).

Victor Schlegel, *Hermann Grassmann: Sein Leben und seine Werke* (Leipzig: Brockhaus, 1878).

Victor Schlegel, "Die Grassmannsche Ausdehnungslehre. Ein Beitrag zur Geschichte der Mathematik in den letzten fünfzig Jahren," *Zeitschrift für Mathematik und Physik*, 1896, *41*: 1-21, 41-59.

Gert Schubring, "The Conception of Pure Mathematics as an Instrument in the Professionalization of Mathematics," *Social History of Mathematics*, eds. H. Mehrtens, H. Bos, I. Schneider (Basel: Birkhäuser, 1981), pp. 111-134.

Gert Schubring, *Die Entstehung des Mathematiklehrerberufs im 19. Jahrhundert. Studien und Materialien zum Prozeß der Professionalisierung in Preußen (1810-1870)* (Weinheim/Basel: Beltz, 1983; 2nd edition 1991).

R. Steven Turner, "The Growth of Professorial Research in Prussia, 1818-1848---Causes and Contexts," *Historical Studies in the Physical Sciences*, 1971, *3*: 137-182.

ZBYNĚK NÁDENÍK

RECEPTION OF GRASSMANN'S IDEAS IN BOHEMIA

In this paper I will discuss how Czech mathematicians, particularly in the Prague community - became acquainted with Grassmann's *Ausdehnungslehre* in the last century.[1]

Two Czech mathematicians wrote on the *Ausdehnungslehre* one hundred years ago: Václav Láska 1862-1943; professor at the university of Lwow from 1895 and at the Czech University of Prague from 1911, and Antonín Libický, a contemporary of V. Láska, and a teacher at grammar-schools in Roudnice and Hradec Králové (Bohemia).

Much is known about the life of V. Láska, but I was not able to find, in spite of sustained efforts, his unpublished manuscript on the *Ausdehnungslehre*. The existence of his manuscript is certain; it appeared not later than in 1895. Biographical information about Antonín Libický, on the other hand, remains scarce, also in spite of intense investigation. He published an extensive paper in Czech on the *Ausdehnungslehre* in 1896 (Libický 1896).

A brief sketch of what went on before Václav Láska and Antonín Libický published their papers will help to characterize the Prague community of Czech mathematicians in the last decades of the nineteenth century.

At that time, mathematical work was concentrated at the two universities of Prague but not exclusively so. There was first the Karl-Ferdinand University, which from 1882 had been sub-divided into a Czech and a German part. Then there was the Prague Polytechnical School, which had been founded at the begining of the 19th century,[2] and sub-divided into a Czech and a German part in 1864. Until the last decade of the 19th century, there was no more than a single chair for mathematics belonging to the Faculty of Philosopy of the university's Czech division. The situation was similar at the Czech Polytechnical

---

[1] For help in providing materials I am grateful to J. Bečvář, Prague; M. Bur a, Prague; M. Lásková, Professor V. Láska's daughter-in-law, Plzeň; L. Peřinová, Roudnice; V. Šobr, Plzeň.

[2] To be more precise: the Engineering School, founded in 1717, became reorganized in 1806 according to the model of the Paris school.

G. Schubring (ed.), Hermann Günther Graßmann (1809–1877): Visionary Mathematician, Scientist and Neohumanist Scholar, 147–153.

School which boasted only one chair for descriptive geometry. The University in Olomouc (Moravia) which had been founded in 1569 had lost its Philosophical Faculty in the middle of the 19th century and was soon closed down. Thus Moravia was left with its only university, the German Technical University in Brno.

This is why there were so few vacant posts for mathematicians at Czech universities, and why Czech mathematicians looked for work with universities abroad. At least three cases illustrate the point that Czech mathematicians were so frequently forced to seek positions abroad.

Matyá Lerch (1860-1922) had published more than 100 scientific contributions by the age of thirty five, half of them in journals abroad; yet the University of Bohemia had no professorial post for him. Recommended by Charles Hermite, with whom Lerch was in a scientific contact, he became professor at the University of Fribourg (Switzerland) in 1896. In 1887/8 he gave lectures on analytic geometry of the rational curves according to Grassmann's approach.

Karel Zahradník (1848-1916) found work in 1876 at the newly founded University of Zagreb. In 1874 he published a thin volume providing a Czech translation of G. Bellavitis' paper. I will return to it later.

Václav Láska had surprisingly extensive interests. At the University of Prague, he studied mathematics, physics and astronomy, later also geodesy, geographics and meteorology. In Lwow, he served as the director of the astronomical observatory. For V. Láska's 60th birthday, a *Festschrift* (Sborník ... 1923), containing a paper on Bernard Bolzano was published by the Association of Czechoslovak mathematicians and physicists. This paper, by grammar-school teacher Martin Jašek (Plzeň, Bohemia) was entitled "On functions with infinite number of oscillations in manuscripts by Bernard Bolzano." In 1918, M. Jašek had found one of Bolzano´s manuscripts in the Vienna *National Bibliothek* which contained the construction of continuous but nowhere differentiable functions.

Bolzano's very first mathematical publication dates from 1804; it concerns the foundations of geometry. It has been analysed by J. Folta, who showed that B. Bolzano, by studying essentially pairs of points, arrived at founding vector space geometrically (Folta 1966) and thus in fact came to a result similar to that Caspar Wessel had attained in 1799 by geometrically interpreting complex numbers, in a similar vein as H. Grassmann did in 1844 in a considerably more general form in his *Ausdehnungslehre*.

The lives of Bolzano and Grassmann show some interesting common features. Neither ever taught mathematics at a university; both were interested in other disciplines; both worked in isolation from the world of university mathematical research; neither lived to see acceptance of their scientific work. And there are other parallels. In the nineteenth century two comprehensive biographies of H. Grassmann were published by R. Sturm, E. Schroeder and L. Sohncke (1879), and Schlegel (1896). Both works praise Grassmann's grammar-school textbooks of arithmetic and especially trigonometry from a pedagogical-didactical point of view. Schlegel wrote that Grassmann's textbooks, nearly unknown in Germany, were better appreciated in North America (Schlegel 1896, p.11). Hence, the reception of Grassmann's textbooks differs from the reception of his *Ausdehnungslehre*. Schlegel explains this by noting that Grassmann's pedagogical activity in teaching arithmetic and trigonometry gave him the opportunity to find and examine the best demonstrations, but that he had no such pedagogical opportunity to develop and improve his major work, the *Ausdehnungslehre*. Something similar happened to B. Bolzano. Vojtěch Jarník (1897-1970), professor of mathematics at Charles University (Prague), charges that Bolzano is not always intelligible in his *Funktionslehre* (Jarník 1981, p. 40). Jarník's explanation is lack of pedagogical opportunity, similar to the explanation that Schlegel offers for Grassmann. On the other hand, Bolzano's sermons in the St. Salvator Church in Prague were clear and intelligible; many have been printed and we can read them ourselves to check. Perhaps the high attendance at Bolzano's sermons, attested to by contemporaries, provided a beneficial stimulus to clarity.

Grassmann and Bolzano suffered the same disappointment - an absence of students to carry on their work. In Grassmann's biography we may read that apparently Hermann Hankel (1838-1873) was the first who called attention to Grassmann in his own 1867 book on theory of complex systems of numbers (Sturm et al. 1879, p.2). M. Jašek begins his discussion of Bolzano's oscillating function as follows:

> "As is known it was Hankel,..., who in 1871 - twenty years after Bolzano's death - paid attention to Bolzano for the first time" (Jašek 1923).

Grassmann's desire (which remained unfulfilled) is expressed in his preface to his *Ausdehnungslehre* of 1862: "to gather around him a circle of pupils and to stimulate them to develop and enrich his own ideas" (Sturm et al. 1879, p. 3). Bolzano also long hoped that he would find a disciple who would complete his own mathematical work. In his last years he set great hopes in Robert Zimmermann (1824-1898) and willed his mathematical *Nachlaß* to him. From 1850, R. Zimmermann was philosophy professor at the University of Vienna. He took Bolzano's manuscripts along, but unfortunately he did not do anything with them. Towards the end of the 1920's they were rediscovered by M. Jašek.

Now I will sketch the milieu of the Czech mathematicians in the second half of the 19th century. After the Austrian Minister of the Interior Alexandr Bach had resigned in 1854, there was an upsurge in Czech political life. In 1861, four students of mathematics and physics at the Philosophical Faculty in Prague composed the statutes of the "Association for open lectures on mathematics and physics." The government of Prague approved of the statutes, and in March 1862 the association was founded with 15 members. This association later became the "Union of Czech mathematicians and physicists" of our days. Since 1872, this association has published a journal of mathematics and physics - *Časopis pro pěstování matematiky a fysiky*. In this journal, A. Libický published his comprehensive paper on Grassmann's *Ausdehnungslehre*.

The first step to acceptance of Grassmann's ideas was made by K. Zahradník in 1874. In 1873, J. Laisant translated Bellavitis' work written in 1854. Emil Weyr (1848-1894), associate professor at the Prague Polytechnical School, and from 1876 professor at the University of Vienna and František Studnička (1836-1903), professor of mathematics at the University of Prague from 1871 proposed the Czech translation of Bellavitis' work which was done by K. Zahradník.

F. Studnička promoted acceptance of the *Ausdehnungslehre* from another direction. In 1876, he published a series of papers on Hamilton's quaternions, later supplemented by his book (Studnička 1894). This book was elementary and did not refer extensively to the literature. Studnička mentioned H. Grassmann only in passing (p. 20): "...Grassmann, together with Scheffler and Schlegel, proceeded in another direction, dominated by situation calculus." Studnička only quotes H. Grassmann's work on geometric analysis of 1847.

J. Odstrčil (who wrote textbooks on physics together with Ernst Mach during his tenure at the University of Prague from 1867-1895), and J. Merten (teacher at the grammar schools in Těšín (Moravia) and Žatec (Bohemia)) published on quaternions in 1879 and then from 1881 to 1889. Odstrčil's approach is geometrical; first he describes operations with vectors, and then he constructs quaternions with their addition, subtraction and multiplication; he adds many applications in the plane and in the spherical trigonometry and in the geometry of triangles but also on tetrahedra. Merten's paper concentrates on applications in mechanics.

A. Libický was the author of the first Czech paper dealing directly with *Ausdehnungslehre*. In the preface he wrote:

"Two years ago, fifty years had passed since one of the most excellent mathematical works of this century appeared, carrying the title: *Die Ausdehnungslehre, ein neuer Zweig der Mathematik, dargestellt von H. Grassmann*. It is well known how unfavourably this remarkable work was received; it remained unnoticed for many years, and in special journals published during that time, we find no trace of Grassmann's principles having been taken up and cultivated. As Grassmann assumed that the cause for that lay in his own rather more philosophical than mathematical style, he elaborated his theory once more in 1862, in a form more common among mathematicians. But even then the book did not attract the attention it deserved, despite the fact that other branches of mathematics, sprouting from the same tree, such as Möbius's barycentric calculus, Bellavitis' method of equipollences, Saint-Venant's geometrical product and particularly Hamilton's vector theory began to be used ever more extensively. Only recently we have seen a turn for the better, be it that contributions on Grassmann's *Ausdehnungslehre* or on parts of it are being published, be it that special journals increasingly print papers intending to cultivate this theory and to prove its utility" (Libický 1906, preface).

As sources, A. Libický gives the books by H. Grassmann (Geometrische Analyse ...) 1847, H. Hankel 1867, F. Kraft 1893, G. Peano 1891 and V. Schlegel 1872-1875; papers by E. Carvallo 1892, W. Clifford 1864, G. Everett 1868 and E. Mueller 1891; and V. Láska's manuscript: *Ausdehnungslehre*. The list ends with this phrase: "I have also used repeatedly Dr. Láska's manuscript: *Die Ausdehnungslehre*, which the author kindly lent me, for which I am grateful to him."

Books by F. Kraft, Peano, Carvallo and E. Mueller had been published only a few years before Libický's series of 1896. This shows that Libický was well informed on the most recent literature on the *Ausdehnungslehre*. Against that, Libický omits quoting the extensive contribution by R. Sturm et al. (1879) containing a bibliography of Grassmann's writings.

A. Libický studied Grassmann's papers in depth, in particular the development of vector calculus and vector analysis. Ten years later he published a second series of papers (1906/07). These were combined to a book in 1914. This shows how interested Czech mathematicians at that time were in the new theory founded by Grassmann. A book containing mathematical contributions separately published years before was quite exceptional in Prague at that time. In his introduction, Libický again quotes Grassmann:

"Vector analysis, one of the youngest branches of mathematical science, was first done in particular by English and American scientists in particular. In Germany, where the bases of this calculus had been established by A. Möbius and H. Grassmann, only recently a vivid interest in them has developed, especially when it became clear, that they can be used succesfully in the new science of electricity. In recent years, this interest is evident in a large number of papers published on this subject."

There is a substancial difference between Libický's paper of 1896 and his second paper of 1906-7, both in the presentation of the content and in the terminology used. For today's reader his second paper is much more modern than the first, and the difference is much more pronounced than an interval of merely ten years would suggest.

Libický's book was soon out of print. In 1923, the Union of Czech Mathematicians and Physicists published Karel Dusl's textbook. Dusl says in his preface: "The origin's of vector calculus must be sought at the beginning of the nineteenth century. In 1827, A. Moebius published *Der barycentrische Calcuel*, 1843 saw Hamilton's first lecture on quaternions, 1844 Grassmann's *Ausdehnungslehre* appeared, which laid the foundations of vector calculus" (Dusl 1923, preface). Dusl adds a table which shows how W. Hamilton's and Grassmann's basic terms developed until his own time. Current use is nearer to the designation's of Libický's older textbook (1914) than to that of Dusl.

Lectures on vector calculus and vector analysis were held at the Czech Technical University from 1922 onwards. These lectures by K. Dusl are an indication that the Czech mathematical community had accepted the conceptions which Grassmann had had such a notable share in developing.

Another proof that Grassmann's ideas were known among the Czech mathematicians at the begining of our century is Vincenc Jarolímek's (1846-1921, professor of descriptive geometry at the Polytechnical Schools of Brno and Prague) study in projective geometry in Grassmann's sense on a surface of third degree (see esp. Jarolímek 1912, p. 76).

Department of Mathematics, Technical University Praha

REFERENCES

G. Bellavitis, "Exposition de la méthode des équipollences" (translated by J. Laisant), *Nouvelles annales de mathématiques, Série 2*, 1873, *12* : 97-113, 145-161 ,193-207, 241-257, 297-315, 501-519, 529-561; 1874, *13* : 58-59, 138-152, 189-198, 220-236, 257-265.

G. Bellavitis, *Methoda equipollencí čili rovnic geometrických (translated by K. Zahradník)* (Praha: Jednota českých mathematiků, 1874).

E. Carvallo, "La méthode de Grassmann," *Nouvelles annales de mathématiques, Série 3*, 1892, *11:* 8-37.

K.Dusl, *Úvod do vektorového počtu* (Praha: Jednota českých mathematiků, 1923).

H. Hankel, "Grenze," *Allgemeine Encyklopaedie der Wissenschaften und Kuenste*, Band 90, Hrsg. J. Ersch, J. Gruber (Leipzig: Brockhaus, 1871), 185-211, par. 19.

J. Folta, "Bernard Bolzano and the Foundation of Geometry," *Acta historiae rerum naturalium necnon technicarum* (Praha: Českoslov. akademie věd, 1966), 75-104.

V. Jarník, *Bolzano a základy matematické analýzy* (Praha: Jednota československých mathematiků a fysiků, 1981).

V. Jarolímek, *Základové geometrie polohy, vol. II* (Praha: Česká matice technická, 1912).

M. Jašek,"O funkcích s nekonečným počtem oscilací v rukopisech Bernarda Bolzana," *Sborník prací matematických a fysikálních k 60. narozeninám prof. Dr. V. Lásky* (Prague 1923), 102-110.

F. Kraft, *Abriss des geometrischen Kalkuels nach den Werken des Professors Dr. H. Grassmann* (Leipzig: Teubner, 1893).

A. Libický, "Základové geometrického počtu Grassmannova," *Časopis pro pěstování mathematiky a fysiky*, 1896, *25:* 187-198, 265-284, 321-341.

A. Libický, "Úvod do vektorové analyse," *Čas. pěst. math. fys.*, 1906, *35:* 207-219, 297-311, 409-441; 1907, *36:* 121-134, 251-271, 345-353, 480-483.

A. Libický, *Úvod do vektorové analyse* (Praha: Jednota českých mathematiků a fysiků, 1914).

E. Mueller, "Die Liniengeometrie nach den Principien der Grassmannschen Ausdehnungslehre," *Monatshefte für Mathematik und Physik*, 1891, *2:* 267-290.

G. Peano, *Die Grundzüge des geometrischen Calcüls* (Leipzig: Teubner, 1891).

V. Schlegel, "Die Grassmannsche Ausdehnungslehre," *Zeitschrift für Mathematik und Physik*, 1896, *41:* 1-21,41-59.

F. Studnička, "O kvaternionech," *Čas. pěst. math. fys.*, 1876, *5:* 49-57, 97-102, 145-151.

F. Studnička, *O kvaternionech* (Praha: Jednota českých mathematiků a fysiků, 1894).

R. Sturm, E. Schroeder, L. Sohncke, "Hermann Grassmann," *Mathematische Annalen*, 1879, *14:* 1-45

Sborník prací matematických a fysikálních k 60. narozeninám prof. Dr. V. Lásky, ed. Editorial staff of Časopis pěst. mathem. fys. (Praha: Jednota československých mathematiků a fysiků).

ALDO BRIGAGLIA

THE INFLUENCE OF H. GRASSMANN ON ITALIAN PROJECTIVE *N*-DIMENSIONAL GEOMETRY

On May 29, 1883, Corrado Segre took his doctorate in Turin (Torino), under Enrico D'Ovidio's guidance. His thesis (Segre 1884a,b) was published one year later in the Journal of the local Academy of Science, and after a short time it became a fundamental starting point for the development of Italian projective n-dimensional geometry.

The paper is devoted to the general study, by geometrical tools, of n-dimensional vector spaces (projective spaces), and in particular of bilinear forms defined on them (quadrics). At the very beginning of the paper, Segre gratefully acknowledges Grassmann's role as his forerunner. It is very likely that Segre did not utilize many of Grassmann's ideas in this first work, but he had obviously perceived their importance in providing sounder foundations to the new geometrical edifice.

Indeed Segre writes:

"We must note that [...] Grassmann, in his *Ausdehnungslehre*, has already developed, together with many other important ideas, that of n-dimensional geometry in 1844, but his work has been taken up only recently. The following very important passage is found in one of Grassmann's papers of 1845 [...] and it shows that the *Ausdehnungslehre* in the mind of this inventive scholar was indeed what we now name «n-dimensional geometry»: *Meine Ausdehnungslehre bildet die abstrakte Grundlage der Raumlehre (Geometrie), d. h. sie ist die von allen räumlichen Anschauungen gelöste, rein mathematische Wissenschaft, deren specielle Anwendung auf den Raum die Raumlehre ist ... die Raumlehre, die auf etwas in der Natur gegebenes, nämlich den Raum, zurückgeht, ist kein Zweig der reinen Mathematik, sondern eine Anwendung derselben auf die Natur*" (Segre 1884a, p. 26).

Since the paper in which Segre included this quotation belongs to the founding contributions to the Italian school of n-dimensional geometry, I consider it useful to clarify some points, if only briefly. The work of the Italian School, and in particular that of C. Segre and E. Bertini, rests on four correlated pillars: Plücker's so-called line-geometry, which was interpreted as a starting point towards a new idea of multi-dimensional space in which the "points" were able to assume different meanings; Grassmann's point of view with regard to "abstract" linear space as the natural framework for any particular geometrical interpretation; von

G. Schubring (ed.), Hermann Günther Graßmann (1809–1877): Visionary Mathematician, Scientist and Neohumanist Scholar, 155–163.

Staudt's notion of the "coordinatization" of synthetic projective space as a fundamental tool for the systematic translation of geometrical concepts into algebraic ones (and vice versa); and finally, algebraic ideas presented by Weierstraß and Frobenius, but suitably translated into geometrical terms.

To be more specific, whereas Plücker suggested to Segre the basic geometrical intuitions, Grassmann allows him to place these same ideas within a much sounder conceptual framework, a broader philosophical view, a more acute formalism. Grassmann's work indeed clearly suggests that the natural environment where we can study multi-dimensional projective geometry is what we now call an abstract vector space (over the complex field).

A closer scrutiny of this will show how the young Segre's ideas on linear spaces, which in the beginning were rather confused, became - also owing to a more and more careful study of the Stettin scholar's work - clearer and ever more precise, and how they finally gave rise, in particular through the contributions of Segre's students, to some fundamental works on modern axiomatics in geometry.

First, a brief sketch on how the first ideas of multidimensional space emerged. It goes back to the beginning of the 19th century and was at first closely tied to mere analytical fact, just to give a geometrical aspect to the use of objects defined through n-numerical parameters. During the next decades, this idea grew into a clearer and more geometrical one.

The 1840s saw publication of several papers in which the idea of multidimensional space appears clearly stated. Besides Grassmann's *Ausdehnungslehre*, there are the papers (Cayley 1845), (Hamilton 1844) and in particular (Plücker 1846).

It was indeed Plücker's approach which had the greatest impact. In his influential paper (Plücker 1865), and above all in his treatise (Plücker 1868-69), he considered as a "point" in a n-dimensional space an arbitrary object from ordinary (tri-dimensional) space depending on n parameters. Thus, a line may be written as

$$\begin{cases} x = rz + a \\ y = sz + b \end{cases}$$

determining a point (r, a, s, b) in four-dimensional space. The set of points satisfying a given algebraic equation is called a *complex* (a hypersurface in four-dimensional space), the intersection of two complexes a *congruence* and the intersection of three complexes a *configuration*.

In line with Plücker's approach, many mathematicians extended his results during the second half of 19th century. In this trend, Felix Klein (considering the minors of the system of equations which determine a line in space) defined a line as an element of a variety in five-dimensional projective space, while Lie introduced the space of spheres and the related contact transformations.

Plücker's methods, which were somewhat analogous to contemporary Euclidean models of non-Euclidean geometry, had made it possible by the mid-seventies to construct a huge number of *ordinary spatial* models of *n-dimensional geometry*. Clearly, this method acquainted mathematicians with an increasingly liberal use of the term of "point", as space element, which had to take on different meanings according to context.

In 1891, Segre noted that there had been various ways to give meaning to the idea of *n*-dimensional space ten years before. One was a purely "naive" interpretation which considered the space C^n analytically, simply putting it into geometrical language; another, in Plücker's style, saw objects in common space defined by *n* complex parameters (in our language this amounts to saying that space was not defined as C^n, but rather as a space isomorphic to it), and yet another saw space axiomatically in accordance with Euclidean style, like Veronese incompletely in 1882, and more successfully nine years later in an influential treatise (Veronese 1891).

Many foundational problems were left completely unresolved. The first was to find an appropriate "abstract" and general framework within which any *n*-dimensional space could be considered while indicating more precisely how to move from one space to another, assigning different meanings to the names (point, line, etc.) of the objects forming the abstract space (isomorphism). Besides, there was the problem of connecting the various points of view, and finding a set of independent axioms from which the Cartesian representation (coordinatization) could be deduced.

Segre and his school were interested in these problems, and the knowledge of Grassmann's ideas was to be of fundamental importance for them.

Segre had begun to understand Grassmann's work about 1884 by reading Grassmann's synthesis of his own prinicpal work in his author's review published in 1845 in *Archiv der Mathematik*. During the 1880s, Segre attained a deeper understanding of the *Ausdehnungslehre* in its 1844 edition and in the second edition published in 1862. It is interesting to see how Segre defines the concept of linear space in his thesis. The very inaccuracy of his own mathematical language suggests that he developed, after reading Klein's works, a theory of his own within which Grassmann's points of view were soon to find recognition. Reading the *Ausdehnungslehre* was to assist the young Italian in getting some order into his own ingenious, but still confused ideas. Thus, he wrote in his thesis:

"Any continuous set of objects, whose number is m times infinity (i. e. such that among them there is generally a finite number that satisfies m simple arbitrary conditions) is said to be a m-dimensional space, of which these objects are the elements.
Any m-dimensional space is said to be linear when it is possible to attach to each element m numerical values (real or complex) in such a way that, without any exception, to any set of values of them there corresponds one and only element of it. The values of these quantities that correspond to that element are said to be its co-ordinates. If we characterise them by the ratios of m quantities with a (m+1)th, they will constitute the m+1 homogeneous co-ordinates of the element of the space, so that any element of it, without exception, will be characterised by the mutual ratios of these homogeneous co-ordinates and, vice versa, will characterise their ratios." (Segre 1884a, p. 38)

The definition, besides being confused, is clearly wrong, leading Peano to challenge Segre during their well-known controversy of 1891 on the subject of importance of rigour versus intuition in mathematics. Peano stressed the need of "absolute rigour" in mathematics, Segre held that intuition paves the way to mathematical discovery, while rigour only confirms it (cf. Manara-Spoglianti 1977).

But, notwithstanding the evident lack of rigour, Segre's definition was an important working tool, which led him on correctly, allowing him to grasp a large number of the more advanced algebraic results which he was able to interpret always in an arbitrary linear space. This capacity requires an idea, even if still confused, of isomorphism between two linear spaces, an idea he expressed as follows:

"Any linear space having the same dimensions, independent of the nature of its elements, can be considered as identical to any other, because, as we already noted, when we study them we don't regard the kind of these elements, but we only consider the property of linearity and the dimension of the space made by the elements themselves. It follows that the theory of linear forms of the first, second and third kind, for instance, of the line, of the plane and of the space considered as dotted, being already known, may be used for any linear space of dimension 1, 2, 3 contained in the linear space of dimension n-1 that one wants to study in general. Therefore one can make use, for example, of the theory of projectivity, of harmonic groups, of involution, etc., in the forms of first kind." (Segre 1884a, p. 46)

In spite of its lack of rigor, Segre's point of view opened towards further developments, even to those of formal algebra. Even if his focus of interest was rather more on algebraic geometry, many foundational problems resulted quite naturally from his formulation, and he was to propose them as research topics to some of his best students four years later.

The framework within which n-dimensional geometry can be developed was thus defined as "abstract" n-dimensional linear space. It seems evident that Segre found in the *Ausdehnungslehre* (which, I should like to recall, he knew only in a vague way in 1884) the first exposition of the major ideas he needed which was clear enough for his own purposes. It should also be noted that at the same University of Turin, and within the same decades, another young mathematician, Giuseppe Peano, undertook to study Grassmann's work and a few years later was to publish a more profound study on that reasoning (Peano 1888). It is hard to believe that the two young colleagues did not interact and influence one another, even if their approaches were rather different; while Peano sought a rigorous logical setting for the mathematical edifice, Segre was looking for the more general definition in which his many geometrical intuitions could be linked better and make a more comprehensive use of the more recent findings of linear algebra (first of all, Weierstraß's theory of elementary divisors and Frobenius's theory of bilinear forms).

It is interesting to consider the large systematical volume Segre wrote many years later (in 1911) for Klein's *Enzyklopädie* (Segre, 1911). Here the great scholar, now in his fifties, reflects deeply on his initial ideas.

In Segre's opinion the theory of hyperspaces ("*n*-fach ausgedehnte Mannigfaltigkeiten") was initiated precisely by Grassmann's studies:

"If one wishes to link the first foundation of the theory of n-fold extended manifolds to a name, it will be that of H. Grassmann, who in his *Ausdehnungslehre* of 1844 (as later in the 1862 edition) developed this very theory. He takes elements (not further defined), and uses them to create such manifolds. The treatment appears as a branch of pure mathematics which encompasses common geometry (which one obtains if one assumes a point as the generating element)." (Segre 1911, p. 772)

Segre indeed points out Grassmann's merits - according to his own interpretation - in clearly defining the abstract concept of point, a definition urgently required after Plücker's theory. In other words, while geometrical practice had for a long time been deprived of any real meaning of the term of "point" by mid-century, that a sound theoretical basis could be restituted to geometrical practice only by Grassmann's approach. While the word "point" lost its concrete meaning, the new "abstract" geometry of *n*-dimensional spaces took on, as a corresponding basic term, precisely the "element" of a linear variety as defined by Grassmann. Segre was to note that this procedure can be interpreted as the "linear" (i. e. projective) version of Riemann's differential theory of variety.

In a footnote, Segre enphasizes how Grassmann used his ideas concretely:

"In A_1 (*Ausdehnungslehre*, 1844) (§ 13) Grassmann defines, as an extensional domain (or system) of the first order, the totality (M1) of those elements into which a generating element is transformed in case of the continuous change of position of a point along a line. ... In A_2 (*Ausdehnungslehre* 1862) Grassmann abolished the excessive indeterminateness of the element by substituting (complex) quantities. There, in § 14, Grassmann designates the totality of quantities numerically derivable from n units a_1, a_2, ..., a_n (i.e. $\Sigma x_i a_i$, where the x_i are n numbers), as the area derivable from those quantities, namely as an area of *n*th order, if the area is not derivable from less than n units (i.e. if the ai are linearly independent)." (Segre 1911, p. 772)

It is useful to compare the idea of *n*-dimensional space explained here with that used by Segre his thesis, in order to understand how much Grassmann's formulation helped to clarify the ideas of the young Italian scholar.

In this context, Segre by now had a clear understanding of the substantial identity between projective *n*-dimensional geometry and the algebra of linear transformations:

"It has been noted ... that, for instance, the general projective geometry that we build in this way is nothing but the algebra of linear transformations. This is only a difference, not an error. Provided that we do mathematics!" (Segre, 1891, p. 54).

We can draw another useful comparison between the ideas formulated by Segre on the two approaches (Grassmann and Plücker) which he considers to be different, but complementary, and Appendix 4 of Klein's *Erlanger Programm*, where the German mathematician expresses similar ideas, but only rather sketchily.

It may also be of interest to maintain how Segre introduced the idea of *n*-dimensional space in his lectures of "Geometria Superiore" at the University of Turin some years later. My quotations are taken from the unpublished notes of the lectures that he gave in the 1892-93 term (*Introduzione alla geometria sugli enti algebrici sempl[icemente] infin[it]i, Cap.I Preliminari. Iperspazi*) recorded in the *Dipartimento di Matematica* of the University of Turin.[1] In spite of their sketchiness, I think these notes are very interesting; we will remember that among Segre's students there were such scholars as G. Castelnuovo, F. Enriques, F. Severi, G. Fano, B. Levi. The lessons begin in this way:

"General warnings about advanced mathematics, and above all about these lectures.
Sketch of the way to develop geometry. The *postulates* or *axioms* in logic and arithmetic (Grassmann, Dedekind, Peano); in geometry (Euclid etc., Pasch, Peano, Veronese) and in mechanics or mathematical physics. Formal, abstract, pure characteristics and experimental ones. Differences (in rank) towards analysis and geometry.
How from its axioms geometry may construct *co-ordinates*. Sketch on projective homogeneous co-ordinates ... Co-ordinates of more complex objects, represented by an equation $f^n(x) = 0$ [i. e. a polynomial equation of degree n], in a linear system of such objects. In this way we obtain varieties of any *dimension*.
Thanks to co-ordinates, geometry can *purify* itself, can get a more abstract character, assigning to the word "point" (etc.) a purely analytical meaning. In this way, we can introduce points etc. *imaginary*, as we will always do from now on. Even mathematical equivalence of two representable geometrical forms, i.e. such that it is possible to establish an single-valued correspondence, becomes now clear... So plane geometry and line-bundle geometry , sphere geometry and line geometry etc. Naturally, this equivalence

---

1 Here I must thank the department, and above all Prof. Livia Giacardi, for allowing me to obtain a photocopy of the manuscript

must be handled correctly: we will return to such problems later.
From co-ordinates and from the above mentioned reasoning, we can proceed to studying *varieties*, whose *element* is a set of numbers x_0, x_1, ... x_r, ... Instead of variety, we may speak of a *space* (r-dimensional), and instead of element, we may say *point*..."

As I have already said, Segre's major effort in this part - which we consider to be foundational - is to unify Plücker's approach and the analytical approach. Segre clearly understood that it is really possible to do this using the language of linear varieties as introduced by Grassmann. The text continues indeed with an extensive consideration of the relations among subspaces and their dimensions (following largely Grassmann's results), etc. in order to solve another foundational problem, i. e. to get co-ordinates started with axioms of projective geometry, thus unifying, always within Grassmann's views, the analytical approach and the axiomatic approach: Thus, Segre is able to outline, in a significant way the path that will later lead us to consider linear algebra as the more important tool to describe, via coordinatization, the various axiomatic systems of modern geometry. Fano (1892) informs us that the problem of obtaining co-ordinates from a set of independent axioms which he studied in this paper had been suggested to him by Segre during the 1890-91 term. We see how he tackles this problem in his lectures of 1892-93; I will give extensively bibliographical quotations to avoid dwelling too long on the sources:

"The problem to *determine* co-ordinates (instead of taking them as a starting point) ... beginning with some primary properties or *postulates* of S_r [the projective r-dimensional space]. *Staudt* Beiträge, *Klein* Math. Ann. IV and above all VI, *De Paolis* Sui fondamenti della geometria projettiva (Mem. Acc. Lincei ...), *Pasch* 1882, *Klein* Nicht-Eukl. Geometrie I Vorlesung (Wintersemester 1889-1990) ... *Clebsch-Lindemann* ... (die Grundlagen der proj. Geometrie).[2]

Segre later gives some information on papers by his students Fano and Amodeo and describes them briefly but completely. It is significant that such a profound program on the foundations of geometry which was to

---

2 The exact references are:

K. G. von Staudt, *Beiträge zur Geometrie der Lage* (Nürnberg 1856);

Felix Klein, "Über die sogenannte Nicht-Euklidische Geometrie," *Mathematische Annalen*, 1871, *4*: 573-625; and id., Teil II, *Mathematische Annalen*, 1873, 6: 112-145;

R. De Paolis, "Sui fondamenti della geometria projettiva," *Memorie della Reale Accademia Nazionale dei Lincei, 1881*, (2) *9*: 489-503;

Moritz Pasch, *Vorlesungen über neuere Geometrie* (Leipzig: Teubner, 1882);

Felix Klein, *Nicht-Euklidische Geometrie. I.* Vorlesungen (Wintersemester 1889-1890), Göttingen 1889;

Alfred Clebsch - Ferdinand Lindemann, *Vorlesungen über Geometrie*, Band I (Leipzig: Teubner, 1875), Band II (Leipzig: Teubner, 1891).

have such an impact on later developments, emerged in lectures, which like his paper published in the *Annali di matematica* (Segre 1894), did not have any foundational purposes at all, but were actually aimed at training that group of Italian algebraic geometers, whose bold edifice had always been considered unsatisfactory from the foundational point of view.

This review of the influence of the *Ausdehnungslehre* on Corrado Segre and his school, even if very brief should emphasize yet two other sets of problems where this influence has been very profound. First of all I mean Segre's papers on complex geometry and, more generally, on the geometry of a generic algebra (cf e.g. Segre 1912), where the inspiration by Grassmann via Study and his school is very evident.

To close, I must quote at least the results obtained by the Turin school of Pieri and Giambelli (cf. e.g. Pieri 1893 and Giambelli 1903) on enumerative problems in Grassmann's varieties (problems arising from Schubert's work). I have not space enough to examine these last two problems, but I should like to conclude that the influence of the geometer of Stettin, after the publication of *Ausdehnungslehre,* on the Italian mathematical school even outside Peano's circle was very profound and significant.

Dipartimento di Matematica, Universitá degli Studi di Palermo

REFERENCES

Arthur Cayley, "Chapters in the analytical geometry of n dimensions," *Cambridge Mathematical Journal*, 1845, *4*:119-127.

Gino Fano, "Sui postulati fondamentali della geometria proiettiva in uno spazio lineare a un numero qualunque di dimensioni," *Giornale di Matematiche*, 1892, *30*:106-132.

Giovanni Zeno Giambelli, "Risoluzione del problema degli spazi secanti," *Memorie della real Accademia delle Scienze di Torino*, 1903, *52*:171-211.

William Rowan Hamilton, "On quaternions, or on a new system of imaginaries in algebra," *Proceedings of the Royal Irish Academy*, 1844, 2:424-434.

Carlo Felice Manara-Maria Spoglianti, "Le idee di iperspazio. Una dimenticata polemica tra G. Peano, C. Segre e G. Veronese," *Memorie dell'Accademia Nazionale di Scienze Lettere e Arti di Modena*, 1977, *19*:109-129.

Giuseppe Peano, *Calcolo geometrico secondo l'Ausdehnungslehre di H. Grassmann preceduto dalle operazioni della logica deduttiva* (Torino: Fratelli Bocca, 1888).

Mario Pieri, "Sul problema degli spazi secanti," (3 note), *Rendiconti del real Istitituto Lombardo di Scienze e Lettere*, 1893, *26*:534-546; 1894; *27*:257-273; 1895, *28*:441-454.

Julius Plücker, *System der Geometrie des Raumes* (Düsseldorf: W. H. Scheller, 1846).

Julius Plücker, "On a new geometry of space," *Philosophical Transactions of the Royal Society of London*, 1865, *155*, 725-791.

Julius Plücker, *Neue Geometrie des Raumes*, 2 Abteilungen (Leipzig: Teubner, 1868-69).

Corrado Segre, "Studio sulle quadriche in uno spazio lineare ad un numero qualunque di dimensioni," *Mem. della R.'Acc. delle Scienze di Torino*, 1884a, *36*:3-86.

Corrado Segre, "Sulla geometria della retta e delle sue serie quadratiche," *Mem. della R. Acc. delle Scienze di Torino*, 1884b, *36*: 87-157.

Corrado Segre, "Su alcuni indirizzi nelle investigazioni geometriche," *Rivista di Matematica*, 1891, *1*:42-66.

Corrado Segre, "Introduzione alla geometria sopra un ente algebrico semplicemente infinito," *Annali di Matematica pura e applicata*, 1894, (2) 22, 41-142.

Corrado Segre, "Mehrdimensionale Räume," *Encyclopädie der mathematischen Wissenschaften*, vol. 3, (C. 7), (Leipzig: Teubner, 1911), 769-972.

Corrado Segre, "Le geometrie proiettive nel campo dei numeri duali," *Atti della real Accademia delle Scienze di Torino*, 1912, *47*:308-327; 384-405.

Giuseppe Veronese, "Behandlung der projektivischen Verhältnisse der Räume von verschiedenen Dimensionen durch das Princip des Projicirens und Schneidens," *Mathematische Annalen*, 1882, *19*:161-234.

Giuseppe Veronese, *Fondamenti di Geometria a più dimensioni e a più unità rettilinee*, Padova, 1891.

KARL-HEINZ SCHLOTE

HERMANN GÜNTHER GRASSMANN AND THE THEORY OF HYPERCOMPLEX NUMBER SYSTEMS

To ask what Hermann Grassmann contributed to the theory of hypercomplex number systems seems to be an appropriate question concerning his work. Grassmann algebras and Grassmann's outer product are well known and important concepts in mathematics. They support the view that Grassmann attained remarkable progress in algebra, or at least in important parts of algebra. Only a few of his articles, however, have an algebraic content. The titles of these articles, it is true, often lead one to expect an algebraical topic, and reading them confirms that Grassmann created or paved the way for important ideas in the theory of hypercomplex number systems. It is not easy, however, to detect these innovations, for they are frequently concealed within a geometrical context.

Grassmann's main objective was to establish an uniform formal construction of mathematics. To reach this goal, he cooperated with his brother Robert. According to his own view of science, Grassmann defined mathematics as a formal science that examined objects established by pure thought, such as the theory of forms. He divided mathematics into four branches: arithmetic, combinatorics ("Kombinationslehre"), the theory of intensive magnitudes, and the theory of extensive magnitudes. The concept of connection ("Verknüpfung") turned out to be the decisive fundamental concept in Grassmann's work. He emphasized connections and the rules of connection, detaching his reasoning from concrete individual objects. His division of the theory of forms into four branches was preceded by a general part explaining all the rules of connection used in all branches of the theory. This part he called the general theory of forms.[1] Grassmann wanted to summarize the basic approaches and principles common to all

[1] Grassmann wrote in the preface of his *Ausdehnungslehre*: "Es kann der Zerspaltung der Formenlehre in die vier Zweige ein allgemeiner Theil vorausgeschickt werden, welcher die allgemeinen, das heisst für alle Zweige gleich anwendbaren Verknüpfungsgesetze darstellt, und welchen wir die allgemeine Formenlehre nennen können" (Grassmann 1894, vol. 1,1, p. 28).

G. Schubring (ed.), Hermann Günther Graßmann (1809–1877): Visionary Mathematician, Scientist and Neohumanist Scholar, 165–173.

branches of mathematics in his own theory of forms. He abstracted for that purpose the common rules from several connections used in various parts of mathematics. In the end, he strove to grasp the structure of connections, and in this context he developed ideas important for algebra.

What is the mathematical content of Grassmann's studies on different kinds of connections? At first, his reasoning implied the requirement that the connection of two thought-forms ("Denkformen") results in another thought-form. In other words, these forms constitute what today would be regarded as a set, and the problem thus becomes: Is this set closed under the given connection? The property of being closed under a relation is decisive in order to define a mathematical structure. In Grassmann's time this question was not self-evident and answering it was not always trivial. It is sufficient to recall W. R. Hamilton's futile efforts to define a three-dimensional system of quantities satisfying the norm equation, i. e. where the norm of a product equals the product of the norms of the factors. Hamilton failed to construct such a system that was closed under multiplication, and this led him to the discovery of quaternions. Grassmann ensured that his concrete systems were closed by defining new objects, i.e. by constructing magnitudes. He realized that it was an important problem to show, for a specific system of magnitudes, that it is closed. But his treatment of this problem was guided by geometry and not abstract algebraic considerations. If, as in his *lineale Ausdehnungslehre*, a multiplicative connection generated extensions (magnitudes) of a higher degree, it was very difficult for him to master the technicalities posed by this problem. Only when he reduced the system to a finite number of units of first degree did the extensions of higher degree become closed in a natural way. A general argument was given in the *Formenlehre*, written by his brother Robert, but with Hermann Grassmann's cooperation. There one finds the assertion that: "The result of a connection of two magnitudes will again be a magnitude, since it is a matter of reasoning and has one value and not several ones".[2] A similar argument seems to form the basis of Grassmann's statements in his "arithmetic textbook" published in 1861. In all, there are only a few remarks that hint at the problem of ensuring closure. In modern terminology, these remarks could be interpreted as if Grassmann considered each specific set of magnitudes as freely generated by units and the rules of connection. These ideas were very vague and intuitive, of course, but they explain his method. If one adds further equations as

---

2 "Das, was durch die Knüpfung zweier Größen entsteht, ist wieder eine Größe, da es Gegenstand des Denkens ist, und nur einen und nicht mehrere Werthe hat,..." (R. Grassmann 1872, p. 8).

side conditions, the set of magnitudes will be restricted, strictly speaking forming a factor set, but the set remains closed under the operations of connection. So Grassmann did not have to prove again that the reduced set is closed under the operations.

In the further development of his theory, Grassmann declared associativity and commutativity to be the only possible restrictive conditions if one single connection were given.[3] Here Grassmann spoke of a synthetic connection, and he assumed these two properties in all studies that followed. The notions of commutativity and associativity were not used, though they were already known at that time. A more precise designation of the kinds of connection was not possible, and Grassmann went on to consider the dissolution of the connection obtained by what he called the analytical method.[4] As a starting point for the definition of the analytical, or inverse operation, he first assumed a noncommutative synthetic connection. Grassmann distinguished two analytical operations. In accordance with the restrictions on connections just mentioned, he passed immediately over to commutative and associative connections also called simple connections.

In addition, the analytical operation had to be single-valued. After these preliminaries, Grassmann defined addition as a simple synthetic connection with a single-valued analytical operation. This analytical operation was then called subtraction. In this way, the connection, called addition, must always be associative and commutative, and there exists a single-valued operation inverse to it. Finally, Grassmann introduced an indifferent and a purely analytical form. In modern notation, these correspond, respectively, to an identity element and an inverse element. At this stage, Grassmann possessed all the concepts necessary to characterize a system of magnitudes with a connection satisfying the properties of a commutative group. But he never actually realized this step toward abstraction. Grassmann never defined a set of magnitudes having one or several possible connections, and then regard this as a new mathematical object. He was not interested in defining new structural elements, but rather was content analysing the kinds of connections that could arise and their properties. This decisive step toward the creation of modern algebra was failed to take, and this gives a clear idea of the extent to which Grassmann contributed to the genesis of modern algebra as well as the boundaries he was unable to overcome.

---

3 For a detailed description of Grassmann's general theory of forms cf. Flament 1992, pp. 209-219; Schlote 1985, pp. 232-240; Lewis 1977, pp. 139-149.

4 "Eine noch weiter gehende Bestimmung ist nun für die Art der Verknüpfung, wenn man nicht auf die Natur der verknüpften Formen zurückgeht, nicht mehr möglich, und wir schreiten daher zur Auflösung der gewonnenen Verknüpfung, oder zum analytischen Verfahren" (Grassmann,1894, vol. 1,1, p. 36).

If a second synthetic connection was defined on a set of magnitudes, Grassmann proceeded in a very modern way. The essential point for him was the fact that both connections had to be combined in some way. They could not exist completely separate from one another. The relations between the two connections took first priority, and from this Grassmann was led to deduce the distributive laws. He noted that these must take into account that the second connection need not be commutative, so that two possible distributive laws will result. If the second connection satisfies distributivity, it is called a multiplication. As in the case of addition, where the set of magnitudes had to form a group, he spoke of a multiplication only if the set of magnitudes formed, in modern terminology, a ring under two operations. Sometimes Grassmann seemed to proceed as follows: Given a set of magnitudes, with one or two connections, he proved that the rules for connections are fulfilled, and only after that did he call them addition and multiplication.

Grassmann concluded his presentation of the general theory of forms with some brief hints regarding division as the analytical operation corresponding to multiplication, but there are no new aspects in this part. Analysis of the different types of multiplicative connections was not part of the theory of forms. He discussed these types from a geometrical point of view in his theory of extensions. All further properties of multiplication, as associativity, commutativity etc., gave rise to a more detailed description of the system of magnitudes. This was established by so-called equations of determination.

Before analysing Grassmann's ideas with regard to multiplicative connections, various changes in the concept of magnitude must be mentioned. At the beginning, and in his *Lineale Ausdehnungslehre* as well, Grassmann rejected the concept of magnitude. In his opinion, it was too closely connected to the concept of number. Grassmann's aim was to select common features found in several parts of mathematics that utilized combinations of elements, and this required abstract objects which could be combined. He created this necessary abstraction with the concept of a thought-form, which also resulted in a generalization of the concept of magnitude.[5] In later works, he returned to the concept of magnitude instead of thought-form. But the concept had now a much broader range: it contained geometrical objects as well and elements of combinatorics like permutations (cf. Grassmann 1861, p. 1).[6]

5 "Der Name Grössenlehre eignet nicht der gesammten Mathematik, indem derselbe auf einen wesentlichen Zweig derselben, auf die Kombinationslehre, keine Anwendung findet, und auf die Arithmetik auch nur im uneigentlichen Sinne. Dagegen scheint der Ausdruck Form wieder zu weit zu sein, und der Name Denkform angemessener,..." (Grassmann 1894, vol. 1,1, p. 23).

6 "Mathematik ... ist die Wissenschaft von der Verknüpfung der Grössen" (Grassmann,

To realize his general concept of multiplication, Grassmann was led to introduce abstract units, systems of units as a base of a domain of magnitudes, and extensive magnitudes. He distinguished between the absolute unit 1 as the unit for numbers and relative units. If relative units form a basis for a domain of magnitudes, their linear independence is assumed (Grassmann 1855, p. 124), which means there is no algebraic relation between units having nonvanishing coefficients and representing the element zero. An extensive magnitude is then defined as a finite linear combination of units of a basis. All these definitions were given by Grassmann for the first time in "Sur les différents genres de multiplication" published in 1855 in a priority claim made against A. L. Cauchy. Grassmann improved this approach in the second edition of his *Ausdehnungslehre* in 1862. There he defined the connection of two arbitrary extensive magnitudes as well as the multiplication of one such magnitude by a number.

Although Grassmann regarded only real numbers as the domain of coefficients, it is possible to choose any abstract coefficient field without difficulties. If one adds the rules of connection defined for addition, one will get a complete characterization of an *n*-dimensional vector space. But here again, Grassmann neglected to introduce this new mathematical object.

If we examine Grassmann's ideas in the theory of hypercomplex numbers it is necessary to analyse his treatment of products of two extensive magnitudes. His most detailed description is given in the article quoted above. In his very first definition of a multiplicative connection of two extensive magnitudes, he already implicitly assumed the distributive rule

$$[\, S\, a_r e_r * S\, ß_s e_s \,] = S\, a_r ß_s [e_r * e_s].$$

There is no further restriction of the multiplicative connection, which corresponds entirely to Grassmann's general remarks in the *Ausdehnungslehre* of 1844. In the spirit of the latter, Grassmann stressed that the product of two relative units has to be regarded as a new unit of second degree, or of n-th degree if n units are multiplied. He knew, of course, that such a definition of multiplication is too general. But in search of further rules of connection, Grassmann pointed out the possibilities for incorporating supplementary rules in this process: The product of extensive magnitudes must, by definition, again be an extensive magnitude or a number, which implies that it be numerically represented by a system of units. But there are no specifications as to

---

1861, p. 1). It is in this general form that the concept of magnitude is used in this article.

how this system of units should be chosen or how the product $[e_r{*}e_s]$ should be composed out of them. An exact definition of such a product thus requires additional conditions on the system of units and the connection of the products (cf. Grassmann 1894, vol. I, 2, p. 28).

The latter is made by adding side conditions: "equations of determination" ("Bestimmungsgleichungen"), and each such system defines a specific kind of product. These different kinds of products are regarded by Grassmann as the decisive distinguishing feature for systems of magnitudes, and this presents a possibility for classifying those systems. In practice, Grassmann mastered this problem only for those types of products which are important for his theory of extension. The reasons for Grassmann's failure to give an extensive elaboration of this classification are very complex.[7]

Grassmann's choice of equations for a further determination of the products was clearly influenced by geometrical ideas, since nature itself calculates so to speak in a geometrical manner (Grassmann 1911, vol. III, 1, p. 33). He did not take into account other ideas, such as Hamilton's quaternions, but rather restricted the equations of determination to those in which all products have the same number of factors, considering mostly products of only two factors. As a consequence, there was no reduction of a product to a linear combination of units, and equations like Jacobi's identity or the associative law were excluded as well. In the case of the associative law, however, Grassmann did not adhere to his own restrictions. Unfortunately, he gave no explanation for this step, and his reasons for doing so remain a mystery to this day.

To sort out from the infinity of possible special multiplications those which seemed to be useful in science, Grassmann formulated three general properties. These properties can be expressed as transformations of the units of the basis. A multiplication will fulfill such a property if the transformation respecting it does not change the equations of determination. Grassmann's first property assumed that the sign of a relative unit can be altered, and that any relative unit can be substituted for any other, different from the first one. Secondly, he supposed that if any two units are chosen, then there exist extensive magnitudes of them which can be substituted for them. The extensive magnitudes formed by the two chosen units should not be equal to nor a multiple of these units. Thirdly, it is permitted to substitute any unit by any other magnitude. Under these assumptions, one obtains what Grassmann called the symmetrical, circular, and linear multiplications namely those satisfying the

---

7 In this context a detailed analysis of the factors that influenced Grassmann's work is necessary. For several aspects cf. Schlote 1987, pp. 72-77, and Lewis 1977, pp. 104-139, 156f.

first, the first and the second, and all three of the above properties (Grassmann 1855, p. 128).

Grassmann pointed out applications of circular multiplications to geometry and mechanics (Grassmann 1855, p. 134). Despite his restrictions, he was able to distinguish between sixteen different types of symmetrical multiplication, to analyse these, and to demonstrate by these examples how to realize his principles. He obtained the equations of determination for all of the different kinds by means of equations of the following form:

1. $e_r{*}e_s = e_s{*}e_r$

2. $e_r{*}e_s + e_s{*}e_r = 0$

3. $e_1{*}e_1 = e_2{*}e_2 = \ldots = e_n{*}e_n$

4. $e_1{*}e_1 + e_2{*}e_2 + \ldots + e_n{*}e_n = 0,$

$r \neq s$ and $r,s = 1, \ldots, n$ (Grassmann 1855, p. 131).

Some of these products should be emphasized, for example the algebraic product or algebraic multiplication leading to a commutative algebra,[8] and the combinatorial product or outer multiplication, which generates a noncommutative algebra with

$$e_i{*}e_j = -e_j{*}e_i, \quad e_i^2 = 0$$

and which is called Grassmann algebra since W. K. Clifford. Both are among the four kinds of linear multiplication and that Grassmann characterized as being of greatest importance for analysis, geometry, mechanics and physics in general (Grassmann 1855, p. 138). He also noted that one can reduce, for instance, every function of several variables to a function of only one variable by applying algebraic multiplication to extensive magnitudes. This allows one to use immediately most of the theorems on functions of one variable. Additionally, Grassmann discussed two of the circular multiplications which are not linear: the inner multiplication or inner product and the complex multiplication (Grassmann 1855, p. 140f.). These play an important role in geometry and mechanics as well, especially when dealing with problems related to circles, angles, etc. These problems cannot be solved by using linear multiplications. Grassmann referred to his *Ausdehnungslehre* and his *Geometrische Analyse* (Grassmann 1894, p. 345 - 395) for applications of his inner multiplication. The equations of determination are given by

---

8 In general, the algebra is to be assumed associative.

$$e_r * e_s = 0 \text{ and } e_1 * e_1 = e_2 * e_2 = \ldots = e_n * e_n,\ r \neq s,$$

(Grassmann 1855, p. 140). Complex multiplication is defined by the equations:

$$e_r * e_s = e_s * e_r \text{ and } e_1 * e_1 + e_2 * e_2 + \ldots + e_n * e_n = 0,$$

Its name comes from the fact that, when n=2 , it represents the multiplication of complex numbers (Grassmann 1855, p. 140).

What is the importance of Grassmann's work for the theory of hypercomplex numbers and for algebra? It consists in establishing an abstract foundation for a theory of hypercomplex number systems by a comprehensive investigation of connections. In this respect, he broke with traditional ways of thinking. The creation of quaternions is often acknowledged as a revolutionary step in the history of algebra, but the same must be said of Grassmann's work. In striving for a uniform, formal construction of mathematics, Grassmann developed ideas which were much more general than those developed within the English algebraic school or W. R. Hamilton's approach. In many ways, he realized the freedom sought by G. Peacock for determining the rules of connection. Like Peacock, Grassmann used analogies to arithmetic as hints for deriving rules of combination. He characterized the multiplication of numbers as commutative, associative, and as having no zero divisors. But there was no principle of permanence as formulated by Peacock, a fact which restricted his choice of rules of connection. A multiplication for extensive magnitudes only had to obey the distributive law; the commutative and associative laws were possible but not necessary conditions. It should be recalled how long Hamilton struggled before he decided in 1843 to give up the commutative law, which opened the way for his discovery of the quaternions.

A thought-form ("Denkform"), an extensive magnitude, as Grassmann defined it, need not be a generalization of numbers, but only an abstract element omitting a more detailed characterization. So there was no necessity for Grassmann to maintain that the connection rules had to mirror those that held for numbers. Thus, two approaches to hypercomplex numbers could be taken, either composing abstract magnitudes or by extending complex numbers, by generalizing the properties of arithmetic. Both had their pros and cons and stimulated further development towards a theory of hypercomplex number systems. The weakness of Grassmann's presentation consisted in the insufficient explanation of his general abstract approach to the connection of magnitudes as formal mathematical elements. His formulations were in part not exact enough, and he often deviated from his own opinions. It should be taken into account, though, that a system of concepts necessary for an exact description of Grassmann's ideas was not available to him.

Moreover, such ideas that might have proved helpful had not been widely disseminated, and consequently he was unaware of them, so that Grassmann failed to develop such a conceptional basis. On the basis of his general principles, the classification of multiplicative connections - that is the classification of the hypercomplex number systems - was an almost unsolvable task. Grassmann had to make additional restrictions in order to obtain concrete results. From a modern point of view, he unfortunately omitted important parts, however this should not be held against him. This was probably Cartan's opinion, too. E. Cartan was the first to reveal extensively the algebraic aspects of Grassmann's work in his article in the French edition of the Encyclopadia of Mathematical Sciences, showing their place in and importance for mathematics. In 1922, in his studies on the calculus of differential forms, Cartan proposed to adopt Grassmann's concept for the corresponding operations of differential forms (Cartan 1922, p. X, 52). In this way Grassmann's name has become a household word in modern mathematics.

Arbeitsstelle für Geschichte der Naturwissenschaften und der Mathematik, Akademie der Wissenschaften zu Leipzig

REFERENCES

Elie Cartan, *Leçons sur les invariants intégraux* (Paris: Hermann, 1922).

Michael J. Crowe, *A History of Vector Analysis* (Notre Dame: University of Notre Dame Press, 1967).

Dominique Flament, "La 'lineale Ausdehnungslehre' (1844) de Hermann Günther Graßmann," *1830-1930: A Century of Geometry*, ed. Luciano Boi, Dominique Flament and Jean Michel Salanskis (Berlin et al.: Springer-Verlag, 1992), 205-221.

Hermann G. Graßmann, *Gesammelte Mathematische und Physikalische Werke*, 3 vols., ed. Friedrich Engel (Leipzig: Verlag B. G. Teubner, 1894-1911).

Hermann Günther Graßmann. "Sur les différents de multiplication," *[Crelles] Journal für die reine und angewandte Mathematik*, 1855, *49*:123-141.

Hermann Günther Graßmann, *Lehrbuch der Arithmetik für höhere Lehranstalten* (Berlin: Verlag Enslin, 1861).

Robert Graßmann, *Die Formenlehre oder Mathematik* (Stettin: Selbstverlag, 1872).

Albert C. Lewis, "H. Graßmann's 1844 *Ausdehnungslehre* and Schleiermacher's *Dialektik*," *Annals of Science*, 1977, *34*:103-162.

George Peacock, "Report on the Recent Progress and Present State of Certain Branches of Analysis," *Report of the Third Meeting of the British Association for the Advancement of Science for 1833* (London, 1834), 185-352.

Karl-Heinz Schlote, "Hermann Graßmanns Beitrag zur Algebrentheorie," *Janus*, 1985, *72*:225-255.

Karl-Heinz Schlote, *Die Entwicklung der Algebrentheorie bis zu ihrer Formierung als abstrakte algebraische Theorie*, Dissertation B, Universität Leipzig, 1987.

JEAN-LUC DORIER

BASIS AND DIMENSION - FROM GRASSMANN TO VAN DER WAERDEN

Basis and dimension are two elementary notions in the theory of vector spaces. The origin of the term 'basis' comes from the possibility of expressing any element of a given set as a linear combination of the basis elements. Therefore, the origin lies in a question of generation; on the other hand the condition of unicity brings out the question of independence. The connection between generation and dependence is certainly one of the most interesting characteristics of the concept of basis: any maximal set of independent vectors or any minimal set of generators, is a set of independent generators and vice versa and such a set is a basis. Moreover, the dimension, beyond its "natural meaning", is the merging point from which the question of invariance is to be drawn out. Indeed, the fact that all bases have the same number of elements entails two results: there cannot be more than a certain number of independent vectors, and fewer than the same number of generators. With a suitable starting point in the presentation of definitions and first properties on dependence and generation, these different aspects seem quite logically connected and easily explainable, but historically, the development of these two concepts was less straight-forward. For various reasons, in the approach to the concept of basis, the connections between dependence and generation were not always exhibited. Therefore the concept of dimension could only partially be drawn out, and some of its aspects were smothered, or even considered as obvious and assumed to be true without proof. On the other hand, the relation between the dimension of a subspace and the rank of any system of linear equations by which it can be represented, played a role in the history of the concepts of basis and dimension.

This diversity reveals the richness of the concepts and their importance in elementary linear algebra. On the other hand, it reflects the historical complexity of the constitution of the concepts of basis and dimension, whose different aspects have been either privileged or smothered depending on the type of problems and settings in which they were discussed. The model of geometry with its three "natural" dimensions, and also the study of linear equations, were two major

G. Schubring (ed.), Hermann Günther Graßmann (1809–1877): Visionary Mathematician, Scientist and Neohumanist Scholar, 175–196.

sources of development. In this paper, I will examine how the concepts of dimension and basis took shape in three contexts which constitute their epistemological origins: Grassmann's *Ausdehnungslehre*, the study of linear equations, and modern algebra.

HERMANN GRASSMANN'S CONCEPTIONS

In his detailed analysis of the influence of Grassmann's Philosophy on the 1844 *Ausdehnungslehre* [A_1], Lewis (1977) studies several aspects of Grassmann's point of view on mathematics and science and on their mode of presentation. One of the most important for my purpose concerns Grassmann's ideas on what a scientific method should be since they govern the mode of presentation of the A_1.

"Now we characterize a method of treatment as scientific if the reader is thereby on the one hand led necessarily to recognize the individual truths, and on the other hand is placed in position from which he can survey a broader sweep of the development."[1] (Grassmann 1844, introduction, §14).

Lewis states that "the complementarity of rigor and overview provides the final contrast which makes the *Ausdehnungslehre* a consistent whole in its methodology. This contrast illustrates Schleiermacher's original meaning of 'heuristic'(...) He implied that one of the principal advantages of this method was its frequent reference back to the original particular intuition. In the first chapter Grassmann makes it clear that in the case of the theory of extension[2] the rigorous development of the science is the development from the basic concepts, while the overview is provided by constantly tying the deduction of new concepts to geometry 'whose basis our science forms'." (ibid., 133). Moreover, in his analysis, Otte suggests that "Grassmann has crystallized the dialectics of process and form into a definite hierarchical structure" (Otte 1989, 29).

The aim of this chapter is to examine how this particular choice in the mode of presentation made Grassmann's approach to the concepts of basis and dimension in the A_1, particularly rich and original. Indeed, these two concepts are involved, on an ontological level, in the foundation itself of the theory of extension, where generation and independence are naturally connected. By using the formal representation as a "vehicle of exploration" (ibid., 31), Grassmann uses the first paragraphs of his theory to explore both the formal and the real aspects of the objects he has introduced, in a dialectic process based on the

1 Any quotation or terms from the 1844 *Ausdehnungslehre*, in English, are taken from the translation of Lloyd C. Kannenberg (Chicago and LaSalle: Open Court Pub. Co., 1995), whom I thank for kindly sending me a preprint of his book.

2 Literal English translation of *Ausdehnungslehre*.

contrast between the intuitive origin and the rigorous development of his theory. Thus, dimension and basis appear, in this context, as the results of a search for independence of the objects from their initial mode of generation.

Indeed, the theory of extension is based on the description of a mode of generation whose origin lies in an analogy with geometry. This gives the theory its intuitive real aspect. This is a dynamic process of construction, which transplants an initial element according to a fundamental method of evolution ("Grundänderung" or "Erzeugungsweise") and its opposite, and then transplants all the transplanted elements according to another fundamental method of evolution (and its opposite), etc., the result consisting in a system of extension (ibid., §16). In this process of generation, the analogy with geometry is a rapidly generalized starting point, as there is no pre-existing set of elements in which the generators (the fundamental methods of evolution) are to be chosen, thus no limitation in dimension. On the other hand, each fundamental method of evolution is chosen in such a way that none will lead from the initial element to one of the elements of the system generated by the preceding methods of evolution. Therefore, the initial generators of a system are, by nature, independent (Grassmann defines independence in these words and for this purpose). So generation and independence are related right in the roots of Grassmann's theory. As a consequence, dimension is naturally present in the theory of extension: it is the number of independent fundamental methods of evolution necessary to generate the whole system from the initial element. Grassmann names it a system of *n*-th order. Thus the order is the dimension with reference to a primitive basis[3] (the fundamental methods of evolution), so this is attached to the initial way of generating the system.

If geometry is a basis for analogy, the reality of the theory of extension is of a different nature; reality refers to the actual way of generating, as described briefly above, which has intuitive foundations in geometry, but concerns the determination of forms of thought. As Lewis pointed out, one should "note the distinction Grassmann makes between (formal) definitions or axioms made for mere convenience and what he regards as the (real) determination of concepts from their original sources" (ibid., p. 139). This distinction is essential to understand the aim of the first twenty paragraphs of the theory. Indeed, once the intuitive sources - determining the real aspect of the concepts - are given, Grassmann is concerned with the exploration of their formal aspects, in relation with

3 The term *basis* is never used by Grassmann, either in 1844 or in 1862, and the term *dimension* is only used in reference to geometry. I will use these two words only in their modern meaning in reference to the modern theory of vector space.

their real aspect, in order to progress in the (never ending) completion of the concepts. In this process, the introductory chapter on the general theory of forms is used as an architectonic reference. In §18-20, Grassmann develops this method in order to give a formal description of a higher order system, which makes evident its independence from the initial mode of generation. This is where the questions related to the concepts of basis and dimension emerge.

Formally, a system of *n*-th order can be represented as a linear space generated by *n* basis elements, or the set of all linear combinations of *n* given basis elements. This presentation was chosen by Grassmann to introduce his theory in the new version he published in 1862 [A_2]. But in 1844, there is no such formal description in the beginning. Indeed, Grassmann did not want to use the concept of number, because his aim was to make his theory self contained and independent from the rest of mathematics. Therefore, there are no linear combinations until §68 where numerical magnitudes (i.e. numbers) are introduced as the quotients resulting from the division of one extensive magnitude by a similar one (i.e."collinear vectors"). Even though this choice does not facilitate the understanding of some statements, it does prevent the temptation of using a formal isomorphic representation by R^n, through the use of a coordinate system. In this matter, Grassmann appears as the successor of Leibniz in his search for a geometrical analysis, which operates on geometrical entities themselves and not on their representations by coordinates (for more details see Otte 1989).

In the A_1, a method of evolution represents a direction and a displacement is equivalent to a directed line segment, although it is more general as Grassmann's point of view goes beyond geometry. A displacement can be represented by [$\alpha\beta$], where α is an initial element and β the final element. Displacements are said to be similar if they belong to the same (method of) evolution - they are collinear vectors. Starting with similar displacements, and then displacements belonging to the fundamental methods of generations, Grassmann determines the concepts of addition and subtraction of such displacements, with respect to the initial mode of generation (§17 to 19). Moreover, Grassmann uses the General Theory of Forms (presented in an introductory chapter to the theory of extension) as an architectonic framework to discuss the relations between this real aspect of addition, and its formal representation. Indeed the way of generating a system of *m*-th order leads to the representation of a displacement *p* by the sum of *m* displacements belonging to the *m* fundamental evolutions *a, b, c,...*: $p = a_1 + b_1 + c_1 + \ldots$ Then §17 is devoted to the possibility of giving a meaning to a change of order in the terms of this relation. This is explicitly made in a dialectical process between the formal and the real aspects of addition. In §19, the formula:

$$p_1 + p_2 = (a_1 + a_2) + (b_1 + b_2) + (c_1 + c_2) + \ldots,$$

which determines the addition of any two displacements is made possible by the preceding result on the commutativity and associativity of the addition of displacements belonging to the fundamental methods of evolution. On the other hand, it also goes back to the original mode of generation, as it symbolizes the way the sum is generated by the initial methods of evolution. It also marks a step forward in the search for independence of the system from its original way of generation. Indeed, a displacement can now be determined as the sum of other displacements not necessarily belonging to the initial fundamental methods of evolution. §20 presents a refinement of this result as it describes conditions for a replacement of the initial methods of evolution. The introduction of this paragraph is clear on this matter:

"The development presented in the previous paragraph prepares us for an independent description of the higher order system. So far this has been presented as dependent on knowing the fundamental methods of evolution by which it was generated. As we will show, we can remove this dependence since the same system of m-th order is generable by any m methods of evolution belonging to it that are mutually independent (in the sense of §16), that is, that are included in no system of lower order (than the m-th)." (ibid., §20).

In this statement, the order appears as a general characteristic of the system which is not only tied to a particular choice of generators, but is a general indicator of the size of a system. Moreover, some of the essential aspects of the concepts of basis and dimension are drawn out. At this stage of the theory, the concept of independence has been briefly introduced in §16, where it concerned only fundamental evolutions. The generalization to all methods of evolution suggested by Grassmann at the end of the preceding quotation, is somewhat ambiguous. Indeed, what must be understood by: "m methods of evolution that are included in no system of lower order"? Does it mean in any system, or only in systems generated by some of the evolutions whose independence is discussed? The reference to §16 makes it clearer that only the evolutions whose independence is discussed are involved. The fact that independent evolutions belong to no system at all of lower order can be deduced from this definition; yet this logical consequence is closely related to the result discussed in §20. It is therefore essential to make the distinction clear at this stage.

To prove the result quoted above, Grassmann uses a step by step method, showing that one fundamental method of evolution can be replaced by another which is independent of the remaining; this is what is now known as the exchange method ("Austauschsatz")[4]. The proof is quite simple, Grassmann makes it in a formal setting but the relations with the real aspect of addition explained in the previous paragraphs make the intuitive meaning visible at each stage. The key point is that if p is independent of $b, c, \ldots$ then in the decomposition of p according to the initial methods of evolution: $p = a + b + c + \ldots$, a is not zero. Therefore, any displacement a_1 belonging to a corresponds to a displacement p_1 belonging to p:

$$p_1 = a_1 + b_1 + c_1 + \ldots, \text{ hence: } a_1 = p_1 - b_1 - c_1 + \ldots.$$

Therefore in the generating process, any displacement belonging to a can be substituted by a sum of displacements belonging to $p, b, c, \ldots$. The possibility of reorganizing the order of the evolutions and the partition in similar displacements (cf. §17 to 19) ensures the exchangeability of p and a. Thus, the exchange theorem appears as a natural conclusion to the preceding paragraphs. Moreover, the dialectics between the formal and the real aspects is quite tangible as the proof of this result is both rigorous in its deductive form of presentation and meaningful in regard to the intuitive origin of the theory.

The exchange theorem is thus an essential issue in the foundations of the theory of extension, it also potentially includes most of the meaning of the concepts of basis and dimension. Yet, Grassmann did not clearly draw out all the implications from this result. For instance, his conclusion of the proof of the result announced in the beginning of §20 is somewhat vague:

"And since one can continue this process, it follows that one can generate the same system by any m mutually independent methods of evolution." (ibid., §20).

But the iteration of the exchange method requires at each step that the conditions of replacement be fullfilled. This is not that straight-forward; it means that if $n(<m)$ initial methods of evolution $a, b, \ldots, c$ have been replaced by n new ones $p, q, \ldots r$, one of the $(m\text{-}n)$ remaining new methods of evolution (s) is independent of $p, q, \ldots r$ and $(m\text{-}n\text{-}1)$ of the remaining initial methods of evolution. This can be deduced from the independence of $p, q, r, \ldots s$, but even if it is only a technical point there is here a lack in Grassmann's reasoning. Moreover, it is easy to deduce from §20, that no set of less than m methods would generate a system of

4 This term is not used by Grassmann; it was introduced by German algebraists around 1930, in reference to Steinitz; this point will be discussed further on.

m-th order.[5] Yet, Grassmann does not explicitly discuss this point either in §20 or further on, although it is implicitly admitted.

In his A_2, Grassmann dropped his philosophical introduction and the general theory of forms. Moreover, he adopts an Euclidian form of presentation. Extensive magnitudes are given as formal entities and they are defined by their relations, without intuitive or geometrical reference. The first two definitions concern generation and dependence. In opposition to the presentation of the A_1, this is given in a numerical setting; an extensive magnitude a is said to be derived from the magnitudes $b, c, \ldots$ by the numbers $\beta, \gamma, \ldots$ if: $a = \beta b + \gamma c + \ldots$ and two or more magnitudes are said to be in numerical relation if one is numerically derived from the rest (Grassmann 1862, §1, def 1 and 2). In this framework, a system of original basis elements $(e_1, e_2, \ldots)$ (a set of magnitudes standing in no numerical relation) replaces the fundamental methods of evolution of the A_1. A magnitude of first order is any magnitude a derived from the original system: $a = \alpha_1 e_1 + \alpha_2 e_2 + \ldots$; α_1, α_2 are called the numbers of derivation. There is an important change in the ontological foundations of the theory compared to 1844. In 1862, the real aspect was dropped, as only the formal aspect of concepts is present in the bases of the theory. The intuitive aspect is discussed afterwards, through the applications to geometry, which appear only in chapter 5. Moreover, the use of numbers gives the presentation a different form as the isomorphism to R^n is now quite apparent. As a consequence, the definition of addition and of multiplication by a number, the next step in the presentation (§2), is introduced through the operation on the numbers of derivation. Moreover, following the definitions of each operation, Grassmann gives a list of properties which he calls *fundamental formulas* because they suffice for deducing all laws of algebraic addition, subtraction, multiplication and division. This led Peano in 1888 to the first axiomatic definition of a vector space (Peano 1888, 141-142), but Grassmann's approach is not axiomatic. In fact, this section of the A_2 cannot be understood independently of the general theory of forms as presented in the introduction to the A_1, this is where the meaning and the origins of the *fundamental formulas* are to be found. I will quote Lewis again:

> "what is presented [in the general theory of forms] is not unproven statements from which succeeding statements are deduced; rather, principles of connection[6] expressed by means of the general concepts of equality and difference, and connection and separation, are symbolized" (Lewis 1977, p. 140).

---

5 If there were a set of $n<m$ generators, any set of m generators could be replaced by a subset of n generators, but this contradicts the assumption of independence of the generators.

The bare formalism of the A_2 is the artificial result of Grassmann's attempt to satisfy the criticisms received after the edition of 1844; this resulted in the inaccessibility of most of the intuitive discussion, which is to be found in the A_1.

§2 entitled "Connection between magnitudes derived from a system of units" contains most of the results about the concepts of basis and dimension. A domain of *n*-th order is defined as the collection of all magnitudes derived from *n* magnitudes and not less than *n* such magnitudes. One can see here a change in the definition compared to the A_1, as the minimality of the number of generators is the criterion, and not the independence of these generators like in the A_1. Moreover the presentation of the different ways of generating a domain of *n*-th order is more precise and detailed in the A_2 than it was in the A_1. As in the A_1, the exchange theorem (ibid., §2-19) occupies a central position, it is the key point to deduce the fact that a domain of *n*-th order is derivable from any set of *n* independent magnitudes (ibid., § 2-24), but this final result is preceded by two theorems which are also deduced from the exchange theorem. One states that a series of 'independent' magnitudes can be completed into a 'basis' (ibid., §2-20), and the other proves that if *n* magnitudes are derivable from less than *n*, they are dependent (§2-22). From this it is clear that a domain of *n*-th order cannot be derived from less than *n* magnitudes. The concepts of basis and dimension are thus more explicitly drawn out in the A_2 than they were in the A_1.

Yet, in spite of this formal improvement, the origin of these two concepts is to be found in the A_1 where both real and formal aspects are discussed. Moreover, the use of the exchange theorem, which appears as essential, comes naturally within the form of presentation adopted in the A_1. In the A_2, the discussion about the exchange of the original units is more artificial, as the question of the independence of the representation of a domain from the original units is not put forward, even though it appears in a remark at the end of §2-24. On the other hand, in this same remark Grassmann admits that the "theorem can also be derived directly from the theory of elimination". Indeed, this can be done through the use of the numbers of derivation. The use of the exchange theorem, rather than a direct proof in R^n is judged by Grassmann to be not only "elementary, but in addition (to have) the advantage that thereby the essential, simple relation between the extensive magnitudes appears more clearly". Such a statement cannot be fully appreciated without knowing the origin of the exchange theorem, as introduced in the A_1. This proves again the importance of the original method of presentation adopted in

---

6 This is the word used by Grassmann to designate an operation.

the first version of the *Ausdehnungslehre* in order to appreciate Grassmann's contribution to the concepts of dimension and basis. Even though the A_2 brought more explicit statements, the A_1 is essential to understand the origin and the pertinence of the question which brought these two concepts into focus.

DIRECT FOLLOWERS OF GRASSMANN

Giuseppe Peano was one of the first mathematicians to have contributed to making Grassmann's *Ausdehnungslehre* more accessible and better known to his contemporaries. In 1888, he published his *Calcolo Geometrico*, in which he presents a shortened and simplified version of the beginning of Grassmann's theory. Most of this treatise is set within the geometrical space and nothing is said about the concepts of basis and dimension. But in the final chapter, Peano gives an axiomatic definition of what he calls a 'linear system', which is very near to the modern definition of a vector space. As mentioned above, Grassmann's viewpoint on mathematics forbids interpretations of any of his writing as close to an axiomatic approach. Nevertheless, the *fundamental formulas* about the operations on extensive magnitudes drawn out in §1-8 and 12 of the A_2 seem to have inspired Peano in his axiomatic approach.

After defining a linear system, Peano gives the definition of dependence and independence of elements in a linear system, then he gives the following definition:

"The number of dimensions of a linear system is the maximum number of independent elements in the system."[7] (Peano 1888, p. 143)

This definition is not similar to Grassmann's definition of the order, either in the A_1 or in the A_2. The main difference is that Peano's definition is not tied to a collection of original generators, it is presented in the definition directly as a general property of the system. Moreover the aspect of generation is absent. In fact, the following theorem shows the relation between this definition and the problem of generation:

" If the system A has n dimensions, let us take n independent elements $a_1, \ldots, a_n$ and a a new element, we can determine n numbers $x_1, \ldots, x_n$ such that: $a = x_1a_1 + \ldots + x_na_n$. Moreover these are unique." (ibid., p. 143).

This shows that any set of n independent elements in a n dimensional system constitutes a basis,[8] and shows the relationship between independence and generation.

---

7 Quotations from Peano have been translated by me.

8 The word 'basis' is not used by Peano.

Peano's approach to the question of dimension is different from Grassmann's. For Grassmann, it comes from a process going from the particular to the general, through the search of independence of the system from its original mode of generation. For Peano, there is no initial mode of generation, and the number of dimensions determine the degree of independence within the system, which is then related to the general question of generating the system, in a later stage only.

Peano's approach might seem attractive, yet it induces a difficulty. Indeed, in his approach nothing proves explicitly that in an *n* dimensional linear system there could not exist a set of less than *n* independent generators. Or, to state it from a different angle, nothing proves that any maximal set[9] of independent elements has the same number of elements. This problem is not raised by Peano. As was said above, Grassmann did not explicitly raise the problem in the A_1 either (although he did so in the A_2) but the use of the exchange theorem could have provided the answer quite easily, as shown in 1862; in Peano's approach the question is smothered. Adopting a general point of view in the definition of the dimension is partly the source of this weakness. My hypothesis is that the model of geometry, and the presentation of the preceding chapters of the *Calcolo Geometrico* induced, to a great extent, this weakness in the approach of the dimension. On the other hand it is difficult to know what Peano knew exactly of Grassmann's theory, and it is likely that his reading may have been superficial. That could also explain the difference in the definition of the dimension. Nevertheless, there is a proof here of the decisive importance of Grassmann's 1844 method of presentation. In Grassmann's approach the generalization of the geometric intuition is carried out through the process of extension; this provides a point of view in which the mode of generation is a starting point. Therefore the dimension appears as a measure of the extent of the generating process. In Peano's work the question of dimension does not appear as part of the constitution of the generalization. This concept is generalized from the point of view of geometry itself. As a consequence, the dimension is not introduced as a measure of the extension but more as a characteristic of the boundary of the extension. Thus Peano's definition does not put forward the mode of generation but the limit in the number of independent elements. Indeed, from a retrospective point of view on the generalization, the geometrical space is characterized by its limitation to three independent directions, although within a dynamic process of generalization, geometry is generalized by the possibility of finding new independent directions (or methods of evolution). This distinction in the

---

9 By maximal set, I mean that any other element is linearly dependent on the elements of this set. This is different "a priori" from a set with the maximum number of independent elements.

process of generalization plays a decisive role in the different approach to the concept of dimension. As Lewis said:

"[...] (for Grassmann) generalization is not the goal of mathematics but rather one pole of contrast which governs its method. This last point is perhaps the key explanation of the difficulty which readers have had with the *Ausdehnungslehre* since 1844. It is not just progress towards generality that makes the 1844 *Ausdehnungslehre* one of the great mathematical works of the nineteenth century; it is rather that this generality is of a kind that elucidates obscurities and problems on lower levels." (Lewis 1977, p. 161).

Indeed making explicit the different aspects of the concepts of dimension and basis, by exploring the relations between dependence and generation is an example of an elucidation of obscurities and problems on lower levels. In this matter the proof is made that Grassmann's approach has been efficient. Peano's generalization did not take into account the whole ontological richness of Grassmann's theory, he thus only partially grasped the meaning of the number of dimensions, even though it may be considered as an elementary concept.

Peano's work did not have much influence outside Italy, but Cesare Burali-Forti and Roberto Marcolongo carried on his work on vector analysis. They also refer to Grassmann as the initiator of their work. In 1909, after several publications on vector analysis, they published their *Omografie Vettoriali* (Burali-Forti and Marcolongo, 1909), which starts with a brief axiomatic presentation of linear systems and transformations. This introduction is quite similar to Peano's last chapter of the *Calcolo Geometrico*, although their axiomatic definition is less accurate (it is redundant and incomplete). But this is the first publication on vector analysis which starts with an axiomatic presentation. Their definition of the dimension is quite similar to Peano's:

"A linear system U is said to be n-dimensional if it contains n independent elements, and if $n+1$ elements of U are always dependent."(ibid., p. 2)

Burali-Forti and Marcolongo have not been much more influential than Peano except in France where their work was better known.

In 1918, Hermann Weyl published the first version of his *Raum-Zeit-Materie*. The first chapter of this book entitled "Foundations of Affine Geometry" contains an axiomatic definition of linear vector-manifolds, which is quite similar to the modern definition of vector-spaces. Weyl does not give any reference to Peano or Burali-Forti and Marcolongo. On the other hand, he refers to Grassmann's *Ausdehnungslehre* as an "epoch-making work", but mainly underlines his contribution to the generalization to more than three dimensions (Weyl 1918, 17, footnote 4). Moreover, Weyl's axioms are closer to the content of the 'general theory of forms' of the A_1, than to Peano's or Burali-Forti and Marcolongo's axiomatic definitions. It is more likely then, that Weyl reestablished the axiomatic approach independently of his Italian predecessors, although like them he was certainly influenced by Grassmann's theory. Nevertheless, his definition of the dimension is quite similar to the Italian's:

"There are *n* linearly independent vectors, but every *n+1* are dependent on one another." (ibid., p. 17).

Weyl comments on this statement, which he adds to his definition of a linear vector-manifold as the dimensional axiom:

"From this property (which may be deduced from our original definition with the help of elementary results on linear equations) it follows that *n*, the dimensional number, is as such a characteristic of the manifold, and is not dependent on the special vector base by which we map it out." (ibid., p. 17).

The fact that Weyl's definition of the dimension is similar to the Italians' is not proof of an influential relation. In fact, one can make the same analysis of Weyl's interpretation of Grassmann's work as the one made above about Peano. Nevertheless, his remark shows a new aspect. Weyl was aware of the possibility of deducing the dimensional axiom from the theory of linear equations.[10] He may have been influenced by Grassmann's remark made in the A_2 (cf. preceding chapter), but then why not use the exchange theorem? In fact, another more probable source of influence may have been the developments of field theory, which will be discussed below. Moreover, Weyl's remark induces implicitly that any basis has the same number of elements, but again the possibility of the existence of a basis of less that *n* elements is not explicitly discussed.

None of the three works discussed here succeeded in imposing the axiomatic approach in linear algebra. This finally happened between 1920 and 1930, to a great extent in relation to the development of functional analysis, mainly due to Banach and Hahn's works. In this context, the vector spaces were of infinite dimension. It would thus go beyond the purpose of this paper to analyse these contributions to the history of linear algebra (for more details see (Dorier 1990 and 1995)).

THE THEORY OF LINEAR EQUATIONS

In the field of linear equations, the concept of dimension is intrinsically related to the concept of duality. Indeed, in modern terms, the dimension of the space formed by the solutions of a system of homogeneous linear equations is related to the rank of these equations which is the dimension of the subspace they generate in the dual. For a

---

10 However, this statement is not absolutely correct. Indeed, it is necessary to assume the existence of a finite set of generators to be able to deduce the dimensional axiom form the theory of linear equations.

long time, it was implictly admitted that n linear equations determine exactly n values. In his work on Cramer's paradox, Euler (1750) points out that if one equation depends on the others, more than n equations are required to determine exactly n values. The theory of elimination is the only theoretical tool he uses. His treatise can be intrepreted as an empirical approach to the theorem: "p $(>n)$ linear equations determine exactly n values if and only if their rank is equal to n.". Yet in his work, Euler does not use the word dependence, he says that an equation is "comprised" in others, which means that the equation does not restrict the set of solutions of the system constituted by the other equations. Practically, with numerical examples, he proves such a dependence through successive eliminations, therefore some linear relations between the equations are involved, but only as transitory results. As a consequence, the type of dependence involved on the equations does not explicitly appear as a linear dependence, I will call it an "inclusive dependence". This point is quite important for the purpose of this study. Indeed, the use of inclusive dependence on equations while the dependence on the n-tuples was clearly linear was to be, for more than a century, an essential barrier to the emergence of the concept of rank, as this distinction between equations and n-tuples prevented the use of duality.

Also in 1750, Cramer published his treatise on algebraic curves in which he introduces a double index notation for the coefficients of a system of linear equations and solves the general 3 by 3 system using of what will later be called determinants.[11] This work marks the starting point of the theory of determinants which was to be immediately popular and became the theoretical framework for the study of linear equations. In spite of this success, the study of rectangular systems or square systems with an evanescent main determinant did not become a major issue before around 1840.[12] Yet only the study of these cases could produce interesting results on the rank of the system or the dimension of the set of solutions. Moreover, within the theory of determinants, the main goal was to describe practical ways of solving a system, thus the theoretical aspect remained implicit even if the generality of the study could make it easily accessible. On the other hand, the use of

---

11 In a letter of 1693 to the Marquis de l'Hospital, Leibniz had already used a double index notation and solved a system using determinants, but this letter was not published until 1850.

12 Several mathematicians were involved in this study, like: James J. Sylvester, Arthur Cayley, Leopold Kronecker, Richard Baltzer, Nicolo Trudi, Henry J. S. Smith, Charles Dodgson, Eugène Rouché, etc. For more precise information on this chapter see (Dorier 1993).

determinants induced a high level of technicity which was often a screen for an intuitive overview of the problem.

In a first phase, partial results were established, which became theoretical bases for a general approach. One of the first results is the fact that a system of $n+1$ homogeneous equations in n unknowns has a non zero solution, which is formed by the n nxn-minors extracted from the array of the coefficients of the first n equations. On the other hand, the dependence of equations was still interpreted from the inclusive point of view, but it was also connected with the evanescence of all the largest determinants extracted from the array of their coefficients. In this framework, the largest non-evanescent minor of the array of coefficients of a system appeared as a central object. On a practical point of view, once one of the largest non-evanescent minor (of order r) is identified in a system of n equations with p unknowns, the first r equations in the first r unknowns (the $(p-r)$ last ones are put on the other side of the equations) can be solved by Cramer's rule. The last $(n-r)$ equations give conditions for consistency. The general solution, if it exists, is therefore given as a linear function of the $(p-r)$ arbitrary last unknowns. So r, which is an invariant of the system, but not necessarily of its solutions, characterizes the degree of inclusive dependence of the equations as well as a particular way of generating the set of solutions. Between 1840 and 1875, this method was established with use of technical results on determinants, it was also gradually simplified, and clarified, but it remained more a practical rule for solving a system than a theoretical framework for the study of linear equations.

In 1875, in a treatise on the problem of Pfaff, Frobenius adopts a wider point of view. He is the first one to introduce explicitly the concept of linear dependence on equations, and to show that this is equivalent to the inclusive dependence and thus to the evanescence of all the largest minors extracted from the array of their coefficients. He also introduces the linear dependence of n-tuples and points out the equivalence with the dependence of the equations whose coefficients are the components of the n-tuples. This presentation lays the bases for a use of duality. Frobenius[13] considers a system of n homogeneous equations with p unknowns, whose largest non-evanescent minor is of order r. First of all he gives an artful process to exhibit $(p-r)$ independent solutions. Then, he shows, in a very elegant manner, that $(p-r+1)$ solutions are always dependent, which implies that any $(p-r)$ independent solutions generate all the others. At this stage, he uses duality, indeed he defines an *associate* ("zugeordnet" or "adjungirt") system as a system of the $(p-r)$

[13] In fact he starts with a system of m independent equations and he generalizes afterwards.

equations with p unknowns whose coefficients are the components of *(p-r)* independent solutions of the original system. The fundamental result of duality proven by Frobenius is that any associate system of an associate system has the same set of solutions as the original system. The possibility Frobenius gives himself of considering n coefficients either as an n-tuple or as an equation, and the fact he uses the same concept of linear independence are the key points in this approach which therefore introduces duality as a prerequisite for a new enlarged theory of linear equations and linear sets of n-tuples. As a consequence, Frobenius is able to connect all the systems of equations having the same set of solutions, and therefore to show that the value of r is not only an invariant of the system but also of its set of solution. It would have been easy to prove in this framework that not less than *(p-r)* solutions could generate the system or that no equivalent system (i.e. having the same set of solutions) could have less than r equations, but Frobenius does not do it explicitly. Nevertheless, his work is of great importance for the development of linear algebra, as it gave a wider perspective to the study of systems of linear equations. Indeed, solving the system is no longer the unique point of view; Frobenius puts the emphasis also on the set of solutions as an independent object to be studied in itself. In this enlarged approach, the same invariant *(r)* characterizes the dimension of the set of solutions *(p-r)* and the number of independent relations that determines them *(r)*. In 1879, Frobenius introduced the term "rank" to designate the order of the largest non-evanescent minor in an array of numbers. Rapidly, the concept of rank was used in several other works dealing with linear equations, and its efficiency was widely recognized.

In the study of linear equations the concepts of dimension and basis are therefore connected with the concept of rank and duality. They emerged progressively within the framework of the theory of determinants; and the duality played a fundamental role in order to put the emphasis on the set of solutions independently of a specific system of equations and also in order to connect the aspect of dependence naturally present in the equations and the aspect of generation naturally present in the set of solution. In this approach the possibility of having a common definition of dependence was an important factor while the technicality of determinants was a difficulty to overcome. Euler's intuitive approach, using only methods of elimination did not have immediate followers. Yet, in 1886, Alfredo Capelli and Giovanni Garbieri showed that the rank of a system of equation can also be characterized as the number of non zero coefficients in the diagonal of any equivalent triangular system. This result is a theoretical basis for most methods of elimination which became important in applied mathematics, with the use of computers.

MODERN ALGEBRA

In the context of what will be called modern algebra, the concepts of basis and dimension were mainly studied within field theory. In this context, the question was to characterize an extension-field as the linear combinations of the exponents of a primitive element having the order of the extension with respect to the original field. The question of generation was therefore essential.

I cannot give a detailed historical report of the treatment of this question in this paper, but I will examine two major works, one by Richard Dedekind and the other by Ernst Steinitz, which were among the first to have clearly elucidated this question and played a fundamental role in the development of field theory as well as linear algebra.

In the fourth edition of Gustav Peter Lejeune-Dirichlet's *Vorlesungen über Zahlentheorie* (Dirichlet 1893), Dedekind added a new 11th supplement entitled *Allgemeine Zahlentheorie*, in which one paragraph presents a very general approach of finite dimensional linear algebra (ibid., §164).[14] A field A being given, a system T of m numbers $\omega_1, \omega_2, \ldots, \omega_m$, is said to be reducible with respect to A, if there are m numbers $a_1, a_2, \ldots, a_m$, in A, which are not all zero, such that: $a_1\omega_1 + a_2\omega_2 + \ldots + a_m\omega_m = 0$. This is of course the definition of what is now known as linear dependence. Dedekind shows that any subsystem of an irreducible system is irreducible and that if a system contains a reducible subsystem it is reducible. Then he defines what an algebraic number with respect to A is, and says that it means that there exists a number n such that the $n+1$ first powers of the algebraic number constitute a reducible system with respect to A. Then he defines the grade ("Grad") of an algebraic number (the minimal value of n), and he remarks that any reducible system with at least one non zero number, contains an irreducible subsystem of which the remaining numbers can be expressed uniquely as linear combinations. This leads him to the definition of a space Ω ("Schaar"), as the set of all linear combinations of an irreducible set of n numbers. He calls these n elements a basis of Ω and defines the coordinates of an element of Ω, as the coefficients of the unique linear combination of the elements of the basis representing the given element. Immediately after that, he presents three properties that he proves to be characteristic of Ω , i.e. of the set of all linear combinations of n independent elements with respect to A.

" I. The numbers in Ω reproduce themselves through addition and subtraction, i.e. the sum and the difference of any two numbers are in Ω
II. Any product of a number in Ω by a number in A is a number of Ω.

---

[14] I must thank Winfried Scharlau who informed me about this contribution of Dedekind.

III. There are n independent numbers in Ω, but any $n+1$ such numbers are dependent." (ibid., p. 468).

For the deduction of the properties from the definition, as Dedekind points out, only the second part of III needs a proof, and he uses a recurrence. Assuming the property for any space having a basis of less than n elements, he takes $n+1$ elements, $\alpha, \alpha_1, \alpha_2, \ldots, \alpha_n$, in a space Ω, with a basis of n elements.

"If one is zero, for instance $\alpha = 0$, then these elements are dependent; if not, one can assume that for instance the first coordinate of α is not zero; so one can obviously find n numbers $c_1, c_2, \ldots, c_n$, in A, such that the first coordinate of each number

$$\alpha_1 + c_1\alpha, \alpha_2 + c_2\alpha, \ldots, \alpha_n + c_n\alpha,$$

is zero; thus these numbers belong to a space whose basis contains only the n-1 elements $\omega_2, \omega_3, \ldots, \omega_n$, and are therefore dependent on each other; consequently there are n numbers $a_1, a_2, \ldots, a_n$, in A, which are not all zero and such that

$$a_1(\alpha_1 + c_1\alpha) + a_2(\alpha_2 + c_2\alpha) + \ldots + a_n(\alpha_n + c_n\alpha) = 0,$$

and as the sum $a = a_1c_1 + a_2c_2 + \ldots + a_nc_n$ is also in A, it follows that the $n+1$ numbers $\alpha, \alpha_1, \alpha_2, \ldots, \alpha_n$, are dependent on each other." (ibid., p. 468).

This result is equivalent to what Grassmann obtained with the exchange theorem, nevertheless, the approach is quite different. Moreover, this proof does not use the theory of linear equations although it uses representation with coordinates.

To prove that these three properties are characteristic of a space as defined by Dedekind is quite simple. He also immediately deduces that any irreducible system of n numbers is a basis of Ω. Raising the problem of change of basis, he also proves that a system of n numbers is irreducible if and only if the determinant of their coordinates on the original basis is not zero.

The impossibility of finding a system of less than n generators is not explicitly stated, but as said about Grassmann in 1844, this question could be easily solved, as the definition put the emphasis on the original set of n independent generators.

At the end of the paragraph, Dedekind turns to the case of an extension-field, i.e. when the space is also closed under multiplication. In this context, the main results proved by Dedekind are:

- In an extension-field with a basis of n elements, any number is algebraic and its grade is at most n.

- Such an extension-field B is said to be of n-th grade with respect to A, and therefore any system of n independent elements constitutes a basis of B.

- If θ is algebraic with respect to A and of grade *n*, then A(θ) is an extension-field of A of n-th grade and 1, θ, ..., θ^{n-1}, constitute a basis of A(θ).

In this last supplement, Dedekind also studies the properties of modules, and especially of their bases, but this goes beyond the purpose of this paper. In §164, the main results of which have been presented briefly above, Dedekind's approach is very close to a modern presentation of the elementary results of finite dimensional linear algebra. the question of dimension is examined with care, and with use of original tools compared to Grassmann's use of the exchange theorem and what had been done in the theory of equations.

STEINITZ AND THE THEORY OF FIELDS

In 1910, Steinitz published his *Algebraische Theorie der Körper*, which marked an important stage in the history of Modern Algebra and was to be a reference work for at least a quarter of a century. In this major work, Steinitz gives a precise definition of linear dependence over a field *R*, and defines a finite extension of order n:

"Let *R* be a subfield of *L*, *L* will be said to be finite with respect to *R* and of order *n* - which will be designed by: [*L*:*R*] = *n* - if there are in *L*, *n* elements linearly independent over *R*, while any set of more than *n* elements of *L* are linearly dependent over *R*." (Steinitz 1910, p. 198).

This definition is identical to the definition of the number of dimensions of a linear space given by Peano or Burali-Forti and Marcolongo. As for Weyl, any influence is very unlikely to have happened as the Italians' work was practically unknown in Germany. There is no explicit reference to Dedekind either, but there is a natural proximity between the two mathematicians which has to be taken into account.

After this definition, Steinitz shows that in an extension of order *n*, for any set of *n* elements, the linear independence is equivalent to the fact that any element of the extension cannot be expressed in more than one way as a linear combination of these *n* elements. This is quite obviousely a consequence of the preceding definition and indeed there is nothing new compared to his Italian predecessors. In fact, Steinitz pursues a goal that he has not explictly stated. For him, a basis of *L* is a set of elements such that any element of *L* can be expressed uniquely as a linear combination of them. Steinitz's goal is to prove that any basis has *n* elements and that any set of *n* independent elements is a basis of *L*. To achieve this goal, he needs to prove that the order of an extension cannot exceed the number of generators. In other words, he has to connect a result concerning generation to a property of dependence. His definition of a basis puts forwards the coordinate system, so his proof is set up within the context of n-tuples and linear equations:

" If one knows that every element β of L can be expressed as: $\beta = c_1\alpha_1 + \dots + c_n\alpha_n$; then for any $(n+1)$ elements $\beta_0, \beta_1, \dots, \beta_n$, there are expressions like:

$$\beta_i = c_{i1}\alpha_1 + \dots + c_{in}\alpha_n ; \quad (i = 0, \dots, n)$$

and therefore, according to the theorem of §1 about homogeneous linear equations,[15] $(n+1)$ elements $d_0, d_1, \dots, d_n$ can be found such that $d_0\beta_0 + d_1\beta_1 + \dots + d_n\beta_n = 0$. The extension L is thus finite and its order is n at most." (ibid., p. 200).

After Grasmman's A_2, this is the first explicit proof of the fact that a set of generators cannot have fewer elements than the number of dimensions, although this was implicit in Dedekind's work, and very easily deducible. Yet, Steinitz and Grassmann's approach and tools are very different. Grassmann knew that a proof could be given with the use of elimination theory (cf his remark quoted above), but he preferred the exchange method for explicit reasons given in 1862.

In fact, further on in his theory, Steinitz uses something very close to the exchange method, but in the context of transcendental extensions, where algebraic dependence replaces linear dependence:[16] "let S be a system of elements of an extension-field L, an element a of L is said to be algebraically dependent on S (with respect to R), if a is algebraic with respect to the field $R(S)$." (ibid., p. 288). Two systems are said to be equivalent if any element of each system is algebraically dependent on the other system. On the other hand, a system is said to be irreducible if it is not equivalent to any of its subsystems. Transferred into the context of linearity, two equivalent irreducible systems correspond to two bases of the same linear space. After a few preparatory results, Steinitz gives three final theorems which contain, for algebraic dependence, results equivalent to the exchange theorem, the theorem about the completion of an independent system into a basis, the invariance of the number of elements in all the different bases of the same linear space, and properties of the dimension of a subspace of a finite dimensional linear space.

"7. Let S be an irreducible system (with respect to R), a a transcendental element with respect to R, but algebraically dependent on S, then S contains a finite subsystem L with the following properties: a is algebraically dependent on L; any subsystem of S, on which a is algebraically dependent, contains the system L; if any element of L is replaced by a, S

---

15 This therorem proves that a system of less than n homogeneous linear equations with n unknowns has a non- zero solution.

16 I want to thank G. Schubring who put me in contact with W. Purkert, who pointed out to me this passage in Steinitz's work.

is changed into an equivalent irreducible system; none of the other elements of S have the same property." (ibid., p. 290).

"8. Let U and B be finite irreducible systems with respectively m and n elements; let us assume that $n \leq m$ and B is algebraically dependent on U. Then if $m=n$ the systems U and B are equivalent, and if on the contrary $n<m$, U is equivalent to an irreducible system made of B and m-n elements of U." (ibid., pp. 291- 292).

"9. An irreducible system B, which is algebraically dependent on a finite system U, cannot have more elements than this one; in case they have the same number of elements they are equivalent." (ibid., p. 292).

This paragraph is presented in such a progressive, deductive way that it is close to an axiomatic definition of a general concept of dependence (which would include algebraic as well as linear dependence) from which the concepts of dimension and basis could be deduced. Therefore, little needs to be done in order to translate the preceding results into the linear context, but Steinitz does not mention (at least here) this possibility. About twenty years later, two of his successors, Bartel L. van der Waerden in (van der Waerden 1930/31, vol. 1, p. 96) and Emanuel Sperner setting up Otto Schreier's work in (1931/35)[17] used the exchange method (and used this term) for one of the first times, in the context of linear dependence and both of them refer to Steinitz, although I have never found proof that Steinitz did actually use the exchange method in the context of linear dependence. On the other hand; there is no reference to Grassmann, although his work was well-known by German algebraists of this time. Nevertheless, even if an influential filiation could be conjectured, in spite of of the gap of nearly a century, the two approaches are embedded in very different mathematical and philosophical contexts.

CONCLUSION

Van der Waerden's *Moderne Algebra* marked the beginning of a new era in algebra and in mathematics in general. His book which was based on lectures given by Emil Artin and Emmy Noether is also the heritage of the works of many German algebraists during the four or five preceding decades (for more details see (van der Waerden 1975)). The great novelty was to give a unified general approach of the new algebra presented in a textbook accessible to beginners. In this sense, it was used by several students some of whom became leading mathematicians in the next years, such as most of the members of Bourbaki's group for instance.

---

[17] To be found on page 20 of the English translation mentioned in the bibliography (Schreier, Sperner 1951).

In the first edition, the results on bases and dimension are presented, with use of the exchange theorem, within the theory of extension-fields (chapter 5). In chapter 15, about linear algebra, the concepts of linear dependence and basis are introduced for modules. In this context, the concept of dimension is more complex. Van der Waerden examines the conditions under which a module has a basis and establishes several results about modules with use of matrix theory and linear equations theory. When he comes to vector-space, the ring becomes a field, and he proves the invariance of the number of elements in a basis, specifying previous results in this new context. But he also mentions that another proof of the invariance of this result has been given about extension-fields. After that, he presents a general study of systems of linear equations, as an application to module theory. He establishes the connection between the concept of rank and the problem of dimension with use of the exchange theorem. In the following editions the importance of linear algebra will increase and the treatment of vector-space becomes more independent from the general theory of modules.

With this treatise, the concepts of dimension, rank and basis reach their maturity. The study of linear problems is unified within one theoretical approach: vector space theory whose elementary foundations are presented in a form which is still today the standard in most research- and textbooks. If from a technical point of view, Grassmann's approach to the concepts of dimension and basis has been rediscovered and even perfected, his philosophical contribution has been left aside. For this reason, his work is still important even concerning such (now) elementary concepts as basis and dimension.

UFR Sciences Économiques, Université Pierre Mendès-France, Grenoble

REFERENCES

Cesare Burali-Forti and Roberto Marcolongo, *Omografie Vettoriali con Applicazioni alle Derivate rispetto ad un Punto e alla Fisica Matematica* (Torino: G. B. Pretrini di Giovanni Gallazio, 1909).

Alfredo Capelli and Giovanni Garbieri: *Corso di Analisi Algebrica* (Padova, 1886).

Gabriel Cramer, *Introduction à l'Analyse des Courbes Algébriques* (Genève: Cramer et Philibert, 1750).

Gustav Peter Lejeune Dirichlet, *Vorlesungen über Zahlentheorie* , ed. and sup. Richard Dedekind (Braunschweig: Vieweg, 1863 (ed.1), 1871(ed.2), 1879 (ed.3), 1893 (ed.4)).

Jean-Luc Dorier, *Analyse Historique de l'Emergence des Concepts Elémentaires d'Algèbre Linéaire*, Cahier Didirem n°7 (Paris: IREM de Paris VII, 1990).

Jean-Luc Dorier, "L'Emergence du Concept de Rang dans l'Etude des Systèmes d'Equations Linéaires", *Cahiers du Séminaire d'Histoire des Mathématiques* (2e Série), 1993, *3*: 159-190.

Jean-Luc Dorier, "A General Outline of the Genesis of Vector-Space Theory", *Historia Mathematica,* 1995, *22*: 227-261.

Leonhard Euler, "Sur une Contradiction Apparente dans la Doctrine des Lignes Courbes", *Mémoires de l'Académie des Sciences de Berlin*, 1750, *4*: 219-223.

Georg Ferdinand Frobenius, "Über das Pfaffsche Problem", *Journal für die reine und angewandte Mathematik*, 1875, *82*: 230-315.

Georg Ferdinand Frobenius, "Über homogene totale Differentialgleichungen", *J. reine und angewandte Math.*, 1879, *86*: 1-19.

Hermann G. Grassmann, *Die lineale Ausdehnungslehre* (Leipzig: Otto Wigand, 1844)

Hermann G. Grassmann, *Die Ausdehnungslehre, vollständig und in strenger Form* (Berlin: Th Fr. Enslin, 1862)

Albert C. Lewis, "H. Grassmann's 1844 Ausdehnungslehre and Schleiermacher's Dialektik", *Annals of Sciences*, 1877, *34*:103-162.

Michael Otte, "The Ideas of Hermann Grassmann in the Context of the Mathematical and Philosophical Tradition since Leibniz", *Historia Mathematica* , 1989, *16:* 1-35.

Giuseppe Peano, *Calcolo Geometrico secundo l'Ausdehnungslehre di H. Grassmann e precedutto dalle Operazioni della Logica Deduttiva* (Torino: Fratelli Bocca, 1888).

Otto Schreier and Emanuel Sperner, "Einführung in der analytische Geometrie und Algebra", *Hamburger Mathematische Einzelschriften,* in two parts, 1931, Nr. 10 and 1935, Nr. 19; English Trans., *Modern Algebra and Matrix Theory*, trans. Martin Davis and Melvin Hausner (New York: Chelsea Publishing Company, 1951).

Ernst Steinitz: 1910. "Algebraische Theorie der Körper", *J. reine und angewandte Math.,* 1910, *137*: 167-309; or reprint, eds. H. Hasse and R. Baer (Berlin/Leipzig: De Gruyter, 1930).

Bartelt L. van der Waerden, *Moderne Algebra*, 2 vols. (Berlin: Springer, 1930/31 [First edition]).

Bartelt L. van der Waerden, "On the Sources of my Book *Moderne Algebra*", *Historia Mathematica,* 1975, *2:* 31-40.

Hermann Weyl, *Raum - Zeit - Materie* (Berlin: Springer, 1918).

KARIN REICH

THE EMERGENCE OF VECTOR CALCULUS IN PHYSICS: THE EARLY DECADES

1. PRELIMINARIES

The following survey considers only two disciplines of physics: mechanics and electrodynamics.

1.1. Mechanics as a discipline of physics

Mechanics has the problem of not belonging exclusively to mathematics, physics, or technology. During the second half of the nineteenth century and even later, mechanics was mostly taught by mathematicians who were either professors of mathematics and mechanics or, more rarely, of geometry and mechanics. At the same time, however, there were chairs of mechanics only or, more precisely, of technical or rational mechanics. Mechanics sometimes was treated as a discipline belonging to applied mathematics. In 1898, however, August Föppl pronounced himself in favor of assigning mechanics to physics; this approach will be adopted here. According to Föppl, mechanics is based on experiments and experience, a typical property of all experimental sciences and particularly of physics (Föppl 1898, p. 5).

1.2. The demand for a new calculus within mechanics

Within mechanics, an urgent demand for a new calculus seems to have risen, as forces, pairs of forces and the operations with them had to be named and represented. Before Grassmann and Hamilton, many authors of mechanical textbooks chose something similar to vector calculus. Very often there was no clear definition of magnitudes, and only a simplified representation. Examples are Domenico Chelini (Chelini 1860), Henri Résal who published many textbooks during the second half of the nineteenth century, and Osip I. Somoff, whose textbook on kinematics in Russian was translated into German by Alexander Ziwet (Somoff 1878).

G. Schubring (ed.), Hermann Günther Graßmann (1809–1877): Visionary Mathematician, Scientist and Neohumanist Scholar, 197–210.

1.3. Möbius's barycentric calculus and its reception

The first author who really introduced a new calculus was August Ferdinand Möbius, who published his so-called barycentric calculus in 1827 (Möbius 1827). Ten years later, in his textbook on statics, Möbius applied his new calculus to mechanics (Möbius 1837).

Quite a number of experts in mechanics were familiar with Möbius's barycentric calculus, for example Luigi Cremona (Cremona 1875, pp. 3, 8, 12, 15), Rudolf Sturm (Sturm 1875/76), Wilhelm Schell (Schell 1879, p. 11-72) and Otto Mohr (Mohr 1888).

1.4. Some details of Grassmann's and Hamilton's theories

In 1844 and in 1862, Hermann Grassmann published his main works on his theory of extension. He introduced so-called extensive magnitudes of arbitrary degree, i.e. dimensions, a term Grassmann himself never used, defining the basic operations with these magnitudes. In his new calculus, he used the so-called outer product and the inner product (Grassmann 1862, p. 56f, 114). In the case of outer multiplication, the result of multipying two magnitudes of the first degree was a quantity of the second degree, i.e. an oriented plane:

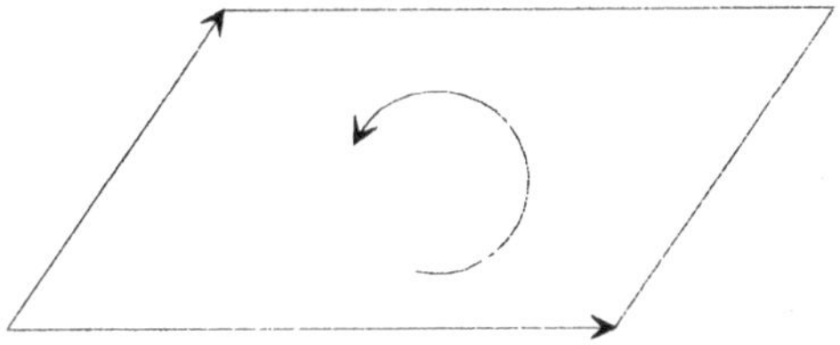

A historical remark is appropriate here. Equivalents of the outer multiplication were introduced much earlier. In 1838, Domenico Chelini used "moments" which were similar to oriented planes (Chelini 1838, p. 70f). In particular with regard to mechanics, Barré de Saint-Venant had initiated a theory of lines with length, direction and orientation as early as 1845. Saint-Venant besides the operations of geometrical sum and geometrical difference, defined geometrical multiplication $\overline{a}\overline{b}$ as an oriented plane, with the properties $\overline{a}\overline{b} = -\overline{b}\overline{a}$ and $\overline{a}\overline{a} = 0$, with $\overline{a}\overline{b}\overline{c}$ being the volume of a parallelepipedon (Saint-Venant 1845). This representation, as well as Hamilton's quaternions, became the basis for Cauchy's theory of "algebraical keys" (Cauchy 1853).

On October 16, 1843, William Rowan Hamilton discovered quaternions. His so-called vector magnitudes v = ai + bj + ck obeyed the laws of complex numbers and were three-dimensional. Hamilton's vector product

$$Vvv' = i(bc'-cb') + j(ca'-ac') + k(ab'-ba')$$

was defined as a vector, and not as a plane. This vector was perpendicular to the plane of the two original vectors and had a length equal to the area of the parallelogram:

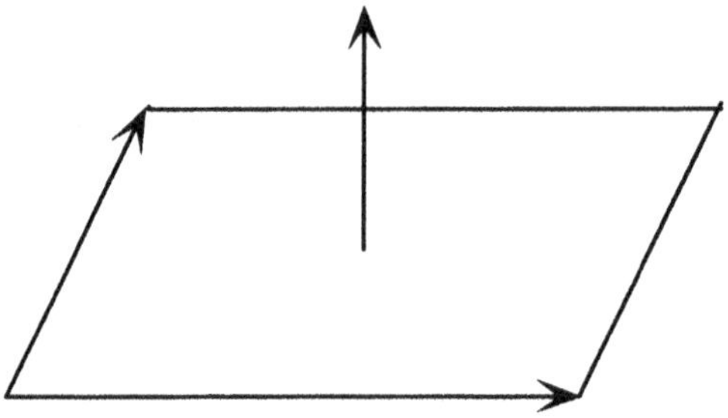

In 1888, Giuseppe Peano called the results of Grassmann's outer multiplication 'bivectors', 'trivectors' and so on (Peano 1888, p. 55, 65). In accordance with Peano, the results of Hamilton's vector multiplication were called monovectors. At this time three-dimensional calculus was sufficient for most cases in physics. Therefore Grassmann's adherents, the Grassmannians, were often called the "bivectorians," and the Hamiltonians were termed the "monovectorians" (Schouten 1914, p. 1).

For the future, to define the outer product and the vector product became the crucial point, especially in the subsequent quarrel between Grassmannians and Hamiltonians, or bivectorians and monovectorians. There were essentially two possible ways to introduce vector calculus, either Grassmann's or Hamilton's, the latter of which was transformed into an even easier form by Josiah Willard Gibbs and Oliver Heaviside by dropping the complex structure.

2. MECHANICS

As for mechanics, Grassmann himself had applied his calculus to it as early as 1867. In 1877, he published his article on "Mechanics according to the principles of extension theory" which began with the following essential sentences:

"There is no field in which it could be better proved that calculus according to my own extension theory is indispensable than mechanics. One can say that every simple mechanical notion is at the same time a simple notion of connection of this calculus" (Grassmann 1877a, p. 46).[1]

Hamilton's theory, however, was applied to mechanics as well. To mention just one case: Peter Guthrie Tait's textbook *Elementary Treatise on Quaternions* was first published in 1867 followed by several subsequent editions, and was even translated into other languages. Tait used his own vector calculus in kinematics and dynamics (Crowe 1967, p. 123f).

2.1. Early reception of Grassmann's calculus

Very soon, mathematicians and physicists working on mechanics became familiar with Grassmann's new calculus. Rudolf Sturm published an article on forces in equilibrium. Besides using Möbius, he also made use of Grassmann's extension theory of 1862 (Sturm 1875/76, p. 234-238).

During the winter term of 1876/77, Ernst Abbe gave lectures on the movement of solid bodies, pointing out the advantages of Grassmann's extension theory (Kircher 1883, p. 3; Jena 1876/77, p. 20). This was probably the first lecture in history given on Grassmannian calculus.

Ernst Mach quoted Möbius, Hamilton and Grassmann in his *Science of Mechanics in its Development* (Mach 1883, p. 452). Mach praised the new presentations which allowed mechanics to use a mathematical language that was closer to geometry than ordinary analytical geometry, combining the advantages of both greater generality and geometrical visualisation.[2] In later editions, Mach referred in detail to Grassmann's extension theory of 1844 (for example Mach 1904, p. 562f).

1881 saw publication of the first textbook on mechanics based on vector calculus, or more precisely on a line segment calculus. Jakob Lüroth, the author, was later among the scientists who edited Grassmann's works. As the foundation of his textbook, Lüroth had used Grassmann's 1867 article "Outline of mechanics," praising it highly. Lüroth, however, also used Hamiltonian terminology (Lüroth 1881,

---

1 Grassmann: "Es giebt wohl kaum ein Gebiet, auf welchem sich die Unentbehrlichkeit des in meiner Ausdehnungslehre (von 1844 und 1862) dargestellten Kalküls so schlagend erwiese wie in der Mechanik. Man kann sagen, jeder einfache mechanische Begriff sei zugleich ein einfacher Verknüpfungsbegriff jenes Kalküls."

2 Mach: "Erwähnt muß werden, daß durch Möbius, Hamilton, Grassmann u.a. eine neue Formwandlung der Mechanik eingeleitet ist, indem die genannten Forscher mathematische Begriffe entwickelt haben, welche sich genauer und unmittelbarer den geometrischen Vorstellungen anschließen, als jene der gewöhnlichen analytischen Geometrie, wodurch die Vorteile analytischer Allgemeinheit und geometrischer Anschaulichkeit vereinigt werden."

pp. III, 1-41). Lüroth's textbook prompted Ferdinand Kraft to write a *Compendium of geometrical calculus* (Kraft 1893). As early as 1887, Georg Emil Kircher, grammar school teacher in Meiningen, published his *Contribution to the motion of invariant plane systems*, depending slightly on Möbius, but primarily on Grassmann's extension theory (Kircher 1887).

2.2. 'Vectors' among Grassmannians

There was no sharp dividing line between Grassmannians and Hamiltonians; they knew and respected one another. Thus, use of the term 'vector', in the beginning reserved for Hamilton and his followers, spread to Grassmannians, although Grassmann himself had never used the term vector, speaking of line segments instead.

This transition took place first in England and in the USA where both theories, Grassmann's and Hamilton's, were widespread.

In his article "Applications of Grassmann's extensive algebra," William Kingdon Clifford compared Hamilton's theory to Grassmann's, as a result vectors started being used everywhere (Clifford 1878). Homersham Cox shared Clifford's objective: in 1883 he concluded that "Calculations with strokes ["Strecken"] will be identical with the calculations with vectors as far as only the product of two quantities is concerned" (Cox 1883, pp. 142f). And in 1884, the American Edward Willys Hyde, a convinced Grassmannian, gave an account of "Grassmann's method of treating points and vectors" (Hyde 1884, p. 5).

Alexander Ziwet, born in Wrocław in 1853, had studied in Karlsruhe, where he graduated as a civil engineer. In 1880, he emigrated to the US, where he became professor of mathematics at the University of Michigan in Ann Arbor and authored several textbooks on mechanics. In 1893/4 Ziwet published his article "A brief account of Grassmann's geometrical theories" expressing his hope:

"It is, however, in the science of rational mechanics that the application of Grassmann's theories and methods, in combination with those of Sir Wm. R. Hamilton, will probably prove of the greatest importance" (Ziwet 1885/6, p. 34).

For Ziwet, straight lines, i.e. segments of straight lines, in the sense of Grassmann were identical with vectors.

A general breakthrough for 'vectors' was achieved by Giuseppe Peano in his *Geometrical Calculus according to Hermann Grassmann's Ausdehnungslehre*. Although he gave a detailed interpretation of Grassmann's extension theory of 1844, Peano used the term 'vector' instead. Arguing that every line can be considered as the product of its origin and its vector, he gave the definition: "The vector of a line AB will be called the vector B-A" (Peano 1888, p. 49).[3]

Peano's friend Filippo Castellano, whose special field was mechanics, adhered exactly to Peano's terminology. Castellano's textbook *Lectures on Rational Mechanics* was the first to be completely based on a Grassmannian type of vector calculus, which was presented in the introductory chapter (Castellano 1894; pp. 1-8); there was a second edition in 1911.

During the 1890s, the term 'vector' became generally used by Grassmannians, though of course there were some authors who only used Grassmann's terminology.

2.3. Monovectorians

From the 1870s to the 1890s, there were many scientists working on mechanics who were adherents of Grassmann, while others were adherents of Hamilton and, of course, there were some who did not introduce any kind of calculus at all.

The years 1897-98, however, were a turning point in the presentation of mechanics. In 1897 Felix Klein and Arnold Sommerfeld published the first part of their theory of the gyroscope and they chose Hamilton's approach (Klein/Sommerfeld 1897, pp. 55-68). This is surprising for Felix Klein, because he later favored Grassmann's approach in vector calculus.

Until then, only the terms 'outer' and 'inner' product were used following Grassmann, while 'vector' and 'scalar' product were used by Hamiltonians. Klein and Sommerfeld, however, dropped this distinction, using Grassmann's outer product as a synonym for vector product (Klein/Sommerfeld 1897, p. 60).

In a way similar to that of Klein and Sommerfeld, August Föppl introduced the Hamiltonian vector calculus in his textbook *Lectures on Technical Mechanics* (Föppl 1898, p. 50-92). One of the reasons was that Föppl had used the same kind of vector calculus earlier in his textbook on electrodynamics. Föppl, too, mixed terminologies: his vector product could also be called outer product. Föppl justified this by mentioning that Grassmann had originally used outer product only to designate the paralellogram. In mechanics, however, only the vector product, i.e. the vector perpendicular to the plane of the two original vectors, is of importance. Hence, Föppl also used the term 'outer product' in case of a vector product. This usage had been criticised, but Föppl considered the

---

3 Peano: "Ogni linea si può considerare come il prodotto della sua origine pel suo vettore". "Def. Dicemo vettore d'una linea AB il vettore B-A."

'outer product' to be such a good term that it should not be abandoned merely for historical reasons (Föppl 1898, p. 80).[4]

Klein, Sommerfeld and Föppl were very influential. After this turning point, most of the textbooks on mechanics adopted the vector calculus according to Hamilton, such as Alexander Wernicke who quoted Föppl (Wernicke 1900 vol.1, p. XI), Gustav Jaumann (Jaumann 1905), Karl Heun (Heun 1906), Roberto Marcolongo (Marcolongo 1911) and Ludwig Silberstein (Silberstein 1913).

2.4. The Encyclopedia of Mathematical Sciences

The fourth volume of the "Encyclopedia of Mathematical Sciences," subdivided into four parts, is dedicated to mechanics; it was edited by Felix Klein and Conrad Müller (Encyklopädie vol. IV). It contained two contributions on vector calculus of which the first came from Max Abraham (Abraham 1901). He mentioned all possible ways of presentation and focused more or less on a special vector calculus (monovectors) which had been introduced by his Göttingen colleague Woldemar Voigt. Voigt was the first to distinguish between polar and axial vectors (Voigt 1895/96, vol.2, p. 801) which he visualized by:

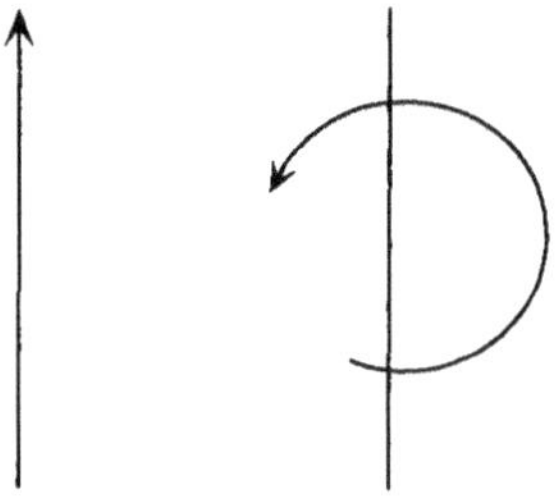

(Voigt 1910, p. 131). The second contribution "Geometrical foundation of the mechanics of a rigid body" came from Heinrich Emil Timerding

---

4 Föppl: "Dann bemerke ich noch, daß die Bezeichnung "äußeres" Produkt von Grassmann, der sie zuerst gebraucht hat, ursprünglich auf das ...Parallelogramm bezogen wurde und nicht auf den <orthogonalen> Vektor, durch dessen Richtung und Größe die Stellung und der Flächeninhalt des Parallelogramms gekennzeichnet wird. Daran halten auch jetzt noch viele fest und unterscheiden daher zwischen dem äußeren Produkt und dem "Vektor-Produkte"...In der Mechanik kommt es aber stets nur auf dieses Vektorprodukt an und der Anschaulichkeit wegen habe ich den Namen "äußeres Produkt" auf dieses Produkt bezogen. Das ist von Kritikern wiederholt beanstandet worden; aber ich werde mich dadurch nicht stören lassen. Eine so gut passende Bezeichnung wie "äußeres Produkt" ist zu wertvoll für eine klare Darstellung, als daß man sie aus solchen historischen Gründen opfern dürfte."

(Timerding 1902), a stout believer in Grassmann who used exclusively the latter's terminology. Later, in his textbook *Geometry of Forces*, he still adhered to Grassmann, but conceded a modification, interpreting the outer product as a vector on the plane ("Flächenvektor"):

"This orientation is an integral part of the geometrical presentation of the vector product, besides the parallelogram's area and the position of its plane, and we may interpret this product as a surface having a certain area and a certain orientation in a plane of a certain position. Such a quantity is in some way analogous to a vector: the length of the vector corresponds to the content of the surface, the vector's direction to the position of the plane and the vector's direction to the area's orientation. Therefore we shall term the geometrical quantity representing the outer product of two vectors as a plane vector" (Timerding 1908, p. 20).[5]

3. ELECTROMAGNETISM

As early as 1845, Grassmann had attempted to apply his new calculus to electrodynamics (Grassmann 1845). Since nobody took any notice, Grassmann published his improved results once more in 1877 (Grassmann 1877b). His second attempt, however, was just as much before its time and was no success either.

The most important developments in electromagnetism took place in England. A new era began when James Clerk Maxwell published his *Treatise on Electricity* in 1873. As far as vector calculus was concerned, Maxwell only pointed out its uses in electrodynamics without going into details. Maxwell distinguished between scalars, vectors, and linear vector functions; there were two kinds of vectors, one referring to lines and one referring to areas (Maxwell 1883, vol. 1, p. 11f). But Maxwell did not introduce a vector calculus.

Oliver Heaviside went much further. He began his *Electromagnetic Theory* with a voluminous chapter on vector calculus (Heaviside 1893, vol. 1, p. 132-305). This was the first physics treatise to start with an entire chapter on mathematics; afterwards this kind of presentation became rather common. Like Maxwell, Heaviside had been influenced by Hamilton's quaternion calculus, but he was also familiar with Gibbs's

5 Timerding: "Dieser Umlaufsinn gehört so, neben dem Flächeninhalte des Parallelogramms und der Stellung seiner Ebene, zur geometrischen Darstellung des Vektorproduks hinzu, und wir können dieses Produkt interpretieren als eine Fläche von bestimmtem Inhalte und bestimmtem Umlaufsinne in einer Ebene von bestimmter Stellung. Eine solche Größe ist dem Vektor in gewisser Weise analog: die Länge des Vektors wird durch den Inhalt der Fläche, die Linienrichtung des Vektors durch die Ebenenstellung und der Richtungssinn des Vektors durch den Umlaufsinn der Fläche ersetzt. Wir wollen deshalb die das äußere Produkt zweier Vektoren repräsentierende geometrische Größe einen Flächenvektor nennen...."

vector analysis. Heaviside used a modified form, a special blend of Hamilton and Gibbs.

Heaviside's *Electromagnetic Theory* formed a basis for August Föppl. He published his own version on the same subject merely one year later (Föppl 1894). Like Heaviside, Föppl began with a chapter on vector calculus, using the same terms as Heaviside. Föppl's *Introduction to Maxwell's Theory of Electricity* was elaborated on by Max Abraham (Abraham 1904) and in this form, it was reprinted many times. Abraham used the same type of vector calculus as Föppl, but changed and simplified several terms. As had become common in mechanics, Abraham used inner and scalar product as synonyms, as well as outer and vector product.

Abraham's and Föppl's textbooks were very successful and became widely disseminated over a long period, also spreading their special kind of vector calculus. This vector calculus became standard and in time prevailed over other kinds of vector calculus used in other disciplines, as in mechanics.

3.1. The Encyclopedia of Mathematical Sciences

The editor of the fifth volume on physics was Arnold Sommerfeld (Encyklopädie vol. V). From the very outset Sommerfeld planned carefully how to handle vector calculus. He got in touch with the principal contributors to be and exchanged views. Finally, all agreed on using vector calculus in only one form: the monovectorial type using scalar products $(\mathfrak{A}\ \mathfrak{B})$ and vector products $[\mathfrak{A}\ \mathfrak{B}]$ (Sommerfeld 1904). By contrast, the volume on "Mechanics" had contained all known kinds of vector calculus.

4. TEXTBOOKS ON VECTOR CALCULUS, 1903-1921

This development within vector calculus towards the monovectorial approach and away from that of Grassmann, can also be recognized in the textbook literature. Figure 1, showing the textbooks in chronological order, will illustrate this new situation.

Only Mehmke remained a steadfast believer in Grassmann; Jahnke and Runge were Grassmannians in principle, but they also made use of Hamilton's specialities. Burali-Forti had changed sides: as one of Peano's students he had begun his career as a Grassmannian, but practice made him see the advantages of the monovectorial type. Schouten's textbook was exceptional: after discussing all existing types of vector calculus, he created a new one. All the others,

Henrici and Turner, Bucherer, Fischer, Gans, Valentiner, Coffin,

Mech/Math	TC	O. Henrici: *Vectors and rotors*. London 1903
Math/Phys	TC	G. Ch. Turner (co-author with Henrici): *Vectors and Rotors*.
Phys	Uni	A. Bucherer: *Elemente der Vektor-Analysis*. Leipzig 1903 (2nd ed.1905).
Tech	-	V. Fischer: *Vektordifferentiation und Vektorintegration*. Leipzig 1904.
Phys	Uni	R. Gans: *Einführung in die Vektoranalysis*. Leipzig 1905 (2nd ed.1909, 3rd 1913).
Phys	TC	Valentiner, Siegfried: *Vektoranalysis*. Leipzig 1907 (many later ed.).
Math	TC	E.Jahnke: *Vorlesungen über die Vektorrechnung*. Leipzig 1905.
Phys/Tech	TC	J. G. Coffin: *Vector Analysis*. New York 1909 (several later ed.).
Math	TC/Uni	C. Burali-Forti: *Elementi di calcolo vettoriale*. Bologna 1909.
Mech	Uni	R. Marcolongo (co-author with Burali-Forti), *Elementi di calcolo vettoriale*
Phys	Uni/TC	W. Ignatowski: *Die Vektoranalysis*. 2 parts. Leipzig, Berlin 1909, 1910.
Math	TC	R. Mehmke: *Vorlesungen über Punkt- und Vektorrechnung*. Vol.1, Leipzig, Berlin 1914.
Tech/Math	Uni	J. A. Schouten: *Grundlagen der Vektor- und Affinoranalysis*. Leipzig, Berlin 1914.
Tech/Phys	TC	J. Spielrein: *Lehrbuch der Vektorrechnung*. Stuttgart 1916 (2nd ed.1926).
Math	TC/Uni	C. Runge: *Vektoranalysis*. Vol.1, Leipzig 1919.
?	?	R. Leveugle: *Précis de calcul géométrique*. Paris 1920.
Mech	Uni	P. Appell: *Éléments de la théorie des vecteurs et de la géometrie analytique*. Paris 1921.

Fig. 1 (Math: professor of mathematics, Phys: professor of physics, Tech: engineer, Uni: teaching at a university, TC: teaching at a technical college or a technical university).

Burali-Forti and Marcolongo, Ignatowski, Spielrein, Leveugle, and Appell
were monovectorians!

Although the list is certainly not complete, the tendency is evident: the vast majority were adherents of the monovectorial approach.

The list also shows something else: Vector calculus was propagated mostly by physicists and by mathematicians working at technical colleges. As can be seen from the list of the textbooks, vector calculus very soon became a standard discipline within physics, and also for engineers.

One category of scientists is nearly totally absent in this list: the mathematicians teaching at universities. In fact, vector analysis then did not belong to the disciplines obligatory for mathematics students. With the exception of mathematicians teaching at technical colleges, most were not fond of vector calculus. University mathematicians normally did not believe that vector calculus was useful in any mathematical discipline, not even in geometry.

5. RESULTS

During the last three decades of the nineteenth century, vector calculus became widely disseminated in several forms. At the end of the century, there were almost no textbooks on mechanics or electrodynamics which did not make use of vector calculus. Vector calculus had become an indispensable tool to physics. Gustav Jaumann, for example, wrote in 1905:

> "Success in theoretical physics depends in part on the development of mathematical forms... Vector calculus was absolutly indispensible" (Jaumann 1905, p. IV).[6]

And Rudolf Rothe pleaded in favor of a "citizenship" ("Bürgerrecht") of vector calculus in mathematics because it had proved its usefulness, in particular in mechanics and in electrical theory (Rothe 1912, p. 249).[7]

One of the reasons why physicists, in particular, favored vector calculus was that it brought together physics and geometry, or at least made them closer. It was quite common to emphasize the geometrical character of vector calculus.

In the beginnings of mechanics, Grassmann's vector calculus was used just as often as that of Hamilton. In electrodynamics, however, the monovectorial kind was applied almost exclusively. At the turn of the

---

6 Jaumann: "Die Fortschritte der theoretischen Physik hängen teilweise von der Entwicklung mathematischer Begriffe ab...Es konnte deshalb auf die Anwendung der Vektoranalysis nicht verzichtet werden."

7 Rothe: "...diese Vektoranalysis hat sich mindestens in der Mechanik, in der Elektrizitätstheorie und vielen andern Zweigen der mathematischen Physik ein Bürgerrecht erworben, daß ihr so leicht nicht wieder genommen werden wird."

century electrodynamics became a major subdiscipline within physics. With the rise of electrodynamics, the latter's particular vector calculus prevailed, and it was later used generally in mechanics and in other branches of physics. Grassmann's vector calculus became extinct in physics.

Institut für Geschichte der Naturwissenschaften, Universität Hamburg

REFERENCES

Max Abraham, "Geometrische Grundbegriffe" (finished in 1901), *Encyklopädie der mathematischen Wissenschaften*, Band IV, 3 (Leipzig: Teubner 1901-1908), 3-47, [1901].

Max Abraham, *Einführung in die Maxwellsche Theorie der Elektrizität, mit einem einleitenden Abschnitte über das Rechnen mit Vektorgrößen in der Physik von Dr. A. Föppl*. 2. Auflage (Leipzig: Teubner, 1904).

Filippo Castellano, *Lezioni di meccanica razionale* (Torino: G. Candeletti, 1894) 2nd ed. 1911.

Augustin Cauchy,"Sur les clefs algébriques," *Comptes rendus hebdomadaires de l'Acad.des sciences*, 1853, *36*, 70-75, 129-136 = *Oeuvres* (1) 11, (Paris: Gauthier-Villars 1909), 439-445 (Nr.514).

Domenico Chelini,"Teoria de' valori delle proiezioni," *Giornale Arcadico de scienze lettere ed arti*, 1838, *74*, 47-73.

Domenico Chelini, *Elementi di meccanica razionale con appendice sui principi fondamentali delle matematiche* (Bologna: G.Legnani 1860).

William Kingdon Clifford, "Applications of Grassmann's extensive algebra," *American Journal of Mathematics*, 1878, *1*, 350-358 = *Mathematical Papers*, ed. Robert Tucker, (London: Macmillan and co.1882), 266-276.

Homersham Cox, "On the application of quaternions and Grassmann's Ausdehnungslehre to different kinds of uniform space", *Transactions of the Cambridge Philosophical Society*, 1883, *13*, 69-143.

Luigi Cremona, *Elemente des graphischen Calculs*. Übersetzt von Maximilian Curtze (Leipzig: Quandt u. Händel 1875).

Michael Crowe, *A History of Vector Analysis. The evolution of the idea of a vectorial system,* (Notre Dame, London: University of Notre Dame Press 1967).

Encyklopädie Band IV. *Encyklopädie der mathematischen Wissenschaften*. Band IV, Hrsg. von Felix Klein und Conrad Müller. Vierter Band in vier Teilbänden (Leipzig: Teubner 1901-1908, 1904-1935, 1901-1908, 1907-1914).

Encyklopädie Band V. *Encyklopädie der mathematischen Wissenschaften*. Fünfter Band in drei Teilen, Hrsg. von Arnold Sommerfeld. (Leipzig: Teubner 1903-1921, 1904-1922, 1909-1926).

August Föppl, *Einführung in die Maxwellsche Theorie der Electricität. Mit einem einleitenden Abschnitte über das Rechnen mit Vectorgrössen in der Physik* (Leipzig: Teubner 1894).

August Föppl, *Vorlesungen über technische Mechanik.* Band 1: Einführung in die Mechanik (Leipzig: Teubner 1898).

Hermann G. Grassmann, *Die lineale Ausdehnungslehre ein neuer Zweig der Mathematik*, (Leipzig: O. Wigand 1844) = *Gesammelte mathematische und physikalische Werke*, Band 1, 1 (Leipzig: Teubner 1894), Reprint (Johnson New York, London 1972), 1-319.

Hermann G. Grassmann, "Neue Theorie der Elektrodynamik", *Poggendorffs Annalen der Physik und Chemie*, 1845, *64*, 1-18 = *Gesammelte Werke* Band 2, 2 (Leipzig: Teubner 1902), 147-160.

Hermann G. Grassmann, *Die Ausdehnungslehre. Vollständig und in strenger Form* (Berlin: Th. Chr. Fr. Enslin 1862) = *Gesammelte Werke Band* 1, 2 (Leipzig: Teubner 1896).

Hermann G. Grassmann, "Grundriss der Mechanik," *Programm des Gymnasiums zu Stettin*, 1862 = *Gesammelte Werke* Band 2, 2 (Leipzig: Teubner 1902), 3-45.

Hermann G. Grassmann, "Die Mechanik nach den Principien der Ausdehnungslehre," *Mathematische Annalen*, 1877, *12*, 222-240 = *Gesammelte Werke* Band 2, 2 (Leipzig: Teubner 1902), 46-72. [1877a]

Hermann G. Grassmann, "Zur Elektrodynamik," *Journal für die reine und angewandte Mathematik*, 1877, *83*, 57-64 = *Gesammelte Werke* Band 2, 2, 203-210. [1877b]

Oliver Heaviside, *Electromagnetic theory*. 3 vols. 1893, 1899, 1912 (London: The "Electrician" printing and publishing Company 1893), Reprint (New York: Chelsea 1971).

Karl Heun, *Lehrbuch der Mechanik. 1. Teil: Kinematik mit einer Einleitung in die elementare Vektorrechnung* (Leipzig: G. J. Göschen 1893).

Edward Willys Hyde, "Calculus of direction and position," *American Journal of Mathematics* , 1884, *6*, 1-13.

Jena. *Index scholarum hibernarum publice et privatim in universitate litterarum Ienensi habendarum* (Iena: Ed. Fromann 1876/77).

Gustav Jaumann, *Die Grundlagen der Bewegungslehre von einem modernen Standpunkte aus* (Leipzig: J. A. Barth 1905).

Georg Emil Kircher, "Ein Beitrag zur Bewegung unveränderlicher ebener Systeme, dargestellt nach den Prinzipien der Graßmannschen Ausdehnungslehre," *Schulprogramm Realgymnasium Meiningen*, 1887, 3-36, 2 plates.

Felix Klein, Arnold Sommerfeld, *Theorie des Kreisels*. In vier Heften (Leipzig: Teubner 1897, 1898, 1903, 1910).

Ferdinand Kraft *Abriß des geometrischen Kalküls. Nach den Werken des Professors Dr. Hermann Günther Grassmann* (Leipzig: Teubner 1897).

Jakob Lüroth, *Grundriss der Mechanik* (München: Th. Ackermann 1881).

Ernst Mach, *Die Mechanik in ihrer Entwicklung historisch kritisch dargestellt* (Leipzig: F. A. Brockhaus 1883).

Ernst Mach, *Die Mechanik in ihrer Entwicklung historisch kritisch dargestellt*. 5. Auflage (Leipzig: F. A. Brockhaus1904).

Roberto Marcolongo, *Theoretische Mechanik*. Autorisierte deutsche Bearbeitung von H. E. Timerding. Band 1: Kinematik und Statik (Leipzig und Berlin: Teubner 1911).

James Clerk Maxwell, *Lehrbuch der Electrizität und des Magnetismus*. Autorisierte deutsche Übersetzung von Dr. B.Weinstein, 2 Bände (Berlin: Springer 1883).

August Ferdinand Möbius, *Der barycentrische Calcul, ein neues Hülfsmittel zur analytischen Behandlung der Geometrie* (Leipzig: J. A. Barth 1827).

August Ferdinand Möbius, *Lehrbuch der Statik*. 1. Teil (Leipzig: G. J. Göschen 1837).

Otto Mohr, "Die Theorie der Streckensysteme," *Der Civilingenieur*, Neue Folge, 1888, *34*, 691-736.

Giuseppe Peano, *Calcolo geometrico secondo l'Ausdehnungslehre di H. Grassmann* (Torino: Frat. Bocca 1888).

Rudolf Rothe, "Anwendungen der Vektoranalysis auf Differentialgeometrie," *Jahresbericht der Deutschen Mathematiker-Vereinigung* , 1912, *21*, 249-274.

Barré de Saint-Venant, Jean Claude Adhémar, "Mémoire sur les sommes et les différences géométriques, et sur leur usage pour simplifier la mécanique," *Comptes rendus hebdomadaires de l'Acad.des Sciences*, 1845, *21*, 620-625.

Wilhelm Schell, *Theorie der Bewegung und der Kräfte. Ein Lehrbuch der theoretischen Mechanik*. 2. Auflage, Band 1 (Leipzig: Teubner 1879).

Jan Arnoldus Schouten, *Grundlagen der Vektor- und Affinoranalysis* (Leipzig und Berlin: Teubner 1914).

Ludwig Silberstein, *Vectorial Mechanics* (London: Macmillan 1913).

Osip Ivanovič Somoff, *Kinematik*. Aus dem Russischen übersetzt von Alexander Ziwet (Leipzig: Teubner 1878).

Arnold Sommerfeld, "Bezeichnung und Benennung der electromagnetischen Grössen in der Enzyklopädie der mathematischen Wissenschaften V," *Physikalische Zeitschrift*, 1904, 467-470 = *Gesammelte Schriften* Band 1 (Braunschweig: Vieweg 1968), 671-674.

Rudolf Sturm, "Sulle forze in equilibrio", *Annali di matematica pura ed applicata*, 1875/76, (2) *7*, 217-246.

Heinrich Emil Timerding, "Geometrische Grundlegung der Mechanik eines starren Körpers" [finished in 1902], *Encyklopädie der mathematischen Wissenschaften*, 1902, IV, *1*, (Leipzig 1901-1908), 125-189.

Heinrich Emil Timerding, *Geometrie der Kräfte* (Leipzig: Teubner 1908).

Woldemar Voigt, *Kompendium der theoretischen Physik*. 2 Bände (Leipzig: von Veit u. Co 1895/96).

Woldemar Voigt, *Lehrbuch der Kristallphysik* (Leipzig und Berlin: Teubner 1910).

Alexander Wernicke, *Adolf Wernickes Lehrbuch der Mechanik in elementarer Darstellung mit Anwendungen und Übungen aus den Gebieten der Physik und Technik*. 2 Teile 1900, 1901. 4. Auflage (Braunschweig: Vieweg 1900/01).

Alexander Ziwet, "A brief account of Grassmann's geometrical theories," *Annals of Mathematics*, 1885/86, 2, 1-11, 25-34.

IVOR GRATTAN-GUINNESS

WHERE DOES GRASSMANN FIT IN THE HISTORY OF LOGIC?

1. THE STATUS OF THE QUESTION

Grassmann's *Ausdehnungslehre* is basically a remarkable geometrical algebraic way of handling geometric objects, and so does not bear directly upon logic. However, the question posed in this title is worth consideration, for the following four reasons.

a) Both Grassmann and logicians of that time were involved in general philosophical questions about mathematics.

b) They both treated collections of things, in the part-whole theory.

c) They both drew upon this form F (including the equation) of linear combination:

$$F := ax + by + cz + ... (= 0), \tag{1}$$

in some reading or another of its symbols. F was in fact rather ubiquitous from the 1840s onwards (Grattan-Guinness 1995): not only in versions already established such as common algebra, equations of hyperplanes, Fourier series and first-order differential equations, but also in new ones such as Grassmann's and Boole's own expansions, quaternions, vectors, matrix multiplication, and group characters.

d) They both paid attention not only to notations in general but also to semiotic aspects, such as the role of brackets and the utility of nesting.

So the question of the title deserves some airing. In fact, it breaks into four, because of two ambiguities which it embodies.

2. TWO TRADITIONS IN SYMBOLIC LOGIC

From the mid 19th century, two quite different lines of symbolising logic developed (Grattan-Guinness 1988). During the 1840s, an algebraic line was launched initially by Augustus De Morgan, with his semiotic schemes for syllogistic logic, and also his logic of relations (the latter closely imitating functional equations, which he had studied in detail). Then soon came George Boole's algebra of logic grounded in his novel

G. Schubring (ed.), Hermann Günther Graßmann (1809–1877): Visionary Mathematician, Scientist and Neohumanist Scholar, 211–216.

'index law' $x^2=x$ but otherwise imitating the calculus of differential operators as much as possible (Panteki 1992).

The next main phase, from the 1870s onwards, is due largely to the American C.S. Peirce and the German Ernst Schröder. A major feature was to unify Boole and De Morgan by extending Boolean algebra in various ways and especially with a logic of relations as an extension to several independent variables of a predicate calculus, and also by developing quantification theory (understood to act over classes).

Meanwhile, Karl Weierstraß led a school from the 1860s to improve rigour in the programme of mathematical analysis that had been laid out 40 years earlier by A.L. Cauchy. This activity led Giuseppe Peano to lead his own school from the late 1880s, to launch 'mathematical logic' (his name), a logic including predicate calculus and quantification as well as the propositional logic. Their follower Bertrand Russell went to a logical conclusion (as it were) around 1900 with his 'logicist' thesis that mathematical analysis (indeed, he hoped, "all" mathematics) was epistemologically founded in mathematical logic.

The differences between the two kinds of symbolic logic are very considerable; the algebraists saw logic as applied mathematics, whereas Russell saw mathematics as applied logic. The algebraists (like Grassmann) used part-whole theory, whereas mathematical logicians used Cantorian set theory, which itself had been created within the Weierstrass scheme. This gives us the first ambiguity of the title, in the word 'logic'; and the connections with Grassmann lay more closely with the algebraic tradition. But we have at once the second ambiguity, in the word 'Grassmann'; for, with two exceptions to be noted in section 4, the main interest for logicians lay not with Hermann but with his brother Robert.

3. ROBERT GRASSMANN ON LOGIC

The work of the two brothers was closely linked; but it was Robert who explicitly applied Hermann's *Ausdehnungslehre* to logic. A group of five little books collectively called *Die Formenlehre oder Mathematik* appeared in 1872. In this remarkable work, written 'in pure German' with all foreign words avoided, he went even beyond Hermann in generality. To start, *Formenlehre* laid out the laws of 'strong scientific thought' of 'Groesen' (sic, following his avoidance of 'ss') denoting any 'object of thought'; each could be composed as a sum of basic 'pegs' ('Stifte') **e** - that is, the expression in (1) with the coefficients set to unity. He admitted, Hermann-style, two means of 'connection' between pegs, 'inner' and 'outer', symbolised by '+' and '×', but then he defined four special kinds of *Formenlehre*, of which Hermann's *Ausdehnungslehre* was only as an example of the last. The members of the quartet were distinguished by the basic laws which their pegs obeyed, under suitable interpretations of them and their means of connection:

'Begriffslehre', or logic: $\mathbf{e} + \mathbf{e} = \mathbf{e}, \mathbf{e} \times \mathbf{e} = \mathbf{e}$ (2)

'Bindelehre', or the theory of combinations: $\mathbf{e} + \mathbf{e} = \mathbf{e}, \mathbf{e} \times \mathbf{e} \neq \mathbf{e}$ (3)

'Zahlenlehre', or arithmetic: $\mathbf{e} + \mathbf{e} \neq \mathbf{e}, \mathbf{e} \times \mathbf{e} = \mathbf{e}$ (4)

'Ausenlehre' (sic), or exterior objects: $\mathbf{e} + \mathbf{e} \neq \mathbf{e}, \mathbf{e} \times \mathbf{e} \neq \mathbf{e}$ (5)

The logic was presented in the first book of the succeeding quartet, as 'the simplest and also most central' kind of *Formenlehre*, in 43 pages and under the title *Die Begriffslehre oder Logik*. The three parts covered, in turn, the development of concepts, judgements and deductions. The theory itself looks like Boole's in much of its contents; but the latest reference in his historical preface is 1825, and he seems not to have known either of Boole or of Boole's first commentator (in 1863), Stanley Jevons.

Robert's signs '·' and '+' stood respectively for 'and' and 'with' ('mal') between concepts ('Begriffe', rather than their associated classes). As with (but independently of) Jevons and then Peirce, he allowed that a + a = a; Boole had admitted 'a+b' only under the hypothesis that a and b were disjoint. Concepts satisfied the relations of identity, non-identity (for which latter he used the symbol '$\gtrless$'), and subordination ('Unterordnung', written 'a < b' or 'b > a'; if identity was possible, then '$\leq$' or '$\geq$'). Somewhat sloppily, he silently gave '=' the additional role of equivalence between propositions.

Apart from the reading of '+', the ensemble of Robert's results strongly resembles Boole's, even to the same orthogonality relation between pegs (in his own notation, '$e_1 \cdot e_2 = 0$') and the same symbol $\bar{a}$ for the negation concept of a concept with respect to the 'totality' T. But his whole approach was much less radical relative to the Aristotelian tradition than had been Boole, especially in Boole's second book *The Laws of Thought* (1854). The theory of judgements was made centred around solving for x the identity a = xu ('x is a u'), and finding equivalents for given ones, while the treatment of deduction did little more than algebraise Aristotelian syllogisms rather than solve sorites (that is, or sets of premises) for a selected unknown; thus he had no analogies to Boole's expansion theorems which did much of the work of derivation of consequences from premises. Presumably also by ignorance, he did not pick up the logic of relations, which De Morgan had outlined in 1860.

In his later work, Robert did cite some works by most of the logicians mentioned above; he also did a modest amount of solving logical equations. But he did not recognise the significance of quantification theory, although Peirce had given it publicity in the early 1880s in connection with the development of the predicate calculus: he seems always to have followed the tradition of conceiving of logic as independent of language, a view which was breaking down for other logicians under the pressure of novelties such as the predicate calculus.

Overall, then, Robert's work was not quite up with the pack, and so failed not to earn great attention. On the other hand, to have achieved so much in apparent isolation, and within the conception of *Formenlehre*, which had no rivals for generality at that time, is a fine achievement. It is well overdue for a detailed study.

For the quantity and range of Robert's production is amazing, surpassing even Hermann's. For just one example, the revived version of the 1872 booklets appeared as the book *Die Logik und die anderen logischen Wissenschaften* (1890) of around 220 pages; it was reprinted as *Die Logik* ten years later. In this version logic was the 'lower analytic' of a quartet of 'logical sciences' of which *Bindelehre* was its lower synthetic companion. But this volume was the second half of the second Book (*Denklehre*) of the second part of the first section of his 10-section 5000-page *Das Gebaeude des Wissens* (1882-1899), which otherwise handled a companion quartet of mathematical sciences (including the *Ausdehnungslehre* as lower synthetic) and elsewhere treated ethics, physics, chemistry, animal and plant physiology, religion, theology, war sciences, and technology. Even in the Germany of the late 19th century, few readers can have had the heart; it seems that hardly any did. One wonders how publisher Robert managed to cope economically with the demands of author Robert.

4. 'IMPACT'

References among the algebraic logicians to Robert are more frequent (or less rare) than to Hermann; they usually cite the *Formenlehre or Begriffslehre*, or one of his later books, usually the *Die Logik und die anderen logischen Wissenschaften* (1890). In an article on logic for the *Encyclopaedia Britannica* for 1882, Robert Adamson prepared a summary of both Grassmanns, after a short survey of Boole (Adamson 1911, pp. 155-158). Schroeder has nearly a score of references to Robert in his three-volume *Vorlesungen ueber die Algebra der Logik* (1890 - 1905), and only a couple to Hermann. John Venn's *Symbolic Logic*

(1881, 1894) also gave Robert somewhat more attention, though again of a marginal or footnotey kind for both men.

Perhaps the most frustrating case of small impact is C.S. Peirce. He did notice the linear form F in Hermann's *Ausdehnungslehre*, but only in 1876, after his own main innovations in algebraic logic had been effected. By 1880 he had seen Robert's *Begriffslehre*, and he mentioned a few features from time to time in his papers on logic (again, Robert more than Hermann). But both men were for him after the fact - confirmatory instances rather than innovating inspirations. He was the one commentator of that time with sufficient philosophical power to master the brothers' strange but tantalising theories; but understandably, the task never seems to have become sufficiently important to him.

With two exceptions still to come, these three appear to have been the only logicians with historical or at least bibliographical interest who gave the Grassmanns much attention, even in German-speaking lands. Elsewhere the silence was often complete; for some British examples, Hugh MacColl, W.E. Johnson, A.T. Shearman and P.E.B. Jourdain.

Among the mathematical logicians, the case of Frege is typical. He wrote a rather poor paper on Boole in 1881, which was rejected for *Mathematische Annalen* by Felix Klein, who explicitly mentioned that Frege had missed both Grassmanns.[1] Thereafter he seems to have ignored Robert; he mentioned Hermann occasionally, but only in passing.

The two exceptions are notable "transfer" figures, from algebraic to mathematical logic. Peano started his logical programme in 1887 within the algebraic tradition, drawing strongly on Hermann (Bottazzini 1985), but inside of a couple of years he had switched entirely into the other one; as far as I know, he did not encounter Robert's work. Similarly, A.N. Whitehead's *A Treatise on Universal Algebra* (1898) drew heavily on Hermann; but his further work in foundations was to follow the guidance of his former student, Bertrand Russell, the leading mathematical logician of his time. Russell himself had discussed Hermann in *AnEessay on the Foundations of Geometry* (1897), but in his pre-logicist phase; by the time of his collaboration with Whitehead, Hermann had virtually disappeared, and Robert seems never to have arrived. So neither Grassmann gained the attention that he deserved among the logicians of either symbolic tradition.

School of Mathematics and Statistics, Middlesex University, London

1 Frege 1976, 134: presumably the criticism of Robert in the version of the paper now published is a later addition (Frege 1983, 38)

REFERENCES

Robert Anderson, *A short History of Logic* , ed. W. R. Solely (London: Blackwood, 1911) [Full version of an *Encyclopaedia* article of 1882].

Umberto Bottazzini, "Dall' analisi matematica al calcolo geometrico: origini delle prime ricerche di Peano," *History and Philosophy of Logic*, 1985, *6*, 25-52.

Gottlob Frege, *Wissenschaftlicher Briefwechsel,* ed. G. Gabriel and others (Hamburg: Meinel, 1976).

Gottlob Frege, *Nachgelassene Schriften,* ed. H. Hermes and others, 2nd ed. (Hamburg: Meinel, 1983).

Ivor Grattan-Guinness, "Living together and living apart: on the interactions between mathematics and logics from the French Revolution to the First World War," *South African journal of philosophy*, 1988, *7*, no. 2, 73-82.

Ivor Grattan-Guinness, "Forms in algebras, and their interpretations: some historical and philosophical features," *What is a form*?, ed. L. Albertazzi, 1996 (to appear).

M. Panteki, *Relationships between algebra, differential equations and logic in England: 1800-1860*, C.N.A.A. (London), Doctoral dissertation 1992.

VOLKER PECKHAUS

THE INFLUENCE OF HERMANN GÜNTHER GRASSMANN AND ROBERT GRASSMANN ON ERNST SCHRÖDER'S ALGEBRA OF LOGIC

1. INTRODUCTION

Today the achievements of Hermann Günther and at least part of Robert Grassmann's results have finally found the recognition they deserve. This is true for Hermann Günther Grassmann's *Ausdehnungslehre*, also for Robert Grassmann's algebraico-logical *Formenlehre* (1872), works which were partly written in collaboration. Robert Grassmann (1815–1901)[1] is now placed undoubtedly among the representatives of the algebra of logic, and his logic system is positioned close to that of George Boole (1815–1864), although published 25 years after the first logical book of the British founder of this symbolic approach to logic (Boole 1847). It is stated with benevolence that Grassmann obviously created his logic without any knowledge of his British predecessors. In order to appraise this last judgement it has to be regarded, however, that in 1872, when Grassmann published his *Formenlehre oder Mathematik*, the British algebra of logic had not yet made the leap across the channel anyway.

Whereas Robert Grassmann's *achievements* are recognized, almost nothing is known as to whether his logic had any *influence*. One could be driven to the opinion that its possible impulses fizzled out unheard, an opinion which is supported by Robert Grassmann's lamentations on the lack of impact of his and of his family's contributions to science. In his book *Das Weltleben oder die Metaphysik*, published in 1881, he directed, e.g., the "special request" (written in the third person) to mathematicians, physicists and philosophers, "not to wait for his death before they begin

---

[1] For Robert Grassmann's biography see especially the extensive autobiographical note appended to the introduction of his *Das Gebäude des Wissens* (1890, XIX–XXVII, note). For an earlier comparison of Robert Grassmann's logic with Ernst Schröder's algebra of logic see Mehrtens 1979. I would like to thank Christian Thiel (Erlangen) and Ivor Grattan-Guinness (Bengeo) for their critical comments.

G. Schubring (ed.), Hermann Günther Graßmann (1809–1877): Visionary Mathematician, Scientist and Neohumanist Scholar, 217–227.

to study the book, contrary to the case of his father and his brother" (1881, p. IX).

To my knowledge this appeal did not help very much. Nevertheless, I would like to question this claimed inefficacy at least concerning Grassmann's logic. I do not intend to support my arguments with the rather destructive allusion to the fact that most of Grassmann's colleagues in logic shared his fate. Ernst Schröder and Gottlob Frege who are much more famous today had to publish their main logical works at their own cost, and they did not have the chance to produce them in their own publishing house like Robert Grassmann. It is quite evident, however, that the overall influence of this new view of formal logic was rather small; and yet it is possible to identify some sort of inner circle of representatives of the subject with working communication among the members, and exchange of their results. Robert Grassmann was counted in this circle by several of its members.

I will focus on aspects of Robert Grassmann's influence on the then contemporary logic and foundations of mathematics. I will do this by concentrating on Grassmann's relations to the mathematician and logician Ernst Schröder (1841–1902),[2] whose monumental *Vorlesungen über die Algebra der Logik,* published between 1890 and 1905, mark the completion of the Boolean period in formal logic (Bocheński 1956, p. 314), and at the same time its final chord and its legacy.

2. H. GRASSMANN, E. SCHRÖDER AND THE FOUNDATIONS OF ARITHMETIC

There were also close connections between Ernst Schröder and Hermann Grassmann which can be seen from the fact that Schröder signed as a co-author the widely read obituary on Grassmann published in 1879 (Sturm/Schröder/Sohnke 1879). As his friend and biographer Jakob Lüroth presumes (Lüroth 1903, p. 350) he may have contributed two pages to this obituary which was at 45 pages unusually extensive. They concern Hermann Günther Grassmann's achievements in pedagogical and didactic fields, especially his *Lehrbuch der Arithmetik* of 1861. Schröder's esteem for this book had its good reasons, since it was the main source for his own *Lehrbuch der Arithmetik und Algebra* published in 1873 when he was still a school professor at the Pro- and Realgymnasium at Baden-Baden. In his *Lehrbuch* Schröder treated, according to its subtitle, "the seven algebraical operations", i.e., the three "direct" operations of adding, multiplying, and raising to a higher power, and their inverses subtracting, dividing, extracting the roots and forming the logarithms, in the beginning applied only to natural numbers.

---

2 For biographical information on Schröder see Lüroth 1903, and Dipert 1990/91.

Schröder defines (pure) mathematics as the "doctrine of numbers" (Schröder 1873, p. 2), and by this definiton he deviates from the traditional view of mathematics as the doctrine of quantities. In this respect he differs from Hermann Günther Grassmann as well who declares mathematics the "science of the connection of quantities" (H. Grassmann 1861, p. 1).

Schröder defines the notion of number without determining what kind of objects numbers are. The notion of number itself, he states, "undergoes a continuing and not yet completed expansion and development" (Schröder 1873, 2). For him a number is a sign created by us arbitrarily in order to reach the most different aims, which can hardly be united under a common aspect.

For the introduction of natural numbers (in the formal part without 0) Schröder recurs to the ability of men to count things which are distinct in our imagination (ibid., p. 3). Each of these things to be counted is called a unity (ibid., p. 4). Schröder introduces the connecting operations which can be applied to numbers in two different ways, in an "independent" manner, i.e., regarding the combination of units (e.g., $4 = 1 + 1 + 1 + 1$), and, explicitly according to Grassmann's model, in a "recurrent" manner, i.e., by reference to definitions already given (e.g., $4 = 3 + 1 = [2 + 1] + 1 = [\{1 + 1\} + 1] + 1$). As Schröder tells us, the recurrent procedure cannot be recommended for beginners of only average ability, although it "originated in the need of greater rigour in order to simplify the presuppositions taken as starting point as far as possible, and to restrain the means of inferences used" (ibid., p. 51).

3. SCHRÖDER'S ALGEBRA OF LOGIC AND ROBERT GRASSMANN'S DOCTRINE OF FORMS

I will now speak about logic and the influence of Robert Grassmann's *Formenlehre* on Ernst Schröder's first steps towards the algebra of logic. Schröder's algebra of logic is usually, and to some extent justly, regarded as a codification of George Boole's logic, which considers the modifications and extensions introduced by Boole's successors and enlarges it by an algebra and logic of relatives following the example of the American philosopher, mathematician, and logician Charles S. Peirce (1839–1914). Such interpretation misses the fact that Schröder's algebra of logic was an integral part of a superior universal-algebraical programme. The term "algebra of logic" refers thus to the algebraical structure of logic and not to some algebraical way of notation as most historians of logic suggest. It seems to be necessary to begin with a sketch of Schröder's programme.

3.1 Schröder's Abstract Algebra

Schröder presented his programme to create an "absolute algebra" in a section of his *Lehrbuch der Arithmetik und Algebra* (1873) entitled "Die Operationsverknüpfung in formaler Behandlungsweise" ("A formal treatment of the connection by operations," pp. 232–294). It was further developed in his school programme dissertation *Über die formalen Elemente der absoluten Algebra* published in 1874—to mention only the first two relevant writings.[3] The formal, and in its last stage of development, "absolute" algebra is founded on a generalization of the notion of number. Schröder proceeds from the existence of an "unlimited manifold [*Mannigfaltigkeit*] of objects (of any kind)" (Schröder 1874, p. 3) which he calls "domain of numbers" (*Zahlengebiet*). Examples for such "objects constituting a manifold" called "general numbers" are "proper names, concepts, judgements, algorithms, numbers [of pure mathematics], symbols for dimensions or operations, points, systems of points, or any geometrical object, quantities of substances, etc." (ibid.).

For such domains of numbers Schröder postulates the existence of connecting operations "with the help of which two objects can be connected to form a third which is to be counted to the same manifold" (ibid.). He especially investigates the non-commutative "symbolical multiplication",

$$c = a \,.\, b = ab,$$

with its related inverse operations "measurement" (*Messung*),

$$b \,.\, (a : b) = a,$$

and "division" (*Teilung*),

$$a/b \,.\, b = a \,.$$

The application of these operations to the general numbers leads to most simple equations ("fundamental relations," *Fundamentalrelationen*), which are now the object of a four-step programme of "formal," and finally, "absolute" algebra:

(1) Formal algebra compiles all assumptions that can serve for defining connectives for numbers of a domain.

(2) Formal algebra compiles for every premiss or combination of premisses the complete set of inferences ("separation" of resulting formulas from the complete stock of combinatorically possible formulas).

(3) Formal algebra investigates which particular domain of numbers can be constructed by the defined operations.

---

3 For Schröder's programme see Peckhaus 1994. Further aspects of his logic are discussed in Peckhaus 1990/91 and Peckhaus 1993.

(4) Formal algebra has finally to decide "what geometrical, physical, or generally reasonable meaning these numbers and operations can have, which real substratum they can be given" (Schröder 1873, pp. 293f.).

In the framework of this absolute algebra Schröder's algebra of logic appears to be an algebraic "Operationskreis," in his terminology a domain of numbers in which more than one connecting operation is working, here logical addition and logical multiplication with their inverses (later additionally relative addition and multiplication). In Schröder's algebra of logic the general numbers are interpreted as domains, classes, propositions and, later, relatives. Thus, he anticipates the lattice-theoretical interpretation of logic.

3.2 First Appearance of Logical Elements in Schröder's Algebra

In the course of Schröder's considerations on formal algebra there are only marginal references to logic. Nevertheless in an early section of the *Lehrbuch* the close connection between logic and arithmetic in Schröder's programme becomes evident. For Schröder this connection results from the problem to express the relation between an "ambiguous expression" in mathematics and its possible solutions symbolically. He exemplifies this by the relation between the expression $\sqrt{9}$ in the domain of real numbers and both of its bases +3 and –3 (Schröder 1873, p. 27). Schröder interprets this relation as a superposition of the general value (e.g., $\sqrt{9}$) upon its particular values (+3, –3). In order to present all combinatorically possible relations between extensions he introduces five logical relations, namely equality ("="), coordination ("$\ni\!\!\in$"), correlation ("(=)") and the subsumption relations superordination ("$\ni\!\!=$") and subordination ("$=\!\!\in$") (ibid., pp. 28f.), the last of which became fundamental in his later logical writings. Schröder emphasizes (ibid., p. 29) that after having introduced negation additionally, a complete terminology has been created for presenting the relations between the extensions of concepts with the help of formulas, "which join closely to the scheme of the mathematical sign language and fall in line with the whole apparatus of signs harmonically."

3.3 Robert Grassmann's Formenlehre

A first glance at the logical parts of Robert Grassmann's *Formenlehre* (1872a–f) which was delivered in five separate pamphlets gives the impression of congruities between Schröder's programme and Grassmann's ideas.

Grassmann starts with a criticism of Hegel's speculative philosophy, which "caused mischief beyond words and evoked distrust of all of philosophy."[4] He sees a decline of logic since its creation by Aristotle. In regard to the exactness of its explanations Kant's logic of 1800 was by far inferior to Aristotle's logic, but Hegel's logic of 1812 was even less useful. On the contrary, Hegel's fallacies and arbitrary assertions had damaged this science tremendously. "Who wants to gain knowledge of the results of the existing logic," he continues, "will find them in Lambert *Neues Organon* 1764, as in Twesten *Logik* 1825."

This hint is interesting in two respects. First of all it gives information on Grassmann's sources. With Johann Heinrich Lambert (1728–1777) he mentions an author who wanted all sciences to draw upon the methodological (Euclidean) paradigm of mathematics, and who became a precursor of the algebra of logic by creating a symbolical calculus of concepts. Like Lambert, August Twesten (1789–1876) stood in the tradition of Leibniz. Twesten is sometimes called a precursor of mathematical logic although hardly with any evidence. Nevertheless he was a consequent advocate of *formal* logic, in his main logical work *Die Logik, insbesondere die Analytik* (1825) trying to make explicit what is implicitly contained in concepts and judgements (Twesten 1825, p. XXVI).

The bibliographical remark suggests, moreover, that Grassmann had no knowledge of the philosophical discussion on logic of his time, which was marked by a criticism on Hegel's conception of a dialectical logic, and by attempts to overcome it. The discussion was thus quite close to Grassmann's objections. It is astonishing that he did not mention the standard logical works of his time, although he can hardly be accused of not having known British symbolic logic.

With the consequences of his criticism concerning the development of logic since Aristotle, Grassmann goes far beyond most of the contemporary philosophers: The second section of his *Denklehre*,

> "should teach us rigorous scientific reasoning which is for all men of any nation, of any language in the same way valid, rigorous, and has the same proving power. It [the section] has therefore to be free from all bounds of single languages, and has to treat the forms of reasoning themselves. In this way it becomes a *doctrine of forms* [*Formenlehre*] or *mathematics*." (R. Grassmann 1875, p. 121)

Grassmann tried to fulfil this programme in his *Formenlehre oder Mathematik* (1872a–f). In a general part he treats the doctrine of quantities (*Grösenlehre*) as "the science of connecting quantities"

4 Grassmann 1872c, 4. For Grassmann's criticism of Hegel's logic cf. also Grassmann 1875, 115–119.

(1872a, 7; cf. 1872b).[5] He adds as special parts *Begriffslehre oder Logik* (1872c), *Bindelehre oder Combinationslehre* (1872d), *Zahlenlehre oder Arithmetik* (1872e), and *Ausenlehre oder Ausdehnungslehre* (1872f).

In the general doctrine of quantities Grassmann introduces the letters *a, b, c, ...* as syntactical symbols for arbitrary quantities. The letter *e* stands for special quantities: elements or, in the Germanophile terminology of Grassmann "Stifte" ("pins"), i.e., quantities which do not emerge from other quantities by applying connecting operations. Beside braces to indicate the order of connections he introduces the equality sign =, the inequality sign $\gtrless$ and a general sign $\circ$ for connecting operations. As special connecting operations he treats "joining" ("fügen") or adding (symbol +), and "weaving" ("weben") or multiplying (symbol ·). These connections can either occur as inner connections (if $e \circ e = e$), or as outer connections (if $e \circ e \gtrless e$).

The different results of connecting pins with themselves give the criteria for distinguishing the special parts of the doctrine of quantities from one another. The "doctrine of concepts or logic" ("Begriffslehre oder Logik") is the first part, "the most simple and, at the same time, the most inward part", as Grassmann calls it. Inner joining $e + e = e$ and inner weaving $ee = e$ are valid. In the "doctrine of binding or doctrine of combinations" ("Bindelehre oder Combinationslehre") as the second part of the doctrine of forms, inner joining $e + e = e$ and outer weaving $ee \gtrless e$ are valid; in the "doctrine of number or arithmetic" ("Zahlenlehre oder Arithmetik") it is outer joining $e + e \gtrless e$ and inner weaving $ee = e$, respectively $1 \times 1 = 1$ and $1 \times e = e$. In the "doctrine of the exterior or doctrine of extensions" finally, the "most complicated and most exterior" part of the doctrine of forms, outer joining $e + e \gtrless e$ and outer weaving $ee \gtrless e$ are valid (R. Grassmann 1872a, pp. 12f.).

In our context we are most interested in the logical part. Its significance can already be deduced from the distinctions mentioned above. In the inner weaving we find Boole's "Law of Duality" also called "Boole's Law," but we have also the inner joining $e + e = e$, a formula which is not valid in Boole's system of logic because of his exclusive interpretation of the logical "or". It was, however, introduced already with William Stanley Jevons's (1835–1882) modifications of the Boolean calculus published in his *Pure Logic* of 1864.

In the beginning of the "doctrine of concepts or logic" the syntactical elements which had been introduced generally were interpreted, thereby gaining a "way of reading" ("Lesart") (Grassmann 1872c, pp. 7f.). Now "quantity" means everything which can become an object of reasoning in

5 This aspect of *connecting* quantities had already been emphasized by Hermann Günther Grassmann in his *Lehrbuch der Arithmetik* published 11 years earlier: "*Mathematics* [...] is the science of connecting quantities" (H.G. Grassmann 1861, 1).

a definite way. "Pins" are originally set objects of thinking which have not been derived from other objects. Two objects are equal if they can be substituted for each other *salva veritate*, they are unequal if such substitution is impossible. Addition is read "and", meaning, however, adjunction, i.e., the logical "or". "Multiplication" is read "times", meaning conjunction, the logical "and". Grassmann introduces the signs $<$ and $>$ to express sub- or superordination of concepts (ibid., p. 12). The sign $\leq$ means that a concept is equal or subordinated to another (ibid., p. 13); it is equivalent to Schröder's (later) subsumption or inclusion. In the doctrine of judgements Grassmann expresses this relation in a shorter way by using the angle sign $\angle$ (ibid., p. 24). The sign "T" stands for "all" or the totality, the sum of all pins (ibid., p. 15). The following laws are valid: $a + \mathrm{T} = \mathrm{T}$ and $a \cdot \mathrm{T} = a$. Zero is interpreted "as the lowest concept which is subordinated to all other concepts". The laws $a + 0 = a$ and $a \cdot 0 = 0$ hold. Finally Grassmann introduces the "not" or negation as complement by the laws $a + \bar{a} = \mathrm{T}$ and $a \cdot \bar{a} = 0$ (ibid., p. 16).

The similarities between Grassmann's calculus and that of Boole, at least in Jevons's modified version, are indeed striking. They have motivated the following benevolent judgement by John Venn (1834–1923), published in the "Historic Notes" for the second edition of his *Symbolic Logic* of 1894:

"Robert Grassmann's scheme was published in 1872, under the title of *Begriffslehre oder Logik*. Like a number of other writers on this subject he seems to have worked out his results in entire ignorance of all that had already been done in this way by those before him. The work is however systematic; and he seems to have been one of the first, after Boole,—with the exception of Peirce,—to realize what was wanted for a complete scheme of Symbolic Logic." (Venn 1894, p. 489)

3.4 The Discovery of Robert Grassmann by Schröder

It should have become clear that Robert Grassmann—like Schröder—propagated a universal-algebraical programme in which the symbols used got different interpretations according to their fields of application. But was there a further effect beside the similarity of the programmes which could be not more than an affirmation of Schröder's own aims? When Schröder wrote the logical portions of his *Lehrbuch* he was not aware of Grassmann's *Formenlehre*. He got hold of it, however, when he was writing a later section. It provoked a voluminous footnote, running over three pages:

"At this point I may not leave a work unmentioned which, published shortly before, came to my sight, when I was just going to send the manuscript of this sheet to the publisher. Although I do not borrow anything from it, I yet find expressed in it (nevertheless in a completely different form and beside many other things) a part of those convictions which I have formed for myself.

[...]
The author of the work thought of uses in the part devoted to logic the + sign for the collective combination, and regards it downright as an *addition*—one could say as a "logical" addition—which has beside the properties of the usual (numerical) addition the basic property: $a + a = a$ additionally.
[...]
Especially interesting and new for me was [...] the rôle the author assigns to *multiplication* in the domain of logic. Whereas the *sum* of two concepts is interpreted as the whole of the individuals belonging to the one or the other of these concepts, the *product* of these concepts is such concepts which comprises the marks of both. Thus, the real extensional addition is opposed to an intensional addition or *addition* of marks as multiplication. This procedure can indeed not surprise if one takes into account that the basic features of addition and of multiplication are essentially the same, that both operations have an already fixed relation to one another only in usual arithmetic, and that one has therefore in new fields from the beginning the choice between the two conceptions". (Schröder 1873, footnote, pp. 145–147)

Grassmann's considerations are thus the source for Schröder's first acquaintance with the essentials of the algebra of logic. As becomes clear from a remark in the *Programmschrift, Über die formalen Elemente der absoluten Algebra* (1874), he got to know of Boole's calculus only after the publication of the *Lehrbuch*. From Grassmann's book he borrows the symbolization of the logical connecting operations which he did not take into account before. He states the analogy of these operations to the respective operations in arithmetic, particularly the arithmetic of the numbers 0 and 1. At this stage he still criticizes the denotation of logical operations with mathematical signs in such contexts which comprise mathematical and logical considerations. Later he will not hesitate to use the signs + and ·, regarded, however, as signs for logically interpretated algebraical operations. Of decisive importance for Schröder's change of his working field seems to be the discovery of the duality of logical addition and logical multiplication which was elaborated by Schröder to perfection in his *Operationskreis des Logikkalkuls* of 1877. At the same time this discovery opened the scope of his programme of abstract algebra, since now logic could be regarded as a representative of a simple structure of abstract algebra.

4. CONCLUSIONS

I would like to maintain that the main impulses for Schröder's change to logical investigations came from Grassmann's *Formenlehre*. There he found his vision of a symbolical representation of logic partly fulfilled in a practical form, which could serve as the base for further considerations anyway. The deeper Schröder studied the logical literature in subsequent times, the more Grassmann's achievements were relativized, at least concerning priorities. Boole had presented his first logical system already in 1847, and Jevons introduced the conception of duality in the algebra of logic already in 1864. Nevertheless, in his *Vorlesungen* Schröder did never forget to mention Grassmann as one of his predecessors.

One could furthermore speculate whether the tradition from Lambert via Grassmann to Schröder forms the hitherto missing link between the symbolic logic in Leibnizian and Wolffian tradition to the various directions of modern symbolic logic. To prove this hypothesis a deeper inspection of Grassmann's logic for influences of Lambert's *Neues Organon* seems to be necessary, which I can, however, only mention as a desideratum of research.

Institut für Philosophie, Universität Erlangen-Nürnberg

REFERENCES

Joseph Maria Bocheński, *Formale Logik* (= *Orbis Academicus*, III, 2; Freiburg i.Br.: Alber 1956, [4]1978).

George Boole, *The Mathematical Analysis of Logic. Being an Essay Towards a Calculus of Deductive Reasoning* (Cambridge: Macmillan, Barclay, and Macmillan/London: George Bell, 1847; reprinted Oxford: Basil Blackwell 1951).

Randall R. Dipert, "The Life and Work of Ernst Schröder," *Modern Logic*, 1990/91, *1*: 119–139.

Hermann G. Grassmann, *Lehrbuch der Arithmetik für höhere Lehranstalten* (= Grassmann, *Lehrbuch der Mathematik für höhere Lehranstalten*, pt. 1 (Berlin: Th. Chr. Fr. Enslin, 1861).

Robert Grassmann, *Die Formenlehre oder Mathematik* (Stettin: R. Grassmann, 1872); reprinted in Grassmann 1966. [1872a]

Robert Grassmann, *Die Grösenlehre. Erstes Buch der Formenlehre oder Mathematik* (Stettin: R. Grassmann, 1872); reprinted in Grassmann 1966. [1872b]

Robert Grassmann, *Die Begriffslehre oder Logik. Zweites Buch der Formenlehre oder Mathematik* (Stettin: R. Grassmann, 1872); reprinted in Grassmann 1966. [1872c]

Robert Grassmann, *Die Bindelehre oder Combinationslehre. Drittes Buch der Formenlehre oder Mathematik* (Stettin: R. Grassmann, 1872); reprinted in Grassmann 1966. [1872d]

Robert Grassmann, *Die Zahlenlehre oder Arithmetik. Viertes Buch der Formenlehre oder Mathematik* (Stettin: R. Grassmann, 1872); reprinted in Grassmann 1966. [1872e]

Robert Grassmann, *Die Ausenlehre oder Ausdehnungslehre. Fünftes Buch der Formenlehre oder Mathematik* (Stettin: R. Grassmann, 1872); reprinted in Grassmann 1966. [1872f]

Robert Grassmann, *Die Denklehre* (= Grassmann, *Die Wissenschaftslehre oder Philosophie*, pt. 1; Stettin: R. Grassmann, 1875).

Robert Grassmann, *Das Weltleben oder die Metaphysik* (Stettin: R. Grassmann, 1881).

Robert Grassmann, *Das Gebäude des Wissens*, vol. 1: *Die Wissenslehre oder die Philosophie*, pt. 1: *Das Verstandeswissen oder das formale Wissen, umfassend die auf die Philosophie vorbereitenden Wissenschaften* (Stettin: R. Grassmann, 1890).

Robert Grassmann, *Die Formenlehre oder Mathematik*, with an introduction by J.E. Hofmann (Hildesheim: Georg Olms, 1966).

William Stanley Jevons, *Pure Logic or the Logic of Quality apart from Quantity with Remarks on Boole's System and the Relation of Logic and Mathematics* (London: E. Stanford, 1864); again in Jevons *Pure Logic and Other Minor Works*, ed. Robert Adamson and Harriet A. Jevons (London and New York: Macmillan and Co., 1890); reprinted in Bristol: Thoemmes Press 1991.

Johann Heinrich Lambert, *Neues Organon oder Gedanken über die Erforschung und Bezeichnung des Wahren und dessen Unterscheidung vom Irrthum und Schein* (Leipzig: Johann Wendler, 1764); reprint ed. Günter Schenk, 3 vols. (= *Philosophiehistorische Texte*; Berlin: Akademie Verlag 1990).

Jakob Lüroth, "Ernst Schröder †. Mitglied der Deutschen Mathematiker-Vereinigung," *Jahresbericht der Deutschen Mathematiker-Vereinigung*, 1903, *12*: 249–265; again in Schröder 1905, III–XIX; reprinted in Schröder 1966, vol. 3.1, III–XIX.

Herbert Mehrtens, *Die Entstehung der Verbandstheorie* (= *arbor scientiarum*, A.VI; Hildesheim: Gerstenberg, 1979).

Volker Peckhaus, "Ernst Schröder und die 'pasigraphischen Systeme' von Peano und Peirce," *Modern Logic*, 1990/91, *1*:174–205.

Volker Peckhaus, "Ernst Schröder und der Logizismus," in: *Philosophie und Logik. Frege-Kolloquien Jena 1989/1991*, ed. Werner Stelzner (= *Perspektiven der Analytischen Philosophie*, 3; Berlin and New York: Walter De Gruyter, 1993), 108–119.

Volker Peckhaus, "Wozu Algebra der Logik? Ernst Schröders Suche nach einer universalen Theorie der Verknüpfungen," *Modern Logic*, 1994, *4*: 357–381.

Ernst Schröder, *Lehrbuch der Arithmetik und Algebra für Lehrer und Studirende*, vol. 1 [no further volume published]: *Die sieben algebraischen Operationen* (Leipzig: B.G. Teubner, 1873).

Ernst Schröder, *Über die formalen Elemente der absoluten Algebra*, supplement to the programme of the Pro- and Realgymnasium in Baden-Baden for 1873/74 (Stuttgart: Schweizerbart'sche Buchhandlung, 1874).

Ernst Schröder, *Der Operationskreis des Logikkalkuls* (Leipzig: Teubner, 1877; reprinted as "Sonderausgabe" Darmstadt: Wissenschaftliche Buchgesellschaft, 1966).

Ernst Schröder, *Vorlesungen über die Algebra der Logik (exakte Logik)*, vol. 1 (Leipzig: B.G. Teubner, 1890); vol. 2, pt. 1 (1891); vol. 3, pt. 1: *Algebra und Logik der Relative* (1895); vol. 2, pt. 2, ed. Karl Eugen Müller (1905); reprinted in Schröder 1966. [1890 –1905]

Ernst Schröder, *Vorlesungen über die Algebra der Logik (exakte Logik)*, ["second edition"], 3 vols. (Bronx, N.Y.: Chelsea, 1966).

Rudolf Sturm, Ernst Schröder, Leonhard Sohnke "Hermann Grassmann. Sein Leben und seine mathematisch-physikalischen Arbeiten," *Mathematische Annalen*, 1879, *14*: 1–45.

August Detlev Christian Twesten, *Die Logik, insbesondere die Analytik* (Schleswig: Kgl. Taubstummen-Institut, 1825).

John Venn, *Symbolic Logic*, 2nd ed., "revised and rewritten" (London: Macmillan & Co., 1894).

PART IV

INFLUENCES OF GRASSMANN'S WORK ON RECENT DEVELOPMENTS IN SCIENCE

Illustration 9: Grassmann Memorial Table in front of the building of the Faculty for Mathematics and Physics of Szczecin University

ANDREA BRINI and ANTONIO G. B. TEOLIS

GRASSMANN PROGRESSIVE AND REGRESSIVE PRODUCTS AND CG-ALGEBRAS

DEDICATED TO THE MEMORY OF ROSA Q. HUANG
I tuoi amici italiani ti ricorderanno sempre con grandissimo affetto
Giandomenico, Flavio, Marilena, Antonella e Domenico, Antonio, Giorgio, Jole e Fabrizio, Francesco, Maria e Andrea, Gian-Carlo

1. CG-ALGEBRAS: AN INTRODUCTION

Let $\mathcal{V}$ be a finite dimensional vector space, $\dim(\mathcal{V}) = n$, and let $(G(\mathcal{V}), \vee)$ be its *exterior algebra*. We will denote by $\vee$ the *exterior product* (equivalently, the *wedge* or Grassmann's progressive product) in order to stress its close analogy with its geometric lattice companion; we call this operation the *join*. Given two *extensors* (i. e. decomposable antisymmetric tensors) A and B, they represent two subspaces $<A>$ and $<B>$ of $\mathcal{V}$, respectively. It is well known that the basic geometric fact about the exterior product of the extensors A and B is the following:

$$<A> \cap <B> = (0) \Rightarrow <A \vee B> = <A> \oplus <B>,$$

$$<A> \cap <B> \neq (0) \Rightarrow A \vee B = 0.$$

In Grassmann's language, the "wedge" product was called the *progressive product*. Grassmann's idea was to develop a "calculus" for linear varieties in vector (projective) spaces. It would, however, be capricious to limit such a calculus to a *single* operation, just as capricious as limiting the algebra of sets to the single operation of union.

Actually, Grassmann defined a further operation between extensors which he called the *regressive product*. Not unexpectedly, the regressive product of A and B corresponds, under obvious conditions, to the "geometric" operation of taking the *intersection* of the subspaces represented by A and B respectively. In our century's mathematical literature, the regressive product appears to be almost completely

G. Schubring (ed.), Hermann Günther Graßmann (1809–1877): Visionary Mathematician, Scientist and Neohumanist Scholar, 231–242.

forgotten: to the best of our knowledge, some notable exceptions are Peano (1888), Segre (1951), Whitehead (1898), Zaddach (1994).

Grassmann defined the *regressive product* in two steps.

STEP 1. Definition of the progressive product (the wedge);

STEP 2. The vector space $\mathcal{V}$ is regarded as an euclidean space; hence, the notion of the orthogonal complement of a subspace of $\mathcal{V}$ is defined; the *Ergänzung operation*

$$A \to |A$$

maps the extensor A to an extensor |A which represent the orthogonal complement of <A>. In modern language, this operation depends from the choice of an orthonormal basis of $\mathcal{V}$. Letting A and B be extensors, the regressive product is defined in the following way

$$(A, B) \xrightarrow{\text{def}} |(\,|A \wedge |B).$$

Two "good" (?) objections can be considered (see e. g. J. Dieudonné 1979).

FIRST OBJECTION (matter of style?) the regressive product is defined by means of the progressive product and the *Ergänzung* operation; clearly, the *Ergänzung* is invariant only under the *orthogonal group*: this fact clashes against the fact that the regressive product turns out to be invariant under the *special linear group*.

SECOND OBJECTION (substantial!?) the definition of the regressive product is not *intrinsic* (i. e. no "closed forms" are given for it). Hence, it is not possible to compute independently and simultaneously with the two kinds of products.

Let $\mathcal{V}$ be a vector space of dimension n and let $(G(\mathcal{V}), \vee)$ be its exterior algebra. As noticed in recent years by several authors (see e. g. Hestenes's paper in this volume), there is no essential need to appeal either to the *Ergänzung* operator or to vector space duality in order to define the regressive product or *meet*, as we prefer to call it. This fact was intuited (at last in dimension 2 and 3) by G. Peano (1888). The basic idea is to endow the n-dimensional vector space $\mathcal{V}$ with a bracket [, , ...,], that is a non-degenerate alternating n-multilinear form

$$[\ ,\ , \ldots,\ ] : \mathcal{V}^{\oplus^n} \to K \text{(the base field)}.$$

The pair ($\mathcal{V}$, [ ]) will be called a *Peano space*. One can recognize that the choice of a bracket in order to define a Peano space is essentially equivalent to the choice of a *neutral element* E_n in $G^n(\mathcal{V})$ (see Zaddach's paper in this volume). However, since this neutral element, that we call the *integral* (this word is borrowed from the theory of Hopf algebras (Sweedler 1969)) and denote simply by E, can be defined as

$$E = e_1 \vee e_2 \vee \ldots \vee e_n \text{ where}$$

$\{e_1, e_2, \ldots, e_n\}$ is any linearly ordered basis such that $[e_1, e_2, \ldots, e_n] = 1$ and the identity $a_1 \vee a_2 \vee \ldots \vee a_n = [a_1, a_2, \ldots, a_n]\cdot E$ holds; we believe that the choice of fixing a bracket first gives the theory a more natural presentation.

REMARK. It is crucial (even if it immediately follows from the definition) that the bracket can be consistently filled not only with n vectors $a_1, a_2, \ldots, a_n$, but also by an ordered k-tuple $(A_1, A_2, \ldots, A_k)$ of extensors: more precisely we can set

$$[A_1, A_2, \ldots, A_k] = 0 \qquad \text{if } \mathrm{step}(A_1) + \ldots + \mathrm{step}(A_k) \neq n$$

$$= [a_{11}, \ldots, a_{1i_1}, a_{21}, \ldots, a_{2i_2}, \ldots, a_{k1}, \ldots, a_{ki_k}] \qquad \text{otherwise,}$$

where $A_1 = a_{11} \vee \ldots \vee a_{1i_1}$, $A_2 = a_{21} \vee \ldots \vee a_{2i_2}$, $A_k = a_{k1} \vee \ldots \vee a_{ki_k}$, and the value of the bracket is *independent* from the representation of $A_1, A_2, \ldots, A_k$ as join of vectors.

We now exhibit *two different closed forms* for Grassmann's regressive product. Let $\mathcal{V}$ be a n-dimensional vector space and $A, B \in G(\mathcal{V})$ be extensors, step(A) = k, step(B) = p. The *meet* $A \wedge B$ is defined as follows:

$$\text{if } \langle A\rangle + \langle B\rangle \neq \mathcal{V} \Rightarrow A \wedge B = 0$$

$$\text{if } \langle A\rangle + \langle B\rangle = \mathcal{V} \Rightarrow A \wedge B \overset{\text{def}}{=} \sum_{\Delta A} \mathrm{sign}(A_{(1)}, A_{(2)})[A_{(1)}, B]\, A_{(2)} \qquad (1)$$

$$= \sum_{\Delta B} \mathrm{sign}(B_{(1)}, B_{(2)})[A, B_{(2)}]\, B_{(1)} \qquad (2)$$

where
$$\Delta A = \sum \mathrm{sign}(A_{(1)}, A_{(2)})\, A_{(1)} \otimes A_{(2)}$$
$$\Delta B = \sum \mathrm{sign}(B_{(1)}, B_{(2)})\, B_{(1)} \otimes B_{(2)}$$
denote the coproducts in the Hopf algebra (Z_2-cocommutative) $G(\mathcal{V})$, in the Sweedler notation (Sweedler 1969).

EXAMPLE. Let n = 3, and $A = a \vee a'$, $B = b \vee b'$; then

$$A \wedge B = [a, b, b']\, a' - [a', b, b']\, a = -[a, a', b]\, b' + [a, a', b']\, b'.$$

REMARK. The crucial identity (1) = (2) is an instance of the straightening formula for (super)symmetric letterplace algebra (Doubilet et al. 1974, Grosshans et al. 1987).

The main properties of the meet are:

i) The meet coincides with the regressive product; more specifically, let A and B be extensors, then:

$$\text{if } \langle A\rangle + \langle B\rangle \neq \mathcal{V} \Rightarrow A\wedge B = 0,$$

$$\text{if } \langle A\rangle + \langle B\rangle = \mathcal{V} \Rightarrow \langle A\wedge B\rangle = \langle A\rangle \cap \langle B\rangle.$$

ii) The meet defines a bilinear associative product on the exterior algebra $G(\mathcal{V})$.

iii) The meet is $\mathbb{Z}_2$-cograded commutative, that is

$$A\wedge B = B\wedge A\,(-1)^{(\dim(\mathcal{V}) - \text{step}(A))(\dim(\mathcal{V}) - \text{step}(B))}.$$

iv) A *covector* is an extensor of step $\dim(\mathcal{V}) - 1$. In the present notation, it immediately follows that every $(\dim(\mathcal{V}) - 1)$-antisymmetric tensor is a covector, and that every extensor of step k can be represented as the meet of n-k covectors.

Thus $G(\mathcal{V})$ turns out to be a double $\mathbb{K}$-algebra $(G(\mathcal{V}), \vee, \wedge)$ which is called the *(double) CG-algebra* of $\mathcal{V}$ (for more details, see e. g. Barnabei et al. 1985). The "intrinsic duality" of the lattice of subspaces of the vector space $\mathcal{V}$ is actually implemented in the present calculus by the Hodge operator $* : G(\mathcal{V}) \to G(\mathcal{V})$ defined as follows.

Let $\{e_1, e_2, ..., e_n\}$ be a *unimodular* basis of $\mathcal{V}$, that is $[e_1, e_2, ..., e_n] = 1$. For every $S = \{i_1 < i_2 < ... < i_k\} \subseteq \{1, 2, ..., n\}$ let $S^c = \{p_1 < p_2 < ... < p_{n-k}\}$ be the linearly ordered complement of S in $\{1, 2, ..., n\}$ and set

$$*(e_{i_1} \vee e_{i_2} \vee ... \vee e_{i_k}) = (-1)^{i_1 + ... + i_k + \frac{k(k-1)}{2}} e_{p_1} \vee e_{p_2} \vee ... \vee e_{p_{n-k}}.$$

THEOREM. $*$ maps extensors of step k to extensors of step n-k (see Barnabei et al. 1985).

The Hodge $*$-operator is invariant under the orthogonal group and coincides with Grassmann's *Ergänzung*. The Hodge $*$-operator is an (anti)isomorphism of CG-algebra $(G(\mathcal{V}), \vee, \wedge)$ that flips meet and join. More precisely, we have:

THEOREM (see [1]). The following properties hold:

i) $*(A \vee B) = (*A)\wedge(*B)$ for every A and $B \in G(\mathcal{V})$,

ii) $*(A\wedge B) = (*A) \vee (*B)$ for every A and $B \in G(\mathcal{V})$,

iii) $*(*A) = (-1)^{k(n-k)} A$ for every $A \in G^k(\mathcal{V})$,

iii) $*(1) = E$, $*(E) = 1$.

Thus, the algebra $(G(\mathcal{V}), \vee, \wedge, *)$ can be seen as the much-sought-after linear analog of the boolean algebra of sets with union, intersection and complement, an analog which was first intuited by Grassmann and later pursued, with partial success, by Clifford, Peano, Burali-Forti and A. N. Whitehead.

2. THE CG-ALGEBRA $(G(\mathcal{V}), \vee, \wedge)$ AS A GEOMETRIC ALGEBRA

An alternative approach to Grassmann's calculus is that of Clifford algebra, or *geometric algebra* as Hestenes prefers to call it (see e. g. Hestenes 1996, 1991). The languages of CG-algebra and geometric algebra are essentially equivalent, even though - under specific circumstances - one could be preferable to the other.

In this section we sketch, without proofs, the main relations connecting the basic CG-algebra operations and the geometric algebra operations related to them. Let $K = \mathbb{R}$ the real field. Given two 1-extensors (i. e. vectors) $a, b \in G^1(\mathcal{V})$, set

$$a \circ b \overset{\text{def}}{=} (-1)^{n-1} ((*a) \wedge b),$$

the *inner product of vectors*. Set

$$ab \overset{\text{def}}{=} a \circ b + a \vee b = (-1)^{n-1} ((*a) \wedge b) + a \vee b,$$

the *geometric product of vectors*. The geometric product (here defined in terms of join, meet and $*$-operation) makes $G(\mathcal{V})$ into a geometric (Clifford) algebra in the sense of Hestenes (1996, 1991). In the theory of geometric algebras a notion of *inner product of extensors* A, B is given. In the case of the geometric algebra associated to the CG- algebra $(G(\mathcal{V}), \vee, \wedge)$ it takes the following explicit form (also denoted by a $\circ$):

$$A \circ B = (-1)^{r(n-r)} (*A) \wedge B \qquad \text{if } step(A) = r \leq step(B) = s,$$

$$= A \wedge (*B) \qquad \text{otherwise.}$$

Let $A, B \in G(\mathcal{V})$, with $step(A) = r \leq step(B) = s$; the following properties hold.

i) $A \circ B = (-1)^{r(s-r)} B \circ A$,

ii) $A \circ B = (-1)^{r(n-r)} \sum_{\Delta B} [*A, B_{(2)}] B_{(1)}$, which is a generalized Laplace expansion, (see Barnabei et al. 1985);

iii) $A \wedge B = (*A) \circ B = (-1)^{s(n-s)} A \circ (*B)$, (see ibid. and Hestenes/ Ziegler 1991).

3. BOURBAKI'S DUALITY BETWEEN $G(\mathcal{V})$ AND $G(\mathcal{V}^*)$

Let $\mathcal{E} = \{e_1, e_2, \ldots, e_n\}$ be a unimodular basis of $\mathcal{V}$ and let $\mathcal{E}^* = \{\eta_1, \eta_2, \ldots, \eta_n\}$ be its dual basis of covectors in $G^{n-1}(\mathcal{V})$, that is

$$[e_i, \eta_j] = \delta_{ij}, \qquad i, j = 1, 2, \ldots, n.$$

Let $\mathcal{V}^*$ be the dual space of $\mathcal{V}$ and let $\psi_{\mathcal{E}}: \mathcal{V} \to \mathcal{V}^*$ be the linear map defined by setting

$$\psi_{\mathcal{E}}(e_i)(x) = [x, h_i] = x \wedge h_i, \qquad \text{for every } x \in \mathcal{V},\ i = 1, 2, \ldots, n.$$

Let us define a bracket $[\,]^*$ on $\mathcal{V}^*$ by setting:

$$[\psi_{\mathcal{E}}(e_1), \psi_{\mathcal{E}}(e_2), \ldots, \psi_{\mathcal{E}}(e_n)]^* = 1.$$

It immediately follows that

$$\psi_{\mathcal{E}}: (\mathcal{V}, [\,]) \to (\mathcal{V}^*, [\,]^*)$$

is a *Peano space isomorphism* and, hence, it canonically extends to a CG-algebra isomorphism

$$\Psi_{\mathcal{E}}: (G(\mathcal{V}), \vee, \wedge) \to (G(\mathcal{V}^*), \vee^*, \wedge^*).$$

The linear invertible operator Φ: $G(\mathcal{V}) \to G(\mathcal{V}^*)$ is defined as the unique linear operator such that the diagram

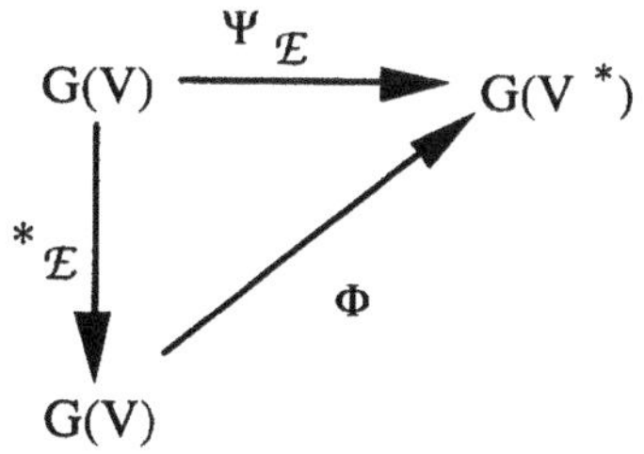

commutes, where $*_{\mathcal{E}}$ is the Hodge operator relative to the basis $\mathcal{E} = \{e_1, e_2, \ldots, e_n\}$ of $\mathcal{V}$. Clearly, the maps $\Psi_{\mathcal{E}}$ and $*_{\mathcal{E}}$ depend on the choice of the basis $\mathcal{E}$ and, hence, are *orthogonal invariants*. To be precise, if $\mathcal{A} = \{a_1,$

$a_2, ..., a_n\}$is any other unimodular basis, then $\Psi_{\mathcal{E}} = \Psi_{\mathcal{A}}$ and $*_{\mathcal{E}} = *_{\mathcal{A}}$ if and only if the transition matrix from $\mathcal{E}$ to $\mathcal{A}$ is an *orthogonal matrix* (Barnabei et al. 1985). However, as we shall see, the map Φ is a *unimodular invariant*, that is, it is the same for any choice of a unimodular basis (informally speaking, this is equivalent to the fact that the construction of the CG-algebra of a Peano space is a unimodular invariant).

PROPOSITION. The map Φ: $(G(\mathcal{V}), \vee, \wedge) \to (G(\mathcal{V}^*), \vee^*, \wedge^*)$ is an (anti)isomorphism of CG-algebras interchanging meet and join.

PROOF. We have

$\Phi((*A)\wedge(*B)) = \Phi(*(A\vee B)) = \Psi(A\vee B) = \Psi(A)\vee^*\Psi(B) = \Phi(*A)\vee^*\Phi(*B)$.

The same argument works for the join.

COROLLARY. The map Φ interchanges the unity and the integral of $G(\mathcal{V})$ and $G(\mathcal{V}^*)$, respectively. In symbols: $\Phi(1) = E^*$, $\Phi(E^*) = 1$.

PROPOSITION. Let $\xi \in G^{n-1}(\mathcal{V})$ be a vector. Then $\Phi(\xi)$ is the linear functional in $\mathcal{V}^* = G^1(\mathcal{V}^*)$ such that

$$\Phi(\xi)(x) = [x, \xi] = x\wedge\xi, \text{ for every } x \in \mathcal{V}.$$

PROOF. Let $\xi = \sum_i c_i\eta_i$. From $*(e_i) = \eta_i$, it is easily seen that $\Phi(\eta_i)(x) = \Psi(e_i)(x) = [x, \eta_i] = x\wedge\eta_i$. Then $\Phi(\xi)(x) = \sum_i c_i\,\Phi(\eta_i)(x) = \sum_i c_i\,(x\wedge\eta_i) = x\wedge\sum_i c_i\,\eta_i = [x, \xi]$.

In the most popular presentations of exterior algebra, a pairing $<,>$: $G(\mathcal{V}) \times G(\mathcal{V}^*) \to K$ is introduced by setting (in the present notation)

$$\begin{aligned} <x_1\vee\ldots\vee x_k, \vee^*\varphi_1\ldots\vee^*\varphi_k> &= 0 && \text{if } h \neq k,\\ &= \det(\varphi_j(x_i)_{i=1,2,\ldots,h=k}) && \text{if } h = k, \end{aligned}$$

where $x_i \in \mathcal{V}$, $\varphi_j \in \mathcal{V}^*$. Hence $G^k(\mathcal{V}^*)$ can be identified with $(G^k(\mathcal{V}))^*$ (see Bourbaki 1979, Marcus 1975, Zaddach 1996). Let us now define a vector space isomorphism g: $G^{n-k}(\mathcal{V}) \to G^k(\mathcal{V}^*) \cong (G^k(\mathcal{V}))^*$ in the following way: for every extensor $A \in G^{n-k}(\mathcal{V})$, set

$$\begin{aligned} g(A)(x_1\vee\ldots\vee x_k) &\overset{\text{def}}{=} [x_1\vee\ldots\vee x_k, A],\\ &= (x_1\vee\ldots\vee x_k)\wedge A. \end{aligned}$$

We explicitly note that g is well defined and linear, thanks to the last identity. Since $A \in G^{n-k}(\mathcal{V})$, it can be written as

$A = \xi_1 \wedge \ldots \wedge \xi_k$, $\xi_j \in G^{n-1}(\mathcal{V})$. It follows that

$$g(A)(x_1 \vee \ldots \vee x_k) = [x_1 \vee \ldots \vee x_k, \xi_1 \wedge \ldots \wedge \xi_k];$$

by a special case of Th. 7.4 of Barnabei et al. 1985, we have

$$g(A)(x_1 \vee \ldots \vee x_k) = \det([x_i, \xi_j]_{i,\, j = 1, 2, \ldots, h = k}). \tag{3}$$

On the other hand, given an extensor A as before, one can compute:

$$\begin{aligned} \langle x_1 \vee \ldots \vee x_k, \Phi(A) \rangle &= \langle x_1 \vee \ldots \vee x_k, \Phi(\xi_1 \wedge \ldots \wedge \xi_k) \rangle \\ &= \langle x_1 \vee \ldots \vee x_k, \Phi(\xi_1) \vee^* \ldots \vee^* \Phi(\xi_k) \rangle \\ &= \det(\Phi(\xi_j)(x_i)_{i,\, j = 1, 2, \ldots, h = k}) \\ &= \det([x_i, \xi_j]_{i,\, j = 1, 2, \ldots, h = k}). \end{aligned} \tag{4}$$

Since (3) = (4), we have proved that $\Phi = g$. More precisely, we have the following

THEOREM. Let $A \in G^{n-k}(\mathcal{V})$, $x_i \in \mathcal{V}$. Then

$$\Phi(A)(x_1 \vee \ldots \vee x_k) = [x_1 \vee \ldots \vee x_k, A] = (x_1 \vee \ldots \vee x_k) \wedge A =$$
$$= \langle x_1 \vee \ldots \vee x_k, \Phi(A) \rangle.$$

REMARK. The last result shows that, even though the operators Ψ and $*$ are *orthogonal invariants*, the operator Φ is the *same* for every choice of a unimodular basis of the Peano space ($\mathcal{V}$, [, ...,]); in other words, as expected, the operator Φ is *unimodular invariant*.

4. A GLIMPSE FROM GEOMETRY

In the following, we keep the notation $\langle A \rangle$ for the subspace of $\mathcal{V}$ represented by the extensor $A \in G(\mathcal{V})$. We recall that, given $\eta \in G^{n-1}(\mathcal{V})$, it follows that $\langle \eta \rangle = \operatorname{Ker} \Phi(\eta)$. This implies that

$$\langle \Phi(\eta) \rangle = \langle \eta \rangle^{\perp} \tag{5}$$

Since Φ is surjective, (5) implies that, given $\varphi \in G^1(\mathcal{V}^*) \cong \mathcal{V}^*$, we have

$$\langle \varphi \rangle^{\perp} = \langle \Phi(\eta) \rangle^{\perp} = \langle \eta \rangle^{\perp\perp} = \langle \eta \rangle = \langle \Phi^{-1}(\varphi) \rangle.$$

Let now $A = \eta_{i_1} \wedge \ldots \wedge \eta_{i_{n-k}} \in G^k(\mathcal{V})$, $\eta_i \in G^{n-k}(\mathcal{V})$; then

$$\begin{aligned} \langle \Phi(A) \rangle &= \langle \Phi(\eta_{i_1} \wedge \ldots \wedge \eta_{i_{n-k}}) \rangle = \langle \Phi(\eta_{i_1}) \vee \ldots \vee \Phi(\eta_{i_{n-k}}) \rangle = \\ &= \langle \Phi(\eta_{i_1}) \rangle + \ldots + \langle \Phi(\eta_{i_{n-k}}) \rangle = \langle \eta_{i_1} \rangle^{\perp} + \ldots + \langle \eta_{i_{n-k}} \rangle^{\perp} = \end{aligned}$$

$$= (<\eta_{i_1}> \cap \ldots \cap <\eta_{i_{n-k}}>)^{\perp} = (\eta_{i_1} \wedge \ldots \wedge \eta_{i_{n-k}})^{\perp} =$$

$$= <A>^{\perp}.$$

Vice versa, let $\underline{\alpha} = \varphi_{i_1} \vee^* .. \vee^* \varphi_{i_k} \in G^k(\mathcal{V}^*)$, $\varphi_i \in G^1(\mathcal{V}^*) = \mathcal{V}^*$; then

$$<\Phi^{-1}(\underline{\alpha})> = <\Phi^{-1}(\varphi_{i_1} \vee^* \ldots \vee^* \varphi_{i_k})> = <\Phi^{-1}(\varphi_{i_1})> \cap \ldots \cap <\Phi^{-1}(\varphi_{i_k})>$$

$$= <\varphi_{i_1}>^{\perp} \cap \ldots \cap <\varphi_{i_k}>^{\perp} = (<\varphi_{i_1}> + \ldots + <\varphi_{i_k}>)^{\perp} =$$

$$= (\varphi_{i_1} \vee^* \ldots \vee^* \varphi_{i_k})^{\perp} =$$

$$= <\underline{\alpha}>^{\perp}.$$

5. INNER PRODUCTS AND INTERIOR PRODUCTS

Some comments about the connection with Zaddach's paper in this volume are in order, especially with Bourbaki's method of introducing the duality between $G(\mathcal{V})$ and $G(\mathcal{V}^*)$ and their "inner products." For the sake of readability, we continue to write the algebra of forms on the right: this amounts to obvious sign changes but preserves the essence of the discussion. The dual integral E^* can be written as

$$E^* = \psi_{\mathcal{E}}(e_1) \vee^* \ldots \vee^* \psi_{\mathcal{E}}(e_n), \text{ with } [\psi_{\mathcal{E}}(e_1), \psi_{\mathcal{E}}(e_2), \ldots, \psi_{\mathcal{E}}(e_n)]^* = 1.$$

This implies that

$$\langle a_1 \vee \ldots \vee a_n, E^* \rangle = \langle [a_1, \ldots, a_n]E, E^* \rangle = [a_1, \ldots, a_n].$$

Now, following Zaddach, we define a mapping φ: $G(\mathcal{V}) \to G(\mathcal{V}^*)$ as follows. Given $A \in G^{n-k}(\mathcal{V})$, $\varphi(A)$ is the element of $G^k(\mathcal{V}^*)$ that satisfies

$$\langle X, \varphi(A) \rangle = \langle X \vee A, E^* \rangle \text{ for every } X \in G^k(\mathcal{V}).$$

PROPOSITION. Since $\langle X \vee A, E^* \rangle = [X, A] = \langle X, \Phi(A) \rangle$, the map φ coincides with our anti-isomorphism Φ.

In closing this section, we briefly describe how Bourbaki's notion of *interior products* (see Bourbaki 1979, Marcus 1975) can be naturally recovered within the theory of CG-algebras.

Let $A \in G^q(\mathcal{V})$; the *interior product* $A \rightharpoondown$: $G^{p+q}(\mathcal{V}^*) \to G^p(\mathcal{V}^*)$ is defined by setting

$$\langle B, A \rightharpoondown \underline{\alpha} \rangle \stackrel{\text{def}}{=} \langle B \vee A, \underline{\alpha} \rangle, \text{ for every } B \in G^p(\mathcal{V}), \underline{\alpha} \in G^{p+q}(\mathcal{V}^*).$$

PROPOSITION. $A \rightharpoondown \underline{\alpha} = \Phi(A) \wedge^* \underline{\alpha}$.

PROOF. Write $\underline{\alpha} \in G^{p+q}(\mathcal{V}^*)$ in the form $\underline{\alpha} = \Phi(C)$, $C \in G^{n-p-q}(\mathcal{V})$; hence

$$\begin{aligned} \langle B, A \rightharpoondown \underline{\alpha} \rangle &= \langle B \vee A, \underline{\alpha} \rangle = [B \vee A, C] = (B \vee A) \wedge C = \\ &= [B, A, C] = \\ &= B \wedge (A \vee C) = [B, A \vee C] = \langle B, \Phi(A \vee C) \rangle = \\ &= \langle B, \Phi(A) \wedge^* \underline{\alpha} \rangle. \end{aligned}$$

Let $\underline{\alpha} \in G^q(\mathcal{V}^*)$; the *interior product* $\underline{\alpha} \leftharpoondown : G^{p+q}(\mathcal{V}) \to G^p(\mathcal{V})$ is defined by setting

$$\langle \underline{\alpha} \leftharpoondown A, \underline{\beta} \rangle = \langle A, \underline{\alpha} \vee^* \underline{\beta} \rangle, \text{ for every } \underline{\beta} \in G^p(\mathcal{V}^*), A \in G^{p+q}(\mathcal{V}).$$

PROPOSITION. $\underline{\alpha} \leftharpoondown A = A \wedge \Phi^{-1}(\underline{\alpha})$.

PROOF. Let $C = \Phi^{-1}(\underline{\alpha}) \in G^{n-q}(\mathcal{V})$; hence

$$\begin{aligned} \langle \underline{\alpha} \leftharpoondown A, \underline{\beta} \rangle &= \langle A, \underline{\alpha} \vee^* \underline{\beta} \rangle = [A, \Phi^{-1}(\underline{\alpha} \vee^* \underline{\beta})] = \\ &= [A, \Phi^{-1}(\underline{\alpha}) \wedge \Phi^{-1}(\underline{\beta})] = \\ &= A \wedge (\Phi^{-1}(\underline{\alpha}) \wedge \Phi^{-1}(\underline{\beta})) = (A \wedge \Phi^{-1}(\underline{\alpha})) \wedge \Phi^{-1}(\underline{\beta}) = \\ &= [A \wedge \Phi^{-1}(\underline{\alpha})), \Phi^{-1}(\underline{\beta})] = \\ &= \langle A \wedge \Phi^{-1}(\underline{\alpha})), \underline{\beta} \rangle. \end{aligned}$$

6. GEOMETRIC CALCULUS, INVARIANT THEORY AND THE CAYLEY FACTORIZATION PROBLEM

A *geometric expression* in $(G(\mathcal{V}), \vee, \wedge)$ is an expression involving extensors, join and meet operations *but not additions*. Clearly a geometric expression in $(G(\mathcal{V}), \vee, \wedge)$ may be immediately translated into a projective geometry statement and viceversa. Furthermore, by expanding the meets by their *closed forms* (1) and (2), any geometric geometric expression in $(G(\mathcal{V}), \vee, \wedge)$ can be translated into a finite set of bracket polynomials, i. e. *invariant polynomial functions* (see, e. g. Weyl 1946). The situation can be pictured as follows:

I) PROJECTIVE GEOMETRY

$\updownarrow$

II) DOUBLE CG-ALGEBRA $(G(\mathcal{V}), \vee, \wedge)$

$\downarrow \uparrow$ (?)

III) BRACKET ALGEBRA (i. e. the algebra of invariants)

The so-called "Cayley factorization problem" (N. White 1991) can be phrased as follows: how can we do the inverse translation III) $\longrightarrow$ II)? Does it always exist?

This problem is crucial in computational geometry, and, more generally, in computer aided geometric reasoning. Usually, computations are efficiently done by symbolic manipulation in the algebra of invariants III), but the last step should be to obtain a geometric interpretation III) $\longrightarrow$ II) of the invariant conditions we got. For example, the infinitesimal rigidity of certain kind of frameworks (White/Whiteley 1983, 1987) and the realizability of certain convex polytopes (Peano 1888) have been characterized in terms of invariant polynomial functions: the Cayley factorization of these results would provide a direct geometric interpretation of them. At present time, the main results about the Cayley factorization problem are - in our opinion - the following ones.

(MULTILINEAR CASE: CONSTRUCTIVE RESULT) For every *multilinear* invariant, there is an effective algorithm to produce its Cayley factorization *if it exists* (that is, to produce an equivalent geometric expression in the CG-algebra) (White 1991).

(GENERAL CASE: EXISTENCE RESULT) *Every* bracket polynomial (with integer coefficients) equals a geometric expression in $(G(\mathcal{V}), \vee, \wedge)$, *up to a bracket monomial factor* (Sturmfels/Whiteley 1991). Informally speaking, it means that the Cayley factorization problem has always a solution, under suitable non-degeneracy conditions (that is every invariant - with integer coefficients - admits a geometric interpretation in the Grassmann algebra).

Dipartimento di Matematica, Universitá degli Studi di Bologna

REFERENCES

M. Barnabei, A. Brini, and G.-C. Rota, "On the exterior calculus of invariant theory," *Journal of Algebra,* 1985, *96:* 120-160.

J. Bokowsky and B. Sturmfels, "Polytopal and non-polytopal spheres: an algorithmic approach," *Israel J. Math.,* 1987, *57:* 257-271.

Nicolas Bourbaki, *Éléments de Mathématique, Algébre Multilinéaire* (Paris: Hermann, 1979).

Andrea Brini and A. G. B. Teolis, "Capelli's method of variabili ausiliarie, superalgebras and geometric calculus," Neil L. White (ed.), *Invariant Methods in Discrete and Computational Geometry* (Dordrecht: Kluwer, 1995), 59-75.

Jean Dieudonné, "The tragedy of Grassmann," *Linear and Multilinear Algebra*, 1979, *8*: 1-14.

P. Doubilet, G.-C. Rota, and J. Stein, "On the foundations of combinatorial theory IX: Combinatorial methods in invariant theory," *Stud. Appl. Math.*, 1974, *53:* 185-216.

Henry G. Forder, *The Calculus of Extension* (Cambridge: Cambridge University Press, 1940). Reprint 1960 by Chelsea Publ. Comp. NY.

F. Grosshans, G.-C. Rota, and J. Stein, "Invariant theory and superalgebras", *American Math. Society C. B. M. S. Regional Conference Series* **69** (1987).

David Hestenes, Grassmann's Vision, 1996, *in this volume* (1996).

David Hestenes and Renatus Ziegler, "Projective geometry with Clifford algebra," *Acta Applic. Math.*, 1991, *23:* 25-63.

M. Marcus, *Finite dimensional Multilinear Algebra*, 2 vols. (New York: Dekker, 1975).

Giuseppe Peano, *Calcolo Geometrico secondo la Ausdehnugslehre di H. Grassmann* (Fratelli Bocca Editori: Torino, 1888).

Günther Pickert, *Analytische Geometrie*, 4. Auflage (Leipzig: Akademische Verlagsgesellschaft Geest Portig, 1961).

B. Sturmfels and W. Whiteley, "On the synthetic factorization of projectively invariant polynomials," *J. Symbolic Computation*, 1991, *11:* 439-453.

B. Segre, *Lezioni sulla teoria delle forme differenziali* (Roma: Docet Editrice Universitaria, 1951).

M. E. Sweedler, *Hopf Algebras* (New York: Benjamin, 1969).

Hermann Weyl, *The Classical Groups* (Princeton: Princeton Univ. Press, 1946).

Neil L. White, "Multilinear Cayley factorization," *J. Symbolic Computation*, 1991, *11:* 421-438.

Neil L. White and W. Whiteley, "The algebraic geometry of stress in frameworks," *SIAM J. Algebraic and discrete methods*, 1983, *4:* 481-511.

Neil L. White and W. Whiteley, "The algebraic geometry of motions of bar-and-body frameworks," *SIAM J. Algebraic and discrete methods*, 1987, *8*: 1-32.

Alfred N. Whitehead, *A Treatise on Universal Algebra with Applications* (Cambridge: Cambridge Univ. Press, 1898).

Arno Zaddach, *Grassmann Algebra in der Geometrie* (Mannheim-Leipzig-Wien-Zürich: B. I. Wissenschaftsverlag, 1994).

Arno Zaddach, "Regressive products and Bourbaki," in this volume (1996).

DAVID HESTENES

GRASSMANN'S VISION

After nearly a century on the brink of obscurity, Hermann Grassmann is widely recognized as the originator of *Grassmann algebra*, an indispensable tool in modern mathematics. Still, in conception and applications, conventional renditions of his *exterior algebra* fall far short of Grassmann's original vision. A fuller realization of his vision is found in other mathematical developments to which his name is not ordinarily attached. This Sesquicentennial Celebration of Grassmann's great book, *Die Lineale Ausdehnungslehre* (Grassmann 1995a), provides the opportunity for a renewed articulation and assessment of Grassmann's vision.

Grassmann is a pivotal figure in the historical evolution of a mathematical language to characterize human understanding of the physical world. Since quantitative concepts of space and time are fundamental to that understanding, the language is fundamentally geometrical and can best be described as a *geometric calculus*. For the most part the evolution of geometric calculus has been tacit and piecemeal, with many individuals achieving isolated results in response to isolated problems. Grassmann is pivotal in this historical process because he made it explicit and programmatic. In no uncertain terms, he declared the goal of his research on *extension theory* (Grassmann 1995a) as no less than the creation of a *universal instrument for geometric research*. Leibniz had clearly articulated that goal long before and blessed it with the name "geometric calculus." However, Grassmann was the first to see clearly how the goal can be reached, and he devoted the whole of his mathematical life to the journey. When Möbius created the opportunity for Grassmann to align his own vision with Leibniz's, Grassmann responded eagerly with an elaboration of his extension theory to show how their common vision can be realized (Grassmann 1995b). Unfortunately, this seminal work failed to attract the recognition and attention he deserved, so Grassmann had to continue his mathematical journey alone, with only an occasional mathematician — Möbius, Hankel, Clebsch — joining him for a few steps along the way.

Most of Grassmann's mathematical discoveries were duplicated independently by others as isolated results. It is fair to say, for example,

G. Schubring (ed.), Hermann Günther Graßmann (1809–1877): Visionary Mathematician, Scientist and Neohumanist Scholar, 243–254.

that Grassmann laid the foundation for linear algebra by himself (Fearnley-Sander 1979). Yet the standard linear algebra of today grew up without his contribution. The frequent duplication of Grassmann's discoveries is not a mark of limited originality but rather a sign that Grassmann was keenly attuned to a powerful thematic force driving mathematical development, namely, the subtle interplay between geometry and algebra. What sets Grassmann ahead of other creative mathematicians is his systemic vision of a universal geometric calculus. This vision marks him as one of the great conceptual synthesizers of all time. Accordingly, the chief presentation of that vision, *Die Lineale Ausdehnungslehre*, deserves a place alongside Euclid's *Elements* and Descartes' *Geometrié* in the library of immortal mathematical masterworks.

Considering the immense power and fertility of Grassmann's vision, we are bound to ask why due recognition has been so long in coming. Several answers have been suggested by commentators and historians. First, it is averred that the original *Ausdehnungslehre* was written in an impenetrable philosophical style. But Grassmann confined his philosophical musings to the first chapter so they could be skipped by those who want to get directly to business. Second, it is suggested that Grassmann was ahead of his time. He was certainly the first person to deal with abstract, multidimensional algebra. But if that was a barrier to understanding during his lifetime, it should have been lesser in subsequent generations. Third, it is noted that Grassmann's unsystematic presentation mixes basic ideas with applications and so places a heavy burden on the reader to disentangle them. But Grassmann's synopsis of the *Ausdehnungslehre* (Grassmann 1995c), written at the behest of a befuddled admirer, did not help much to ease the burden. Moreover, more systematic presentations of Grassmann's calculus by distinguished mathematicians, most notably Peano in 1888 and Whitehead in 1898, also failed to inspire the mathematical community at large.

Grassmann himself identified a more critical condition for comprehending his vision, namely *immersion* in his conceptual system. That system is, after all, a rich mathematical language, so it takes years to develop the proficiency that opens up mathematical insights. Having acquired such proficiency, the superiority of his system in applications throughout geometry and physics was obvious to Grassmann, so it was a great frustration to him that others, even distinguished mathematicians, were unable to recognize this. It was easy for experts to dismiss Grassmann's work as "old wine in new bottles" when he was unable to show them applications that could not be handled with techniques they already knew. Grassmann thus suffered from a credibility gap that prevented experts from investing the time necessary to acquire his perspective.

There is another, more profound reason for the muted impact of Grassmann's extension theory. Something critical was missing. As Grassmann admitted in the preface to his *Ausdehnungslehre* of 1862:[1]

"I am aware that the form which I have given the science is imperfect." And he went on to say "there will come a time when these ideas, perhaps in a new form, will arise anew and will enter into a living communication with contemporary developments."

The purpose of this paper is to point out that the missing ingredient in Grassmann's theory has been found and put in place to create a geometric calculus that fully realizes his vision. Indeed, the seeds for this advance were already present in Grassmann's work. Moreover, Grassmann's bold prophecy is in the process of fulfillment even today, as his insights are revived to enrich modern mathematics and physics.

I. COMPLETING THE VISION

In the Foreword to his *Ausdehnungslehre* of 1844, Grassmann outlined some ideas to be incorporated in a second volume to "complete the work." This included the invention of a new kind of product, the essentials of which can be described as follows. He defined the *quotient* $\mathbf{a/b} = \mathbf{ab}^{-1}$ of vectors **a** and **b** by writing

$$(\mathbf{a} / \mathbf{b})\mathbf{b} = \mathbf{a} . \tag{1}$$

He noted that, if **a** and **b** have the same magnitude, then (1) implies that **a/b** is an operator which rotates **b** into **a**. He expressed this with the exponential form

$$e^{\angle \mathbf{ab}} = \mathbf{a} / \mathbf{b}, \tag{2}$$

where $\angle \mathbf{ab}$ denotes the angle between **a** and **b**. It follows then that $(\mathbf{a/b})^2$ is a rotation through twice the angle. In particular, if **a/b** is a right angle rotation, then $(\mathbf{a/b})^2\mathbf{b} = -\mathbf{b}$, so $(\mathbf{a/b}) = -1$ and

$$\mathbf{a/b} = \sqrt{-1} . \tag{3}$$

If we write $\angle \mathbf{ab} = \mathbf{i}x$, where $\mathbf{i} = \sqrt{-1}$ and x is the radian measure of the angle, then

$$e^{\mathbf{i}x} = \cos x + \mathbf{i} \sin x \tag{4}$$

has a purely geometrical significance denoting the rotation through an arbitrary angle.

---

1 Translation by M. Crowe in (Crowe 1967)

Grassmann concludes:

"From this, all imaginary expressions now acquire a purely geometric meaning and can be described by geometric constructions. . . . it is likewise now evident how, according to the meaning of the imaginaries thus discovered, one can derive the laws of analysis in the plane; however, it is not possible to derive the laws for space as well by means of imaginaries. In addition there are general difficulties in considering the angle in space, for the solution of which I have not yet had sufficient leisure." (Grassmann 1995a, p. 15)

That is the last we hear about this interpretation of the imaginary unit from Grassmann. He never revived and completed the argument to handle general rotations in space and develop his new product into a complete algebraic system. However, a brilliant young English mathematician, William Kingdom Clifford, did just that. The resulting new algebraic system is known as *Clifford algebra* in mathematical circles today.

Ironically, Clifford is seldom mentioned in accounts of Grassmann's influence on other mathematicians, though it may be through Clifford algebra that Grassmann's ideas exert their most profound influence today. We have no direct knowledge that Clifford was influenced by Grassmann's little argument above, but it could hardly be otherwise, because the argument appeared right at the beginning and nothing else in Grassmann's corpus is so obviously pertinent. Indeed, Clifford made no great claim to originality, referring to his algebra as a mere application of Grassmann's extensive algebra (Clifford 1878). Unfortunately, his life was to brief for him to communicate the dimensions of his debt to Grassmann.

Clifford was surely assisted in his adaptation of Grassmann's ideas by his mastery of Hamilton's quaternions. Ironically again, Grassmann was led along the same path in the last years of his life by his endeavor to fit quaternions into extension theory (Grassmann 1995c, 1995d). He was induced to define the *central product* **ab** of vectors **a** and **b** by writing

$$\mathbf{ab} = \lambda[\mathbf{a}|\mathbf{b}] + \mu[\mathbf{ab}], \tag{5}$$

where $[\mathbf{a}|\mathbf{b}]$ is the *inner product*, $[\mathbf{ab}]$ is the *outer product*, and λ, μ are arbitrary non zero constants. With $\lambda = \mu = 1$, we have the alternative notation

$$\mathbf{ab} = \mathbf{a}\cdot\mathbf{b} + \mathbf{a}\wedge\mathbf{b}. \tag{6}$$

This is essentially the basic product of Clifford algebra, and it was published before Clifford's paper (Clifford 1878). Thus, Grassmann could be credited with the invention of Clifford algebra. However, he failed to investigate the central product with the verve that he bestowed on the inner and outer products in bygone years. Had he done so, he would have been pleasantly surprised to find that quaternions drop out without resorting to the deformation of the central product (5) that he employed in his paper (Grassmann 1995d).

Grassmann also failed to notice that the product in (6) can be identified with the quotient in (2). Then, for $\mathbf{a}^2 = \mathbf{b}^2 = 1$, we can write (2) in the form

$$\mathbf{ab} = e^{\mathbf{i}x}. \tag{7}$$

Using (2) and (6) to expand the right and left sides of (7), we infer that $\mathbf{a}\cdot\mathbf{b} = \cos x$ and, more important,

$$\mathbf{a}\wedge\mathbf{b} = \mathbf{i}\sin x. \tag{8}$$

This tells us immediately that the *unit imaginary* **i** must be interpreted as the unit bivector for the plane containing **a** and **b**, something that Grassmann never realized.

There is another easy inference that would have pleased Grassmann mightily, because it reveals a hidden unity and simplicity in his system which he had not anticipated. Using the symmetry of the inner product and the skew symmetry of the outer product, we infer from (6) that

$$\mathbf{ba} = \mathbf{a}\cdot\mathbf{b} - \mathbf{a}\wedge\mathbf{b}. \tag{9}$$

Then, from (6) and (9) we can deduce

$$\mathbf{a}\cdot\mathbf{b} = \frac{1}{2}(\mathbf{ab}+\mathbf{ba}) \tag{10}$$

and

$$\mathbf{a}\wedge\mathbf{b} = \frac{1}{2}(\mathbf{ab}-\mathbf{ba}) \tag{11}$$

This shows that the inner and outer products can be defined in terms of the central product, thus reducing three products to one. Full details of this reduction and simplification of extension theory are given in (Hestenes 1985).

The insight that all multiplicative aspects of extension theory can be reduced to the single central product leads to the conclusion that Clifford algebra is the proper completion of Grassmann's vision for a universal geometric calculus. Ample justification for this conclusion comes from the manifold applications of Clifford algebra mentioned below. Recognizing the geometric origin of his algebra, Clifford called it *geometric algebra.* Considering the universality of this algebra and the fact that so many others besides Grassmann and Clifford have contributed to its development as a mathematical language, it should be regarded as common intellectual property of the mathematical community, so it should not be tagged with the name of a single individual. As no other mathematical system deserves the title more, Clifford's choice should be adopted and, on occasion, expanded to *universal geometric algebra* for emphasis. For the same reasons, the name *geometric product* is to be preferred in place of Grassmann's "central product."

As counterpoint to these unifying developments, we note a final irony in Grassmann's treatment of quaternions in his dismissal of Hamilton's ideas without due credit, just as his own ideas had been summarily dismissed by other mathematicians before. It is likely that Grassmann did not have the opportunity to study Hamilton's work first hand, so he learned about it from secondary sources. That only deepens the irony in his refusal to admit that Hamilton has something to teach him. He could, at least, have graciously acknowledged that the crucial idea of adding a scalar to a bivector in (5) came from Hamilton, for he himself had never taken this step before. Indeed, he had explicitly warned against such mixing of quantities of different kind.

II. SOURCE OF THE VISION

The incredible power and versatility of Grassmann's approach derives from the fact that he systematically incorporated geometric interpretations into the design of his algebraic syntax; then he abstracted the algebraic structure from the geometric interpretation for analysis as a formal algebraic system.

Grassmann introduced two complementary geometric interpretations for vectors and their outer products. Let us refer to them as the *direct* and the *projective* interpretations to keep them distinct. One difficulty in understanding Grassmann's work is the informal fluency with which he moves from one interpretation to the other. His synopsis (Grassmann 1995c) is helpful by bringing the two interpretations together and comparing them. Recently, it has been realized that *geometric algebra* provides a surprising formal connection between the interpretations - as given in Hestenes 1988,[2] Hestenes 1991a, Hestenes/Ziegler 1991 - and thus clarifies their complementary roles to see that we need to review Grassmann's perspective.

We begin with the modern concept of a *vector* as an abstract element of a vector space. The concept of a vector space had not been formalized in Grassmann's time, though it was certainly implicit in his work. To coordinate the two geometric interpretations, we assign them to vectors in different spaces which are distinguished by using *boldface* to denote

---

2 This is a synopsis of Hestenes/Ziegler 1991 and Hestenes 1991a.

vectors with the direct interpretation and *italics* for vectors with a projective interpretation.

Under the *direct interpretation* each vector **a** represents a directed line segment, and the outer product $\mathbf{a} \wedge \mathbf{b}$ represents a directed plane segment. This interpretation is thoroughly discussed and applied to classical mechanics in ref. (Hestenes 1985) using the full geometric algebra.

Under the *projective interpretation* each vector a represents a geometric point. The outer product $a \wedge b$ represents a directed line segment determined by points a and b, while $a \wedge b \wedge c$ represents a directed plane segment determined by three points. This interpretation is thoroughly discussed and applied to projective geometry in (Hestenes/Ziegler 1991). which also explains the advantages of the full geometric algebra in this domain.

The two different interpretations correspond to two different representations of geometric concepts. The relation between the two is elegantly characterized by a *projective split* in geometric algebra, as explained in (Hestenes 1991a) and also in (Hestenes 1988). The projective split relates each vector **a** in an n-dimensional vector space $\mathcal{V}_n$ to a vector a in an $(n+1)$-dimensional vector space $\mathcal{V}_{n+1}$ by the equation

$$a e_0 = a \cdot e_0 + a \wedge e_0 = a_0 (1 + \mathbf{a}), \tag{12a}$$

where e_0 is a distinguished vector in $\mathcal{V}_{n+1}$, and

$$a_0 = a \cdot e_0, \tag{12b}$$

$$\mathbf{a} = \frac{a \wedge e_0}{a \cdot e_0} \tag{12c}$$

According to (12c), vectors in $\mathcal{V}_n$ correspond to bivectors of $\mathcal{V}_{n+1}$ with a common factor e_0. This is a projective relation, because (12c) shows that **a** is unaffected if a is replaced by a non zero scalar multiple of a. The projective split amounts to a mechanism for introducing *homogeneous coordinates.* It is more powerful than the conventional approach to homogeneous coordinates, however, because it prepares the way for drawing on the computational and conceptual advantages of geometric algebra. Since there is insufficient space here to demonstrate those advantages, the reader is left to consult the references.

Grassmann made much ado about separating *extension theory* from *space theory*, in other words, separating formal algebraic structures from geometric interpretation. He was ahead of his time in regarding geometry (space theory) as an empirical discipline derived from human perception and therefore limited to spaces of no more than three dimensions. Extension theory formalizes space theory, and Grassmann emphasized that geometric interpretation is essential to its applications in physics and geometry. On the other hand, he argues that separation from space theory enables the cultivation of extension theory as an abstract science and admits generalization to higher dimensions. He could as well have added that geometric interpretations can also be generalized to give insight to algebraic structures in higher dimension.

One clear benefit of separating geometric algebra from geometric interpretation is that a single algebraic entity can be assigned many different geometric interpretations. We have already seen, for example, that vectors can be assigned either direct or projective interpretations, and this does not exhaust the possibilities. However, abstraction has its dangers, as the following important historical example serves to show.

After Clifford algebra was invented by Clifford, it was cultivated as an abstract discipline for the better part of a century without reference to its geometrical roots. Clifford algebra thus became a minor mathematical subspecialty — just one more example of an algebra among many others. In regard to applications, it was almost sterile, with representations of the orthogonal group as the only example of much interest.

The situation has changed drastically since 1966 when, for the purpose of application to physics, Grassmann's direct interpretation was incorporated into Clifford algebra and a variant of the projective split (now called the *spacetime split*) was found to simplify an clarify relativistic physics (Hestenes 1966). This has reinvigorated Clifford algebra and fueled a growing realization that it provides the structural basis for a truly universal geometric algebra.

III. REALIZING THE VISION

Realizing Grassmann's vision of a universal geometric calculus to serve as a unified language for mathematics and physics is necessarily a community enterprise. Though Grassmann and Clifford have supplied the language with a suitable syntax, enrichment of the language with a broad spectrum of applications is a vast undertaking drawing on the intellectual resources of the entire scientific community. Grassmann understood this perfectly, so he devoted considerable effort to developing applications in geometry and physics, primarily mechanics and electrodynamics.

The main concern of Grassmann's scientific papers is the mathematical form of physical equations. Much of his work on mechanics is standard fare in textbooks today, with only minor differences in notation, though direct influence of Grassmann on textbooks is unlikely. The only mechanics problem of any complexity that Grassmann addressed is the "theory of the tides." While that work has no special scientific importance, a reformulation of its vectorial perturbation approach to tidal theory in modern terms may have pedagogical value. The main deficiency in Grassmann's approach to tidal theory and the rest of mechanics was in his mathematical treatment of rotations and rotational dynamics. Though Grassmann attacked this problem with great ingenuity, the ideal solution escaped him, because it required the geometric product. Grassmann's grand vision for mechanics has recently been utterly fulfilled with a complete reformulation of the subject in terms of geometric algebra (Hestenes 1985). As Grassmann desired, the treatment is completely coordinate-free, it employs his direct geometric interpretation, and it incorporates all of his keen algebraic devices.

At this point I must confess my personal interest and apologize for the extensive self-citations in a tribute to someone else. The fact is that most of my professional life has been devoted to the same grand theme as Grassmann's, though I came to realize this only recently. My work could fairly be described as restoring Grassmann's original insights to Clifford algebra and systematically developing it into a universal language through extensive applications in mathematics and physics. I was guided by Grassmann's ideas from the beginning and referred to them frequently in my published work. However, everything I knew about Grassmann came from secondary sources, so it suffered from the limitations and distortions of such filtering. It was only in the last few years that I learned about the incredible confluence of my own vision with Grassmann's from the translation of Grassmann's work by Lloyd Kannenberg. I was amused to note that I had independently rederived practically every algebraic identity in Grassmann's work (especially in (Hestenes/Sobczyk 1984). This illustrates how a shared vision can stimulate parallel discoveries. That occurs frequently in mathematics, and, as I have emphasized before (Hestenes 1986a), it is especially significant in the history of geometric algebra because of its universal importance.

In my own work on geometric calculus, I have enjoyed the huge advantage over Grassmann of more than a century of developments in mathematics and physics. Even so, I find little in Grassmann's mathematical work that is outdated, and I am confident in asserting that Grassmann's optimistic vision of the future for geometric calculus has been fully vindicated. The references cited herein demonstrate in detail that geometric calculus embraces a greater range of mathematics than any other mathematical system, including linear and multilinear algebra, projective geometry, distance geometry (Dress/Havel 1993), calculus on manifolds (Hestenes 1993a), hypercomplex function theory, differential geometry, Lie groups and Lie algebras (Doran et al. 1993).

Geometric calculus has also fulfilled Grassmann's vision of a universal language for physics. Besides integrating the mathematical formulations of mechanics (Hestenes 1985, 1993b), relativity and electrodynamics (Hestenes 1966), it has revealed a geometric basis for complex numbers in quantum mechanics (Hestenes 1986b, 1993c).

Leading into the twentieth century, Grassmann's algebra was often cast as a competitor with quaternions and vector calculus for the role of a language for physics. Vector calculus was victorious, owing principally to Oliver Heaviside's slick formulation of Maxwell's equations with applications to electromagnetic wave propagation. The victory was only temporary, however. With the completion of Grassmann's vision by geometric calculus, we can now see Heaviside's vectorial equations and Hamilton's quaternions as components of a single, more powerful system. Indeed, the geometric calculus reduces the four Maxwell equations of Heaviside to the single equation (Hestenes 1966, 1986a)

$$\partial F = J\,, \tag{13}$$

where J is the spacetime charge current density, F is the complete electromagnetic field (without separation into electric and magnetic parts), and ∂ is the vector derivative with respect to a spacetime point. With the choice of an inertial reference frame, (13) can be split into the four equations of Heaviside (Hestenes 1966). However, in many applications it is simpler and more informative to solve (13) directly. Equation (13) is typical of the simplifications that geometric calculus brings to every branch of physics. This justifies its claim to be a universal language for physics.

The fusion of Grassmann algebra with Clifford algebra has been the subject of some debate (Hestenes 1991b). The chief point of contention has been the concept of *duality*. Dieudonné summarizes the dominant "modern view" in (Dieudonné 1979) and declares it superior to Grassmann's. In fact, they are just different, and Grassmann's has distinct advantages which are made more evident by using the full geometric algebra (Hestenes/Ziegler 1991). On the other hand, the modern view has unnecessary limitations which can be eliminated with geometric algebra. It works with a "dual pair" of vector spaces, each with its own exterior algebra. As shown in (Doran et al. 1993), a far more powerful algebraic structure is created at no cost by joining the dual pair into a single vector space with twice the dimension of each and then incorporating the duality relations into the geometric algebra of that space. The remarkable result is that *every* linear transformation on the base space can be represented in this algebra as a monomial product of vectors. This promises to open a new chapter in linear algebra.

IV. RECOGNIZING THE VISION

An adequate history of geometric calculus remains to be written. It calls for a historian steeped in Grassmann's vision who can recognize how the vision is played out in the various branches of mathematics and physics. We have seen how easy it is to misconstrue or overlook important historical threads. Even the central theme — one of the great themes of cultural history — has been consistently overlooked by mathematicians and historians. The time is ripe for a historian with the insight and dedication to flesh out the whole story.

Department of Physics and Astronomy, Arizona State University

REFERENCES

William K. Clifford, "Applications of Grassmann's Extensive Algebra," *American Journal of Mathematics* 1878, *1:* 350-358.

Michael Crowe, *A History of Vector Analysis* (Notre Dame: University of Notre Dame Press, 1967).

Jean Dieudonné, "The Tragedy of Grassmann," *Linear and Multilinear Algebra,* 1979, *8:* 1-14.

C. Doran, D. Hestenes, F. Sommen & N. Van Acker, "Lie Groups as Spin Groups," *Journal of Mathematical Physics,* 1993, *34:* 3642-3669.

Andreas Dress & Tim. Havel, "Distance Geometry and Geometric Algebra," *Foundations of Physics,* 1993, *23:* 1357-1374.

D. Fearnley-Sander, "Hermann Grassmann and the Creation of Linear Algebra," *American Mathematical Monthly*, 1979, *86*: 809-817 .

Hermann G. Grassmann, "Linear Extension Theory" (*Die Lineale Ausdehnungslehre*), translated by L. C. Kannenberg in *The "Ausdehnungslehre" of 1844 and Other Works* (Chicago, La Salle: Open Court Publ., 1995). [1995a]

Hermann G. Grassmann, "Geometric Analysis," translated by L. C. Kannenberg in *The "Ausdehnungslehre" of 1844 and Other Works* (Chicago, La Salle: Open Court Publ., 1995). [1995b]

Hermann G. Grassmann, "Survey of the Essentials of Extension Theory," translated by L. C. Kannenberg in *The "Ausdehnungslehre" of 1844 and Other Works* (Chicago, La Salle: Open Court Publ., 1995). [1995c]

Hermann G. Grassmann, "The Position of Hamiltonian Quaternions in Extension Theory," translated by L. C. Kannenberg in *The "Ausdehnungslehre" of 1844 and Other Works* (Chicago, La Salle: Open Court Publ., 1995). [1995d]

S. Gull, A. Lasenby & C. Doran, "Imaginary Numbers are not Real – The Geometric Algebra of Spacetime," *Foundations of Physics,* 1993, *23:* 1175-1201.

David Hestenes, *Space-Time Algebra* (New York: Gordon & Breach, 1966).

David Hestenes, *New Foundations for Classical Mechanics* (Dordrecht/Boston: G. Reidel Publ. Co., 1985. Third printing 1993).

David Hestenes, "A Unified Language for Mathematics and Physics," *Clifford Algebras and their Applications in Mathematical Physics*, eds. J.S.R. Chisholm/A.K. Common, (Dordrecht/Boston: Reidel, 1986), p. 1-23. [1986a]

David Hestenes, "Clifford Algebra and the Interpretation of Quantum Mechanics," *Clifford Algebras and their Applications in Mathematical Physics,* eds. J.S.R. Chisholm/A.K. Common (Dordrecht/Boston: Reidel, 1986), p. 321-346. [1986b]

David Hestenes, "Universal Geometric Algebra," *Simon Stevin,* 1988, *62:* 253-274.

David Hestenes, "The Design of Linear Algebra and Geometry," *Acta Applicandae Mathematicae,* 1991, *23:* 65-93. [1991a]

David Hestenes, "Mathematical Viruses," *Clifford Algebras and their Applications in Mathematical Physics*, eds. A. Micali, R. Boudet, J. Helmstetter (Dordrecht/Boston: Kluwer, 1991), 3-16. [1991b]

David Hestenes, "Differential Forms in Geometric Calculus." *Clifford Algebras and their Applications in Mathematical Physics*, eds. F. Brackx *et al.* (Dordrecht/Boston: Kluwer, 1993), p. 269-285. [1993a]

David Hestenes, "Hamiltonian Mechanics with Geometric Calculus," *Spinors, Twistors and Clifford Algebras*, eds. Z. Oziewicz, A. Borowiec & B. Jancewicz, (Dordrecht/Boston: Kluwer, 1993), p. 203-214. [1993b]

David Hestenes, "The Kinematic Origin of Complex Wave Functions," *Physics and Probability: Essays in Honor of Edwin T. Jaynes*, eds. W. T. Grandy & P. W. Milloni (Cambridge: Cambridge U. Press, 1993), p. 153-160. [1993c]

David Hestenes and G. Sobczyk, *Clifford Algebra to Geometric Calculus, A Unified Language for Mathematics and Physics* (Dordrecht/Boston: G. Reidel Publ. Co., 1984). Third printing 1992.

David Hestenes and Renatus Ziegler, "Projective Geometry with Clifford Algebra," *Acta Applicandae Mathematicae,* 1991, *23,* 25-63.

F. WILLIAM LAWVERE

GRASSMANN'S DIALECTICS AND CATEGORY THEORY

PROGRAMMATIC OUTLINE

In several key connections in his foundations of geometrical algebra, Grassmann makes significant use of the dialectical philosophy of 150 years ago. Now, after fifty years of development of category theory as a means for making explicit some nontrivial general arguments in geometry, it is possible to recover some of Grassmann's insights and to express these in ways comprehensible to the modern geometer. For example, the category $\mathcal{A}$ of affine-linear spaces and maps (a monument to Grassmann) has a canonical adjoint functor to the category of (anti)commutative graded algebras, which as in Grassmann's detailed description yields a sixteen-dimensional algebra when applied to a three-dimensional affine space (unlike the eight-dimensional exterior algebra of a three-dimensional vector space). The natural algebraic structure of these algebras includes a boundary operator ∂ which satisfies the (signed) Leibniz rule; for example, if A, B are points of the affine space then the product AB is the axial vector from A to B which the boundary degrades to the corresponding translation vector: $\partial(AB) = B-A$ (since $\partial A = \partial B = 1$ for points). Grassmann philosophically motivated a notion of a "simple law of change," but his editors in the 1890's found this notion incoherent and decided he must have meant mere translations. However, translations are insufficient for the foundational task of deciding when two formal products are geometrically equal axial vectors. But if "law of change" is understood as an action of the additive monoid of time, "simple" turns out to mean that the action is internal to the category $\mathcal{A}$ at hand, in the following sense: The affine category has a unique closed structure, consisting of an internal hom functor with an adjoint tensor-product functor which neither is the cartesian product nor underlies the usual tensor product of vector spaces, and since addition of times does define an internal monoid structure $R \otimes R \to R$ with respect to this tensor product, it is very natural to interpret "internal law of becoming" on a space E to mean an action (= "flow") $R \otimes E \to E$ with respect to this tensor. Such actions

G. Schubring (ed.), Hermann Günther Graßmann (1809–1877): Visionary Mathematician, Scientist and Neohumanist Scholar, 255–264.

turn out to be determined by shear transformations, of which there are indeed enough to detect equality of axial vectors!

ORIENTATION

Grassmann in his philosophical introduction describes the two-fold division of formal sciences, that is, the science of thinking, into dialectics and mathematics. He briefly describes dialectics as seeking the unity in all things, and he describes mathematics as the art and practice of taking each thought in its particularity and pursuing it to the end. There is a need for an instrument which will guide students to follow in a unified way both of these activities, passing from the general to the particular and from the particular to the general.

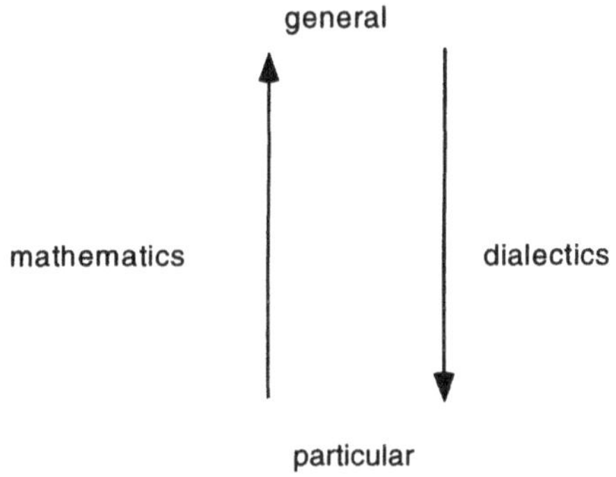

I believe that the theory of mathematical categories (which was made explicit 50 years ago by Eilenberg and Mac Lane, codifying extensive work done by Hurewicz in particular during the 1930's), can serve as such an instrument. It was introduced and designed in response to a very particular question involving passage to the limit in calculating cohomology of certain portions of spheres, but this particular calculation necessitated an explicit recognition of the manner in which these spaces were related to all other spaces and, in particular, how their motion might induce other motions. In other words, category theory was introduced (and still serves) as "a universal geometrical calculus."

Looking more closely into Grassmann, Stephen Schanuel and I found numerous ways in which it could be justified to say that Grassmann was a pre-cursor of category theory. The general algebraic operations which he discussed have become the explicit object of a particular developed theory, and those general concepts, general operations, systems of operations and systems of equations in invariant coordinate free form have been made into a part of category theory. More specifically, we find that in certain cases the famous distinction between analytic and synthetic operations can only be explained in terms of adjoint functors.

This is because, although in simple cases the analytic operation may be the actual inverse of the synthetic one, there are a great many cases (arising in deductive logic and finance and so forth) where there are clearly pairs of operations having this analytic/synthetic relation, but the precise mathematical form of their relationship is not that of inverse, but rather of adjointness, in the sense of category theory. I hope to illustrate the use of category theory in making explicit certain of Grassmann's ideas, which have been misunderstood or neglected, with a few examples. I have written several other papers on the general principles of dialectics and how these can be formalized using category theory, so I think it is appropriate here to describe two particular algebraic examples worked out with Schanuel.

GRASSMANN FUNCTOR MORE STRUCTURED THAN EXTERIOR ALGEBRA

If as modern geometers we look into Grassmann's *Ausdehnungslehre,* we see on page 305 that the Grassmann algebra built on three-dimensional space has itself dimension 16, whereas in all the modern text books, the exterior algebra of a three-dimensional space has dimension 8. This distinction is by no means trivial; in fact a large part of the geometrical content of Grassmann's algebra, as he explains it in detail in the book, can be traced back to this opposition which stems from the opposition between points and vectors. As explained in most modern text books, the space of points has no preferred origin and the space of vectors lives principally because it acts on points; a special property of this action is that with two points there is a unique vector which is their difference and therefore the same translation motion by the vector can be applied to a new starting point.

The Grassmann algebra is the anti-commutative graded algebra which is universally associated to the space. There is the category of "commutative" (which is the modern word for anti-commutative) graded algebras and homomorphisms thereof. In particular, there is the possibility of taking the grade 1 portion of any such algebra which we can consider as an affine space of points; moreover, any homomorphism which preserves the grading and the addition and multiplication between these graded algebras will induce an affine linear map between the grade 1 portions. Now, the algebra Grassmann considered, (and discussed in detail the nature of grade 0, grade 1, grade 2, grade 3, grade 4) is uniquely characterized as the construction which is left adjoint to this forgetting process. This is one of the qualitative improvements of category theory over previous formulations, that a process can have a uniquely determined reverse attempt, which is however not actually an inverse.

Thus the Grassmann construction is uniquely determined by the fact that it assigns to any affine space of points a commutative graded algebra with the original points included into the grade 1 portion of the algebra in a way which is universal with respect to homomorphisms of graded algebras.

⊣ means "left adjoint of"

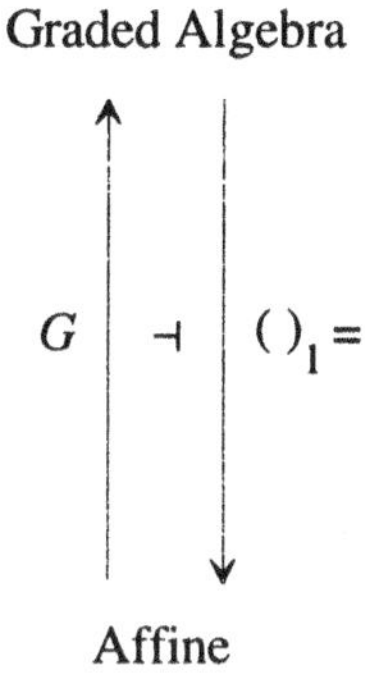

$(\ )_1$ = forgetful functor taking the grade 1 portion of any graded algebra

We can explain this construction with the help of the standard 20th century constructions by noting that this forgetful (throwing away the multiplication and higher grades) functor can actually be factored into two parts, namely the grade 1 portion of an algebra is a vector space and then we can forget the origin. If we take a vector space and forget the origin, what's left is an affine space. Now, it's a general principle of adjoints (proved by Kan in 1958) that the adjoint of a composite is the composite of the adjoints. So the Grassmann adjoint G is actually the composite of two left adjoints: the second is the exterior algebra of a vector space, as described in the text books; in particular, it assigns to a vector space of dimension n a graded commutative algebra of dimension 2^n; for example, to a three-dimensional vector space it associates an 8-dimensional algebra, and to a four-dimensional vector space it associates a 16-dimensional algebra etc.

$\wedge$ = modern exterior algebra

$G = \wedge \circ G_1$

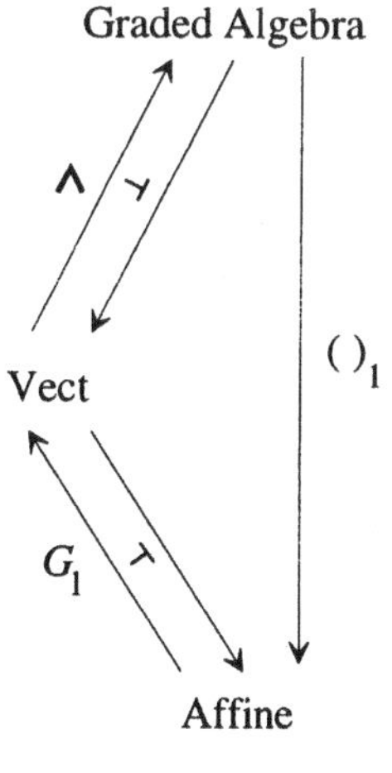

The process of forgetting the origin is not trivial, because the left adjoint to it is the following construction: You have a flat space (another word for affine is "flat") and you must adjoin in a universal way an origin. That origin cannot be in the original space, it must be adjoined outside, so the vector space generated by the original affine space and this new origin will always be of dimension 1 more than the dimension of the affine space. For example, the 3-dimensional affine or flat space by this composite process has at first associated to it a 4-dimensional vector space; so if we then take the exterior algebra in the modern sense of that 4-dimensional vector space, we will obtain the 16-dimensional Grassmann algebra, the one that he describes step by step in his book.

$$\begin{array}{ccccccccc} & & E & & & & & & \\ & & \downarrow & & & & & & \\ R=G_0 & \xleftarrow{\partial} & G_1 & \xleftarrow{\partial} & G_2 & \xleftarrow{\partial} & G_3 & \xleftarrow{\partial} & G_4 \\ & & \uparrow & & & & & & \\ & & V & & & & & & \end{array}$$

$E = \{x \in G_1 \mid \partial x = 1\}$

$V = \{x \in G_1 \mid \partial x = 0\}$

It has a grade 0 portion which is just the scalar field R ; it has as grade 1 portion the 4-dimensional vector space G_1 generated by the 3-dimensional affine space E , a 6-dimensional grade 2 portion and at the end a 1-dimensional grade 4 portion. An extremely important feature of this construction is that the Grassmann algebra has much more *natural structure* than the exterior algebra. The natural structure (in the technical sense of my 1963 doctoral thesis) of any functor consists of all natural operations, where a natural operation is an assignment to every value of the functor of an operation which commutes with all the morphisms which are values of the functor. In this case the natural structure includes a "boundary" operator ∂ which decreases grade by one and commutes with all the maps induced by an arbitrary affine–linear map α .

$$\begin{array}{ccccc} E & & G_k(E) & \xleftarrow{\partial_E} & G_{k+1}(E) \\ \Big\downarrow \alpha & & \Big\downarrow G_k(\alpha) & & \Big\downarrow G_{k+1}(\alpha) \\ F & & G_k(F) & \xleftarrow{\partial_F} & G_{k+1}(F) \end{array}$$

Thus we can consider the richer category of *differential* graded algebras and we find that the Grassmann functor actually lifts up along the forgetful functor (from this richer category back to the graded algebras). The axioms satisfied by the boundary operator in each differential graded algebra are that it is linear and satisfies the graded (signed) Leibniz rule with respect to the product, while of course reducing the grade by 1, and has square zero. By contrast, there is no such natural boundary operator on the exterior algebra of a vector space. Again we can explain this by modern constructions. The specific construction of the first step can be understood in the following way: The $G_1(E)$ is the 4-dimensional vector space generated by the 3-dimensional affine space E. There is the unique affine map from any affine space E to the one point affine space and yet $G_1(1) = R$. Hence, by the functoriality of G_1 there is a canonical linear map from $G_1(E)$ to $G_1(1) = R$; in fact, this map is just the lowest-grade instance of the boundary operator. If the elements of $G_1(E)$ are thought of as weighted points, then this linear functional simply specifies the weight value. The kernel of this linear functional is a 3-dimensional vector space which is actually the space of translations of E acting as follows: The part of $G_1(E)$ where this functional takes the value 1 is just the original affine space E , and obviously the addition of an element of value 0 takes elements of value 1 into elements of value 1 by linearity. Similarly, subtracting two points, that is, two elements of value 1, will give an element of value 0, i.e. a translation vector. The whole chain complex, i.e. the sequence of boundary operators of higher grade, can be described as the 20th century construction of the "Koszul" complex, the unique differential graded algebra structure which extends this lowest-grade part, which for the Koszul construction could be any linear functional on any given R-module.[1]

---

1 For a non–empty affine space the homology of this complex is trivial, since a contracting homotopy is defined by the Grassmann product with any chosen point.

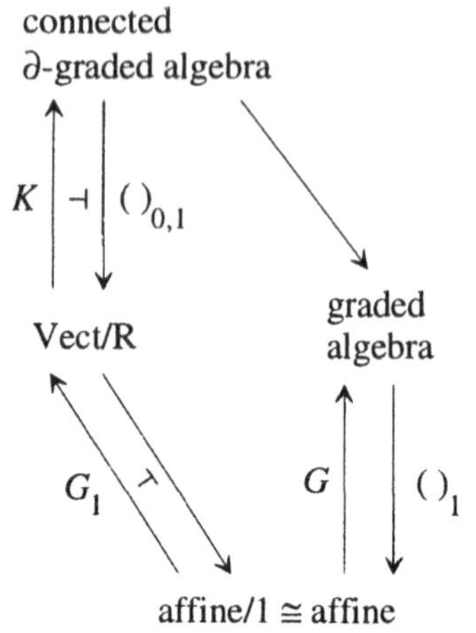

Vect/R = category of vector spaces equipped with a linear functional

K = Koszul

This boundary operator contains much of the geometry which Grassmann was talking about: for example, if we take two points, (i.e. two elements of G_1 which have boundary one) and form their Grassmann product, we obtain an element of G_2 which represents the axial vector from the first point A to the second point B. The boundary of the product is equal to $B - A$, the ordinary vector which translates A to B for

$$\partial (AB) = (\partial A)B \pm A\partial B$$

but if A and B are points, then ∂A and ∂B are both 1 and moreover in this grade it is the minus sign which we must take in the Leibniz rule. Higher products represent triangles, tetrahedra and so on and their boundaries are their usual combinatorial boundaries. Thus not only do the volume elements etc. live at the expected grades, but moreover, their boundaries are connected in the expected way.

Grassmann's term for the boundary operator is "Ausweichung." The exactness (vanishing of homology) he affirms on page 187; the special case, that the top-grade instance of the boundary operator is injective, is used for example on page 198. The fact that the lowest-grade instance of the boundary operator coincides with the weight functional is explicit on page 185. The quantities of boundary zero (which Grassmann called "Ausdehnungsgrösse") constitute the exterior algebra of the translation vector space of the original affine space; through their inclusion into the larger algebra these quantities can act on the "rigid" quantities which contain point factors.

"SIMPLE" LAWS AND EQUALITY OF AXIAL VECTORS

Grassmann apparently intended that his philosophical notion of "Becoming according to a simple law of change"[2] should serve as a foundation for his geometrical algebra. But Engel and Study, the editors of his collected works, found his discussion incoherent and decided that "in fact" (page 405) his changes were merely parallel translations. However, let us consider the foundational problem in the following sense. While indeed the description of the Grassmann algebra construction as the left adjoint into graded "commutative" algebras is very effective for calculation, we can ask why that particular category is chosen as geometrically natural. The crucial ingredient in presenting some algebra generated by points is the relation stating when two formal products of generating points are to be considered equal; there should be a criterion for $AB = CD$ in terms of the basic affine category itself. One idea for solving the foundational problem of providing such a criterion is as follows. Assuming that "simple law of change" is somehow interpreted as a special kind of affine–linear morphism, we can associate, with each pair of points A, B in a given affine space E, the set of all simple endomorphisms S of E for which $SA = B$; then define an equivalence relation on the set of pairs of points by the requirement that (for B distinct from A, and D distinct from C), these sets should be equal, in other words (A,B) is equivalent to (C, D) if and only if for every simple endomorphism S, $SA = B$ iff $SC = D$. If "simple" merely meant translations, this idea would collapse, because the equivalence would just say $B - A = D - C$, equality of the mere boundaries of the axial vectors. But it seems that Grassmann's description on the cited pages supports a broader interpretation of "simple" which both narrows the above equivalence idea enough to separate distinct bound or axial vectors and is also extremely natural in view of the particular nature of the affine category.

For an affine–linear endomorphism S of E, the geometrical property that for every point A, the product $A(SA)$ is the same as the product $(SA)(SSA)$, is equivalent to the algebraic property $S^2 = 2S - 1$. That equivalence can be proved in one direction by applying the boundary operator and in the other direction by using (anti)commutativity and the fact that the square of any point, such as SA, is zero. The foundational significance of this equivalence is that our "algebraic" condition does not presuppose the Grassmann product, but only involves composition and the fact that affine combinations of such maps (specifically the one with coefficients 2, –1) can be formed. Let us call such endomorphisms S "simple." Translation by a given vector is indeed simple, but there are

2 A_1, Einleitung, Nr. 9-10, and: Abschnitt 1, Kap. 1, §13 and §14, and: Anhang III, I. 5.

more examples such as any linear transformation which differs from the identity by a linear transformation of square zero, and certain combinations of these two types.

The simple laws of change in the above sense do indeed determine relations for presenting the Grassmann algebra. More precisely, not only does $AB = CD$ imply that any simple change taking A to B also takes C to D, but also that the converse is true. That converse can be proved as follows: the two sets of simple maps assumed equal include the translation and so, in particular, $D = C + (B - A)$, but more, if C were not on the line spanned by A and B, then there would be a simple S for which $SA = B$ but SC is not D (for example we could even have $SC = C$, a shearing transformation with C as origin.)

The apparently bizarre equation $S^2 = 2S - 1$ is in fact determined by applying to the affine category $\mathcal{A}$ the following very general consideration. It is a closed category in a unique way. This means that there is an internal hom functor with a left adjoint tensor product functor and unit object. For example, the category of sets has the exponential function-set of any two sets, the left adjoint to that is the ordinary cartesian product of two sets and the unit object is the one point set. The category of vector spaces has the well-known internal hom and tensor product whose unit is the ground field. More complicated categories may have more than one closed structure, but when the unit object is faithful, the closed structure is unique (but only rarely exists). For the category of affine spaces the unit object is the one point space, which is indeed faithful; the tensor product (uniquely determined by representing bi–affine maps) is bigger than the corresponding one for the category of vector spaces, since the functor (discussed previously) of adjoining a generic origin does transform the affine tensor product into the linear tensor product and hence $nm + n + m$ is the dimension of the affine tensor product of two spaces of respective dimensions n, m. In any closed category there is the notion of a monoid structure on an object M and of M–action on any object, both with respect to the tensor product. For example, in vector spaces a "monoid" is really any associative linear algebra and an "action" of such turns out to mean any module over the algebra. The affine tensor product of the scalar field R with itself is three-dimensional, and ordinary multiplication is an example of a monoid structure on R; however, addition gives a different monoid structure on R in the affine category, which is the one that interests us here because actions of the additive monoid of time translations are considered in many categories as a model of "laws of Becoming" (or in modern terminology "dynamical systems"). Of course, when the monoid structure is written additively, the associative law required for any action looks like an exponential law (also sometimes known as the "semigroup property"). A bi–affine action of the additive monoid R of all real numbers on a given affine space E is determined by its instance at $t = 1$ because for other times it just affinely follows the straight line through its state A at $t = 0$ and its state SA at $t = 1$. The associative (or exponential) law is a strong restriction on which affine–linear endomorphisms S can occur within such a flow: they are exactly those satisfying $S^2 = 2S - 1$ which we interpreted above as Grassmann's simple laws of change. In this sense the relation imposed on the Grassmann multiplication has a foundation in the way in which the affine category particularizes the very general notion of internal law of change.

Department of Mathematics, State University of New York at Buffalo

REFERENCES

Hermann G. Grassmann, *Die Ausdehnungslehre, ein neuer Zweig der Mathematik* [A_1], id., Gesammelte Werke, Ersten Bandes erster Theil (Leipzig: B. G. Teubner 1894) (Reprint Bronx, N. Y.: Chelsea 1969).

Gregory M. Kelly, *Basic Concepts of Enriched Category Theory*. London Mathematical Society Lecture Note Series 64, (Cambridge: Cambridge University Press, 1982).

F.William Lawvere, *Functorial Semantics of Algebraic Theories*, Thesis Columbia University, 1963.

F.William Lawvere, "Some Thoughts on the Future of Category Theory," Lecture Notes in Mathematics 1488 (Berlin, etc.: Springer, 1991) 1–13.

F.William Lawvere, "Categories of Space and of Quantity," *The Space of Mathematics. Philosophical, Epistemological and Historical Explorations*, eds. Javier Echeverria et al. (Berlin: De Gruyter, 1992), 14–30.

F.William Lawvere, "Tools for the Advancement of Objective Logic: Closed Categories and Toposes," *The Logical Foundations of Cognition,* eds. J. Macnamara & G.E. Reyes (Oxford: Oxford Univ Press, 1994, 43–56.

F.William Lawvere, "Cohesive Toposes and Cantor's 'lauter Einsen',"*Philosophia Matematica,* The Canadian Society for History and Philosophy of Mathematics, Series III, Vol. 2 (1994), 5–15.

F.William Lawvere, "Unity and Identity of Opposites in Calculus and Physics," Proceedings of the European Colloquium on Category Theory, Tours 1994, to appear in *Applied Categorical Structures.*

F.E.J. Linton, "Autonomous Categories and Duality of Functors" *Journal of Algebra,* 1965, *2*: 315-349.

ALVIN SWIMMER

THE COMPLETION OF GRASSMANN'S *NATUR-WISSENSCHAFTLICHE METHODE*

INTRODUCTION

In (Lewis 1977) Albert C. Lewis deduces, by a brilliant analysis, that the scientific method or '*wissenschaftliche* unity' displayed in Grassmann's A_1 is based mainly upon the philosophy of his father Justus Grassmann and that of the famous theologian F.E.M. Schleiermacher. Schleiermacher was one of Grassmann's teachers during his student years at the University of Berlin.

An important aim of Grassmann's scientific method, as expressed in the Forward of the A_1, was that

> "... each step from one formula to another appears at once as just *the symbolic expression of a parallel act of abstract reasoning*. The methods formerly used require the introduction of arbitrary coordinates that have nothing to do with the problem and completely obscure the basic idea, leaving the calculation as a mechanical generation of formulae, inaccessible and thus deadening to the intellect. *Here, however, where the idea is no longer strangely obscured, but radiates through the formulae in complete clarity, the intellect grasps the progressive development of the idea with each formal development of the mathematics.*
>
> Thus I feel entitled to hope that in this new analysis I have found the only natural way in which mathematics should be applied to nature and likewise that in which geometry should be treated, if it is to lead to general and fruitful results." (Grassmann 1894, p. 9)

While generally speaking, we may say that Grassmann was successful in his aim of making the mathematical symbolism act as a mirror for the related mental activity, his fascination with the concept of a *Strecke*, (i.e. a 'displacement' or 'vector'), acted as a deterrent to achieving this aim completely in investigations where points play a more important role than the vectors.

The modern definition of a linear space evolved directly from the A_1, principally through the following works: (Hankel 1867), (Schlegel 1872, 1875), (Peano 1888), (Whitehead 1898), (Burali-Forti, Marcolongo 1909), (Burali-Forti, Marcolongo 1909), (Weyl 1918). All of these authors explicitly acknowledge Grassmann's A_1 as their source and all of them,

G. Schubring (ed.), Hermann Günther Graßmann (1809–1877): Visionary Mathematician, Scientist and Neohumanist Scholar, 265–280.

with the exception of Whitehead, (Whitehead 1898), followed Grassmann's lead in geometrical applications, by regarding the vector as the fundamental entity to be algebraically manipulated.

From a geometrical point of view, this was indeed unfortunate and reintroduces a complication that precludes the possibility of using the mathematics as a symbolic mirror of the abstract thought. Indeed, it has led to the modern day impression that the only 'proper' ways to deal with geometrical and physical notions and theorems in which points occur as principle entities is either (i) by the convoluted process of treating these points as the endpoints of vectors, or (ii) to utilize affine spaces.

The usual process in case (i) is to combine appropriate vectors by algebraic manipulation and then extract the desired conclusion by looking at the *endpoints* of the resulting vector combinations. As an example of this procedure and because it is important for our subsequent developments, we consider the center of mass concept.

The center of mass of a system of mass particles at points P_i having mass μ^i is, since Grassmann's time, usually defined as the *end point* M of the vector $\overrightarrow{PM} = \dfrac{\Sigma \mu^i \overrightarrow{PP_i}}{\Sigma \mu^i}$ where the vectors $\overrightarrow{PP_i}$ all have the same arbitrarily chosen initial point P and terminate in the particle locations P_i. Of course, in order to use this definition, one must show that the location of M is independent of the arbitrarily chosen initial point P.

For more sophisticated applications, where the points impose themselves more obtrusively (such as occur in computer aided geometric design, applied mathematics or theoretical physics) one is often led to utilize alternative (ii): the affine spaces. Now these are rather peculiar objects consisting of two different types of entities called 'vectors' and 'points'. The vectors (as one would expect) form a 'vector space' and thus can be manipulated by addition and scalar multiplication according to the well known rules.

The points on the other hand cannot be added to each other but, by a process that could descriptively be called *miscegenated addition*, they can be combined with or 'added to' vectors to produce new points. Consequently, points can be 'subtracted' from each other to form vectors but *there is no sense in which points may be added together*. This is indeed a strange prohibition, resulting in the circumstance that we have an inverse operation for the points but no direct operation, utilizing only points, of which it is the inverse!

The reason for the indirect method in case (i) and the rules and prohibition in case (ii) is, no doubt, the belief that only these allowable operations are easy to interpret geometrically. However, the general misimpression (leading to the above prohibition and consequent occlusion of the mathematical mirror) that there is no direct geometric interpretation for the 'sum' of two points, is wrong.

My purpose here is to correct that misimpression. We lose absolutely nothing and gain a much richer tripartisan (physical, algebraic and geometric) structure when, instead of regarding the vectors as fundamental, we follow Möbius (Möbius 1827) and regard the fundamental entities of a linear space as symbolizing the mathematical generalization of the mass particles and magnetic and electric charges so useful in physical applications.

Only in this way does the exact relationship between points and vectors become evident. And only in this way can we truly understand why homogeneous coordinates are so useful in projective geometry and its applications.

1. POINT ADDITION

To be specific, let capital Latin letters, with or without subscripts, such as A, B, P_0, P_1,... represent weighted points. These are characterized by both their weight and location in space.

For the weighted point A, its weight, which can be any real number, will be denoted by $[A]$ and its location or associated geometric point by $\bar{A}$. Thus two weighted points, A and B, are the same, symbolized by $A = B$, if and only if (i) they have the same weight: $[A] = [B]$ and (ii) they coincide (i.e. have the same location) which we denote by $\bar{A} = \bar{B}$.

We use the symbol $[A,B]$ to denote the oriented line segment beginning at A and terminating at B and $[AB]$ to signify the oriented length of $[A,B]$.

As usual, we use '+', to denote the binary operation of the abelian group of weighted points determining our linear space. As unusual, we interpret it as the process of finding the center of mass (henceforth abbreviated to 'cm') so that $C = A + B$ iff C is the point on the line $\overline{AB}$ satisfying:

(*i*) $$[C]=[A]+[B]$$

and

(*iia*) $$[A][AC]=[CB][B]$$

(*iia*) is Archimedes' Law of Moments and it is equivalent to:

(*iib*) $$\frac{[AC]}{[CB]}=\frac{[B]}{[A]}$$

In this geometric form, the left side is called **'the ratio in which C divides $[A,B]$'**. Since we use oriented lengths, $[AC] + [CB] = [AB]$, no matter what the relative positions of these three collinear points are. Hence (*iia*) and (*iib*) are also equivalent to:

(*iic*) $$\frac{[AC]}{[AB]} = \frac{[B]}{[A]+[B]}$$

This gives us a geometric construction for $\overline{C}$, when the ratio is a rational number. Thus let $\frac{\lambda}{\mu} \doteq \frac{[B]}{[A]+[B]}$, where λ is an integer and μ is a natural number. Divide $[A,B]$ up into μ parts and locate C at λ of those parts from A. (For convenience, we use here and below, the symbol $\doteq$ as an abbreviation for "is defined as.")

We also have a means for locating C by coordinates. If we set up a cartesian coordinate system on $\overline{AB}$ in which A has coordinate α and B has coordinate β then the coordinate for C will be the weighted average $\frac{[A]a+[B]b}{[A]+[B]}$.

The equivalence of these various forms means that each of our equations can be interpreted both physically and geometrically.

It also means that one can find the cm either physically, geometrically or algebraically. For computer graphics applications, it is also important to note that computers can be hard wired to find the cm electronically (Axelrod et al. 1989, Cameron et al. 1992).

The relation between the ratio $\rho \doteq \frac{[AC]}{[CB]}$ in which C divides $[A,B]$ and the location $\overline{C}$ of C is given in Figure 1. where M is the midpoint of $[A,B]$.

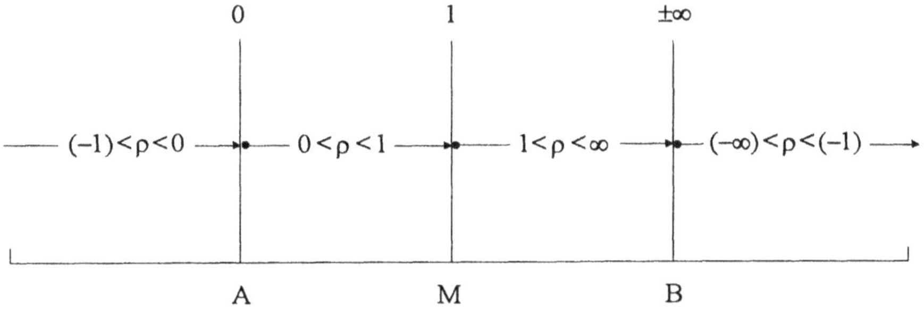

Figure 1.

Notice the special role played by -1 in *Figure 1*. There is no ordinary point corresponding to the ratio -1 since if $\frac{[AC]}{[CB]} = -1$ then C must be outside of $[A,B]$ (because of the $-$ sign) and yet equidistant from *both* endpoints!

However, writing ρ in the form $\frac{[AB]+[BC]}{[CB]} = \frac{[AB]}{[CB]} - 1$, we see that -1 is the limiting value to which ρ approaches as C recedes further and further away from the basic interval, *on either side of that interval*. In order to have closure in the abelian group structure that the weighted points determine we must allow the ratio -1 to correspond to some explicitly invented extraordinary weighted points which we call **points at infinity**.

When $\frac{[B]}{[A]} = -1$, of necessity $[C] = [A] + [B] = 0$. Thus *all the points at infinity have weight* 0.

However these points at infinity (which we shall investigate in detail in §5.) must be distinguished from the ordinary points (i.e. those located in the accessible part of the ambient geometric space) at weight 0.

These latter points have no effect, when combined with other points, in determining the cm. To see this suppose $[B] \doteq 0$, $[A] \neq 0$ and $C \doteq A + B$.

Then $[C] = [A] + [B] = [A]$ and $\frac{[AC]}{[CB]} = \frac{[B]}{[A]} = 0$ implies that $\overline{C} = \overline{A}$.

Thus $C = A + B = A$, and this is true *no matter where B is located*.
For this reason we often use the symbol $\mathit{0}$ to denote any ordinary point of weight 0. (This is the same algebraic device used when we say that for any numbers α and β, $\alpha - \alpha = \beta - \beta = 0 \cdot \alpha = 0 \cdot \beta = 0$ in ordinary numerical algebra.) (For points at infinity see §5.) Thus $\mathit{0}$ acts as the identity element of our group.

The reason we use the symbol + to denote the process of finding the cm and regard it algebraically as an 'addition' is that, for all points A, B, C, it satisfies both:

(a) The Commutative Law: $A + B = B + A$

(b) The Associative Law: $(A + B) + C = A + (B + C)$

From the physical point of view these laws obviously hold and it is not hard to supply geometric verification and analytical proofs for them also.

To complete the group structure of these points, we define, for each weighted point A, its additive inverse $-A$, as the new weighted point satisfying:

(iii) $[-A] \doteq -[A]$ and

(iv) $\overline{-A} \doteq \overline{A}$. Then, of necessity,

$$A + (-A) = 0$$

With the usual definition of subtraction as the inverse operation of addition, it follows that $A - B = A + (-B)$.

Thus all the manipulations of ordinary numerical addition are available for weighted points just as with numbers. Further, each of these manipulations can be immediately interpreted, both geometrically, in terms of the ratio in which a point divides an interval and physically, in terms of the cm concept. In this way we complete Grassmann's *wissenschaftliche* unity for the addition operation.

2. WEIGHT MULTIPLICATION

The second law of composition for our linear space we call "weight multiplication" and it concerns expressions of the form αA where α is a real number and A is a weighted point. By definition αA is the weighted point satisfying:

(v) $[\alpha A] \doteq \alpha[A]$

and

(vi) $\overline{\alpha A} \doteq \overline{A}$

Again, it is easy to prove either physically, geometrically or algebraically that the following laws hold for all real numbers α, β and all weighted points A, B:

(c) The Point Distributive Law: $(\alpha + \beta)A = \alpha A + \beta A$;

(d) The Weight Distributive Law: $\alpha(A + B) = \alpha A + \alpha B$;

(e) The Weight Associative Law: $\alpha(\beta A) = (\alpha\beta)A$

(Note that the '+' sign on the left of (*c*) refers to numerical addition, while in the other three instances in which it occurs, it means point addition, i.e. finding the cm.)

3. COORDINATES

We can now show the relationship between this way of representing points and the usual cartesian coordinate method.

Let $[P_0] \doteq [P_1] \doteq 1$ and let A have cartesian coordinate α in the system determined by $[P_0, P_1]$ (By this we mean P_0 is the origin and $[P_0 P_1]$ is the oriented unit of length.)

Then we have

$$\frac{[P_0 A]}{[P_0 P_1]} = \alpha$$

so that (since $[P_0 P_1] = [P_0 A] + [A P_1]$)

$$\frac{[P_0A]}{[AP_1]} = \frac{\alpha}{1-\alpha}.$$

Thus if $[A] \doteq \beta$ then we can write $A = \beta(1-\alpha)P_0 + \beta\alpha P_1$.

In other words, when α is the cartesian coordinate of $\overline{A}$, $((1-\alpha), \alpha)$ are its barycentric coordinates and for each non zero real number β, $(\beta(1-\alpha), \beta\alpha)$ is a set of homogeneous coordinates for $\overline{A}$.

The projective point $\overline{A}$ can now be recognized as the seat of the whole family of weighted points of the form βA. This makes clear why homogeneous coordinates work so well in projective geometry and gives geometric meaning to each of the possible choices for the homogeneous coordinates of a given projective point.

If we take $\beta = 1$, and write the equation in the form

$$A = P_0 + \alpha(P_1 - P_0)$$

we see that the coordinate α is also the coefficient of the point at infinity $V_1 \doteq P_1 - P_0$ when we use $[P_0, V_1]$ instead of $[P_0, P_1]$ for the coordinate basis.

4. CEVA-MENELAUS THEOREM

As an example of the use of point addition and weight multiplication we now prove the very important *Ceva-Menelaus* Theorem.

(In what follows we employ a very convenient *cyclic notation* wherein we use i, j, k as ordered in $[i,j,k]$ to denote, in turn, the three even permutations [1,2,3], [2,3,1] and [3,1,2].)

Let P_1, P_2, P_3 denote three non collinear points and let D_i divide $[P_j, P_k]$ in the ratio

$$\beta_i \doteq \frac{[P_jD_i]}{[D_iP_k]}$$

then

1. $\beta_1\beta_2\beta_3 = 1$ iff the lines $\overline{P_iD_i}$ are concurrent. (*Ceva's* Theorem)
2. $\beta_1\beta_2\beta_3 = -1$ iff the points D_i are collinear. (*Menelaus*'s Theorem)

Proof:

Let the masses $[P_i]$ be chosen so that $D_1 \doteq P_2 + P_3$ and $D_2 \doteq P_3 + P_1$. Thus

$$\beta_1 = \frac{[P_3]}{[P_2]} \text{ and } \beta_2 = \frac{[P_1]}{[P_3]}.$$

1. If $\beta_1\beta_2\beta_3 = 1$ then

$$\beta_3 = \frac{1}{\beta_1\beta_2} = \frac{[P_2]}{[P_1]}.$$

Hence we may set $D_{3c} \doteq P_1 + P_2$ and then

$$P_0 \doteq P_1 + P_2 + P_3 = P_1 + D_1 = P_2 + D_2 = P_3 + D_{3c}$$

is the point of concurrency of the lines $\overline{P_iD_i}$.

2. If however, $\beta_1\beta_2\beta_3 = -1$ then

$$\beta_3 = \frac{-1}{\beta_1\beta_2} = \frac{-[P_2]}{[P_1]}.$$

In this case we may satisfy the hypothesis by setting $D_{3m} \doteq P_1 - P_2$. It then follows that $D_{3m} = (D_2 - P_3) - (D_1 - P_3) = D_2 - D_1$ also. Thus D_{3m} is also on $\overline{D_1D_2}$. Indeed it is located at $\overline{P_1P_2} \cap \overline{D_1D_2}$.

Conversely suppose

1. that the lines $\overline{P_iD_i}$ are known to be concurrent at the point P_0. By choosing the weights $[P_i]$ appropriately, we may set $P_0 \doteq P_1 + P_2 + P_3$. It then follows that $P_0 - P_i = P_j + P_k$.

The expression on the left of this last equation shows that this point is on $\overline{P_0P_i}$ and the one an the right shows that it is on $\overline{P_jP_k}$. Hence it may be set equal to D_i since $\overline{D_i} \doteq \overline{P_0P_i} \cap \overline{P_jP_k}$.

From the above equations, it follows that

$$\beta_i = \frac{[P_k]}{[P_j]} \text{ and } \beta_1\beta_2\beta_3 = \frac{[P_3][P_1][P_2]}{[P_2][P_3][P_1]} = 1.$$

For the converse part of *Menelaus's* Theorem we assume

2. that the points D_i are collinear.

Then for appropriate weights λ and μ we have $D_3 = \lambda D_1 + \mu D_2$. Hence

$$D_3 = \lambda(P_2 + P_3) + \mu(P_3 + P_1) = \mu P_1 + \lambda P_2 + (\lambda + \mu)P_3.$$

Since, by hypothesis, D_3 is on $\overline{P_1P_2}$ we must have $\lambda + \mu = 0$ and so $D_3 = \mu(P_1 - P_2)$. Hence

$$\beta_3 = \frac{-\mu[P_2]}{\mu[P_1]} \text{ and } \beta_1\beta_2\beta_3 = -1.$$

5. POINTS AT INFINITY

An apparent exception to *Menelaus's* Theorem occurs when the *Menelaus* line is parallel to one side of the triangle $[P_1, P_2, P_3]$. For then we don't have a third intersection point. Suppose for definiteness that $\overline{D_1D_2}$ is parallel to $\overline{P_1P_2}$. This will be true if and only if

$$\beta_2 \doteq \frac{[P_3D_2]}{[D_2P_1]} = \frac{[P_3D_1]}{[D_1P_2]} = \frac{[D_1P_3]}{[P_2D_1]} = \frac{1}{\beta_1}$$

that is if and only if $\beta_1 \beta_2 = 1$. But then by *Menelaus's* Theorem $\beta_3 = -1$.

Now $\beta_3 = -1$ means that D_{3m} is a point at infinity on $\overline{P_1P_2}$ and the expression $D_{3m} = \mu\,(P_1 - P_2)$ derived above in this case means that $[P_1] = [P_2]$. But then $[D_2] = [D_1]$ and $D_{3m} = \mu\,(D_2 - D_1)$ shows that D_{3m} is also a point at infinity on $\overline{D_1D_2}$. Thus (just as Kepler first conceived the notion) parallel lines do meet at infinity and the apparent exception to *Menelaus's* Theorem vanishes, like the Chesire Cat Alice discovered in Wonderland, leaving only a complacent smile at the realization that these points at infinity give us a means whereby we can unify geometric notions concerning intersecting and parallel lines.

Before using these weighted points at infinity in combination with the ordinary weighted points, there is one more matter that must be attended to.

Since points at infinity - which always occur as the difference of two ordinary points having equal masses - divide the interval determined by these two points in the ratio –1, of necessity they all have mass 0. In order to distinguish between *different* points at infinity on the *same* line (which will then also be on the whole family of lines parallel to that line), it is necessary to find a new numerical magnitude to characterize these points at infinity.

This can be done as follows. Let A, B, C be three collinear points each having mass 1. Then for an appropriate number α, we can write

$$A = (1 - \alpha)\, B + \alpha C = B + \alpha\,(C - B)$$

or

$$A - B = \alpha\,(C - B).$$

Now $A - B$ and $C - B$ are both in the same location, corresponding to the ratio –1, both have the same mass 0, and yet the last equation tells us that the former is α times the latter. From the first equation we have

$$\frac{[BA]}{[AC]} = \frac{\alpha}{1-\alpha}$$

which is true if and only if

$$\frac{[BA]}{[BA]+[AC]} = \frac{\alpha}{\alpha+(1-\alpha)} \text{ or } \frac{[BA]}{[BC]} = \alpha\,.$$

Thus, α, which is the numerical factor relating $A - B$ and $C - B$, is also the ratio of the *lengths* of the intervals $[B, A]$ and $[B, C]$. In other words, α

is the cartesian coordinate of A in the system having B as origin and $[BC]$ as unit of length.

Based upon this observation and taking into account the points of arbitrary weight, we make the following definitions. If A and B are two mass points satisfying $[A] = [B]$ then $[A][AB] = [AB][B]$ is the **magnitude** (henceforth symbolized by $[[B\text{-}A]]$) of the point at infinity $B - A$. Two points at infinity are **equal** if and only if (i) they are at the same location (i.e. they are on the same or parallel lines) and (ii) have the same magnitude.

With these definitions we have a means of comparing different points at infinity on the same family of parallel lines. We also favorably resolve the apparent inconsistencies about these points at infinity. Furthermore, these definitions also allow us to declare that the set of all mass points, in a geometric space of any dimension, satisfy the axioms of a linear space.[1]

Note that because of the points at infinity that are in our linear space, geometrically speaking, our space is now projective rather than Euclidean.

Furthermore, each algebraic operation, in concurrence with Grassmann's *wissenschaftliche* unity, can now be directly interpreted in a geometrical or physical fashion. To the explanation of these interpretations we now turn.

6. VECTORS

When $[A] = [B] = 1$ the magnitude of $B - A$ is the same as the length of $[A,B]$, i.e. $[[B\text{-}A]] = [AB]$. Also $B - A$ is associated with the family of lines parallel to $\overline{AB}$ or, what comes to the same thing, the *direction* of that family of parallel lines. Thus, in this case $B–A$ is characterized by the same two qualities that define the *Strecke* or vector usually denoted by $\overrightarrow{AB}$.

Thus 'vector' and 'point at infinity' are simply two different ways of looking at the same concept. And now we can see that **by regarding weighted points as the fundamental entity in our linear space, the vectors are automatically included as a special species of extraordinary points, but subject to the same basic operations.**

With this interpretation, there is no problem in adding two vectors, two points or a point and a vector. No inconsistencies will arise and, in accordance with Grassmann's *wissenschaftliche* unity, each algebraic manipulation is directly interpretable in geometric fashion. All the familiar

---

1 Indeed, it was this particular example (with one essential difference), first introduced by Möbius (Möbius 1827) which led to the abstract definition in the first place.

manipulations associated with vector addition are still available for use, but we can now also manipulate the points by algebraic manipulations and immediately perceive their geometric meaning.

Further corroboration of the identity between these points at infinity of weight zero and the vectors of calculus and physics is given by the 'obvious' equation:

$$(A - B) + (B - C) = (A - C)$$

This equation holds no matter which points we are talking about, but when each of A, B, C has weight 1, each parenthesis denotes a point at infinity and the equation expresses the triangle law of vector addition. Note that, in this context, vector addition need not be separately defined. *It is just a special case of the point addition that applies universally.*

We can also now see that the coordinate form:

$$A = (1 - \alpha)P_0 + \alpha P_1 = P_0 + \alpha V_1$$

shows that adding a vector (i.e. point at infinity) to an ordinary point has the effect of translating that point by the amount and direction determined by the vector. Thus the idea of a vector being a translation or displacement operator is also built into our system. This was actually the genesis of the 'addition' defined between points and vectors in the affine spaces.

To elaborate on this idea let us now consider four noncoplanar points P_0, P_1, P_2, P_3 each having weight 1. These points determine a cartesian (but not necessarily a rectangular) coordinate system. To find the coordinates of any point A in the 3 dimensional space $\overline{P_0P_1P_2P_3}$ determined by these four points, we draw a line through A parallel to $\overline{P_0P_3}$ intersecting $\overline{P_0P_1P_2}$ in the point A_3. Next draw a line through A_3 parallel to $\overline{P_0P_2}$ intersecting $\overline{P_0P_1}$ in the point A_1 and another line through A_3 parallel to $\overline{P_0P_1}$ intersecting $\overline{P_0P_2}$ in the point A_2.

Assigning the weight 1 to each of the points A_1, A_2 we can, just as above, find numbers α^1, α^2 satisfying $A_1 = (1 - \alpha^1)P_0 + \alpha^1 P_1$ and $A_2 = (1 - \alpha^2)P_0 + \alpha^2 P_2$. Then

$$A_3 = A_1 + \alpha^2(P_2 - P_0) = A_2 + \alpha^1(P_1 - P_0) =$$
$$(1 - \alpha^1 - \alpha^2)P_0 + \alpha^1 P_1 + \alpha^2 P_2.$$

If $[A_3A] \doteq \alpha_3[P_0P_3]$ then, assigning weight 1 to both A_3 and A we have $A - A_3 = \alpha^3 (P_3 - P_0)$. Thus if $V_i = P_i - P_0$ for $i = 1, 2, 3$ then

$$A = A_3 + a^3V_3 = (1 - \sum a^i)P_0 + \sum a^i P_i = P_0 + \sum a^i V_i \ .$$

From these equations it is seen that it is very easy to convert between the cartesian coordinate representation and the weighted point representation of a given geometric point.

7. POINT MULTIPLICATION

As mentioned above, with the definitions of point addition and weight multiplication, it is easy to verify that the set of all weighted points on a line, in a plane or in any higher dimensional geometric space satisfy the axioms of a linear space. To make this set an algebra we must introduce a third operation, a multiplication, compatible with the other two operations.

Compatibility with the point addition means that the (using juxtaposition to denote multiplication)

(f) Distributive Laws: $A(B + C) = AB + AC$ and $(A + B)C = AC + BC$

should be true.

Compatibility with the weight multiplication means that, for any weight β, the

(g) Osmosis Law: $A(\beta B) = (\beta A)B = \beta(AB)$

should also hold.

Now none of the right sides of these equations make any sense unless the set of all products of the form AB form a new linear space. We shall thus make that assumption.

Any other conditions imposed upon these point products are somewhat arbitrary and depend upon the uses that are to be made of the multiplication. If $[P_0,P_1,...,P_n]$ is a basis for the linear space of points, $A \doteq \sum \alpha^i P_i$ and $B \doteq \sum \beta^i P_i$ then, assuming the distributive and osmosis laws, their product is $AB = \sum_{i,j} \alpha^i \beta^j P_i P_j$.

Thus we need only determine the relationship between the $(n+1)^2$ basis products to define the multiplication. In order that the multiplication be geometrically interpretable, the form of any further conditions must be independent of the basis chosen. That is, they must be *invariant equations of condition*, satisfied by all points of the space.

Grassmann proved that there are essentially only two types of invariant equations of condition on product pairs. These are $AB = BA$ and $AA = \mathbf{0}$. (Here, of course we are using the symbol $\mathbf{0}$ as the additive identity of the *new* linear space generated by the products P_iP_j). The latter equation, which distinguishes the Grassmann Algebras from other possibilities, is the one we henceforth postulate. It corresponds to the principle that products of points should, in some sense, represent the geometric objects those points determine.

Setting $A \doteq B + C$ in our invariant equation of condition

(vii) $AA = \mathbf{0}$,

we find a most important consequence: $BB + BC + CB + CC = \mathbf{0}$. Thus, since the first and last terms are also equal to $\mathbf{0}$, CB and BC are additive inverses of each other:

$$CB = -BC.$$

Conversely this last equation implies that $AA = 0$ which was our original postulate.

Another important consequence of our invariant condition is that if $[C] \doteq [B] \doteq 1$ and $E \doteq C + \mu B$ is any point on $\overline{BC}$, then

$$BE = B(C + \mu B) = BC.$$

Now the definition of E implies $(1 + \mu)\,[BE] = [BC]$.

These results lead us to define the numerical measure or **magnitude** of a product XY of two points to be $[[XY]] \doteq [X]\,[Y]\,[XY]$. Then the above consequences imply $[[CB]] = -[[BC]]$ and that

$$[[BE]] = 1 \times (1 + m)\,[BE] = 1 \times 1 \times [BC] = [[BC]]$$

These consequences also give credence to the idea that the product of two points geometrically represents a *linebound* vector on the line determined by the two points, whose length is equal to the magnitude of that product.

8. FORCES

To discover the geometric meaning of the sum of two point products, we consider three unit mass points A, B, C and the midpoint $M \doteq B + C$ of $[B,C]$. $AM = A(B + C) = AB + AC$ is a linebound vector of magnitude $1 \cdot 2 \cdot [AM] = [AD]$ where $D \doteq B + C - A = C + B - A$ is the fourth vertex of the parallelogram determined by A, B and C.

Thus we see that the geometric interpretation of point products as *linebound* vectors is quite different from the *free* vector interpretation of the points at infinity encountered earlier. The latter are zero-dimensional and thus are of the same nature as points. Even when they are interpreted as vectors (i.e. one-dimensional objects), they combine by a triangle law, while the former, which truly are one-dimensional, combine by the parallelogram law.

It is just because they combine by the parallelogram law of addition that these linebound vectors perfectly represent the forces of classical Newtonian physics. This representation is much superior to the usual representation of forces by free vectors, since by its nature, a force is associated with a line of action whereas free vectors (i.e. points at infinity) are associated with families of lines or directions rather than individual lines.

Furthermore, two linebound vectors combine by the parallelogram law of vector addition, and thus their sum physically represents the resultant of the two forces simultaneously in magnitude, direction, *and line of application*.

To find the resultant of a system of n forces f_i, all emanating from the same position $\overline{A}$, let these forces be represented by the linebound vectors

AP_i, where the mass points A, P_i all have mass 1, the lengths $[AP_i]$ represent the magnitudes of the f_i and $\overline{AP_i}$ are the respective lines of action. Then the resultant is $\sum AP_i = A\sum P_i$.

The right side of this equation tells us that we can construct the resultant by first finding the cm of the endpoints, $P \doteq \sum P_i$ which has mass n, and then finding the unit mass point

$$Q \doteq (1-n)A + P = A + (P - nA) \text{ on } \overline{AP}$$

which satisfies $[AQ] = n[AP]$, so that $AQ = AP$.

In this way the relationship between centers of mass and resultants of forces becomes completely obvious. The latter is exactly the one dimensional counterpart of the former.

If we use linebound vectors instead of free vectors to represent the forces we find a marvelous unification of the two basic laws of Newtonian mechanics:

I. Force equals the time derivative of (linear) momentum.

II. Torque equals the time derivative of angular momentum.

To see this clearly, assume $x^i = x^i(t)$ are the parametric equations of the path of motion in a rectangular system determined by the unit mass point basis $[P_0, P_1, P_2, P_3]$, and that the $V_i \doteq P_i - P_0$ are the three free vectors of unit length in the coordinate axes directions. Then

$$P = P_0 + \sum x^i V_i$$

is the general point of the trajectory and λP is the trajectory point at mass λ.

$$\dot{P} = \sum \dot{x}^i V_i$$

is the (free) velocity vector and

$$\lambda\dot{P} = \sum \lambda\dot{x}^i V_i$$

is the momentum vector (also free). The total momentum at P is

$$P(l\dot{P}) = l(P\dot{P}) = l\left(P_0\dot{P} + \left(\sum x^i V_i\right)\dot{P}\right)$$

$$= \sum l\dot{x}^i P_0 V_i + l\sum_{j<k} \begin{vmatrix} x^j & \dot{x}^j \\ x^k & \dot{x}^k \end{vmatrix} V_j V_k \doteq M_L + M_A$$

M_L is the linear momentum and M_A is the angular momentum. The determinant coefficients occurring in the latter sum are the **moments of momentum.**

If the force acting on the mass particle is described by the (free) vector $F \doteq \sum f^i V_i$ then the **total force** at P is the linebound vector

$$PF = P_0F + \sum x^i V_i F = \sum f^i P_0 V_i + \sum \begin{vmatrix} x^j & \dot{x}^j \\ x^k & \dot{x}^k \end{vmatrix} V_j V_k = F_L + F_A$$

Here F_L is the linear force and F_A is the angular force or torque. The components of F_A are the force moments.

The basic Newtonian Laws, I and II now become component parts of a single *Unified Law Of Motion:*

The time derivative of the total momentum is the total force:

$$\frac{d}{dt}\left(\lambda P\dot{P}\right) = PF \,.$$

In the equations leading to this unified law the right sides naturally split up into two components. One of these involves products of the origin P_0 with the coordinate direction vectors: P_0V_i. These are all ordinary linebound vectors $P_0(P_i - P_0) = P_0P_i$ in the finite portion of three space.

The other part involves products of two free vectors and thus is a linebound vector at infinity. It is the one-dimensional counterpart of the points at infinity and physically represents a couple determined by two parallel forces equal in magnitude but opposite in direction, as follows from

$$V_jV_k = (P_j - P_0)\;(P_k - P_0) = P_jV_k - P_0V_k = P_0V_j - P_kV_j.$$

The linebound vector part gives the translational aspect of the motion and the part at infinity gives the rotational part of the motion.

We hope that these ideas amply demonstrate the advantages of interpreting the fundamental entities as weighted points rather than vectors. Particularly since it allows us to manipulate both points and vectors algebraically, either individually or together. Besides this we achieve the complete symbiosis between the algebraic manipulation and the geometric and physical interpretation in accordance with Grassmann's *wissenschaftliche Methode*.

Department of Mathematics, Arizona State University

REFERENCES

T. Axelrod, T. Tassinari, G. Barnes, K. Cameron, "A VLSI Centroid Extractor For Real Time Target Tracking Applications," Proceedings of the *SPIE, The International Society of Optical Engineering*, vol. 1154 (San Diego, California August 1989), 306 - 312.

C. Burali-Forti and R. Marcolongo, *Elementi di Calcolo Vettoriale* (Zanichelli: Bologna 1909).

C. Burali-Forti and R. Marcolongo, *Omografie Vettoriali con applicazioni alle derivate* (G. B. Pretrini di Giovanni Gallazio: Torino, 1909).

K. Cameron, J. Canaris, G. Barnes, S. Whittaker, N. Liu, K. Liu, "An Automatic Centroid Extractor For Video Image Processing," *ASIC (Application Specific Integrated Circuits) '92*, Rochester, N.Y. (September 1992).

Hermann G. Grassmann, *Die lineale Ausdehnungslehre, ein neuer Zweig der Mathematik*. Gesammelte mathematische und physikalische Werke, Band I, 1 (Leipzig: Teubner, 1894).

Hermann Hankel, *Theorie der Complexen Zahlensysteme* (Leipzig: L. Voss 1867).

Albert C. Lewis, 'H. Grassmann's *Ausdehnungslehre* and Schleiermacher's *Dialektik*," *Annals of Science*, 1977, *34* :103-162.

August Möbius, *Der Barycentrische Calcül* (Leipzig: J.A. Barth, 1827)

Giuseppe Peano, *Calculo Geometrico secondo l'Ausdehnungslehre di H. Grassmann* (Torino: Fratelli Bocca, 1888).

Victor Schlegel, *System der Raumlehre nach den Principien der Grassmann'schen Ausdehnungslehre und als Einleitung in dieselbe dargestellt* (Leipzig: Teubner 1872,1875).

Hermann Weyl, *Raum-Zeit-Materie* (Berlin: Springer, 1918).

Alfred North Whitehead, *A Treatise on Universal Algebra with Applications* (Cambridge: The University Press, 1898).

WILHELM KLINGENBERG

GRASSMANNIAN MANIFOLDS IN GEOMETRY

Nowadays, every student of modern geometry has heard of Grassmann manifolds, and also he will have a more or less clear idea what they are. But it is highly unlikely that the student will know something about the man whose name is attached to this basic concept.

It seems that the name Grassmann manifold was first used in the mid-1930s by Charles Ehresmann in the school of geometers around Élie Cartan. And what now is a Grassmann manifold or, as one also says, a Grassmannian?

Fix integers $0 < k < n$. Consider the n-dimensional euclidean space $\mathbb{R}^n$. The elements of the Grassmannian $G(k,n\text{-}k)$ are the k-dimensional subspaces U of $\mathbb{R}^n$. Sometimes, one also writes $G(k, n)$ instead. The first important observation is that $G(k, n\text{-}k)$ is a compact manifold of dimension $k(n - k)$. Indeed, we have a natural atlas of charts. Each element $U_0 \in G(k,n\text{-}k)$ determines a chart as follows. Let $U_0^\perp$ be the space orthogonal to U_0. Denote by $\mathcal{U}(U_0)$ the set of $U \in G(k,n\text{-}k)$ which intersect $U_0^\perp$ only in the origin $0 \in \mathbb{R}^n$. That is, under the orthogonal projection $\mathbb{R}^n \to U_0$, the image of U shall be all of U_0. Such an U is the graph of a linear mapping L_U: $U_0 \to U_0^\perp$. Associate to $x_0 \in U_0$ first its counter image x in U and then the projection of x into $U_0^\perp$.

Conversely, the graph of a linear mapping U_0 into $U_0^\perp$ is a subspace U with
$U \cap U_0 = 0$. Thus, we get the chart

$$\varphi_{U_0}:\mathcal{U}(U_0) \to L(U_0,U_0^\perp) \cong M(n-k,k)$$

On the right hand side stands the vector space of $(n\text{-}k, k)$- matrices.

Clearly, the atlas formed by the charts $(\varphi_{U_0},\mathcal{U}(U_0))$ is smooth, even analytic. Actually, $G(k,n\text{-}k)$ is a symmetric space. To see this observe that on $G(k,n\text{-}k)$ the special orthogonal group $SO(n)$ is operating, and that this operation is transitive. The isotropy group of the standard subspace $\mathbb{R}^k$ of $\mathbb{R}^n$ is $SO(k) \times SO(n\text{-}k)$. Thus,

$$G(k,n\text{-}k) = SO\,(n) \,/\, SO\,(k) \times SO\,(n\text{-}k).$$

G. Schubring (ed.), Hermann Günther Graßmann (1809–1877): Visionary Mathematician, Scientist and Neohumanist Scholar, 281–283.

The symmetry at $U_0 \in G(k,n\text{-}k)$ is given by $(id_{U_0}) + (-id_{U_0^\perp})$.

If $k = 1$, $G(k,n\text{-}k)$ is nothing but the projective space P^{n-1}, something quite familiar to geometers in Grassmann's time. The boldness of Grassmann's idea to consider k-dimensional subspaces in $\mathbb{R}^n$ as "points" can hardly be overestimated. Here he used an idea which belongs in the twentieth century rather than the nineteenth. It is clear that Grassmann was not familiar with the modern concept of a smooth manifold, given by an atlas of charts where the change of charts is smooth.

Next we want to explain why Grassmannians are so important in modern geometry. For that purpose we give the canonical decomposition of $G(k,n\text{-}k)$ into cells, first described by Hermann Cäsar Hannibal Schubert (1848-1911). These cells, called Schubert cycles, are determined by an index of the form

$$\sigma = (\sigma_1, \cdots, \sigma_k) \text{ with } 1 \le \sigma_1 < \cdots < \sigma_k \le n.$$

The cycle $e(\sigma)$ consists of all $U \in G(k, n\text{-}k)$ having the property

$dim\,(U \cap \mathbb{R}^{\sigma_i}) = i$, $\dim(U \cap \mathbb{R}^{\sigma_i - 1}) = i\text{-}1$, $i = 1,\ldots,k$.

One easily calculates

$$dim\, e(\sigma) = \textstyle\sum_1^k (\sigma_i - i). \quad \partial e(\sigma) = 0 \bmod 2.$$

Thus, the Schubert cycles form a base for the Z_2-homology of $G(k, n\text{-}k)$.

For the following it is convenient to get rid of the parameter n by letting n go to infinity. E.g., one may take the infinite dimensional Hilbert space H instead of $\mathbb{R}^n$. $G(k)$ denotes the set of all k dimensional subspaces of H. The orthogonal complement $U^\perp$ of U again is an infinite dimensional Hilbert space. The dimension of the Schubert cycles $e(\sigma)$ now is unbounded. $G(k,n\text{-}k)$ is canonically embedded into $G(k)$ as the Schubert cycle $e(\sigma)$ with $\sigma = (n\text{-}k\text{+}1,\ldots, n)$.

We now come to the concept which is at the source of the fame of the Grassmann manifolds, i.e., the k-vector bundle. This is a mapping

$$\beta\colon E \to B$$

where B, the so-called base space, is a CW-complex. We don't go into the details of the definition of a CW-complex. For us, it suffices to observe that it is a topological space built up by cells, just as $G(k)$ is built up by the Schubert cycles, each of which is a cell. E is called the total space of the bundle and β is the projection mapping. It is a surjective mapping such that the pre-image $\beta^{-1}(b)$ of each $b \in B$ is a k-dimensional vector space, called the fibre over b. One last property holds: Every $b \in B$ has an open neighborhood $O = O(b)$ such that the restricted bundle $\beta\colon \beta^{-1}(O) \to O$ is bundle-isomorphic to the product bundle $\mathbb{R}^k \times O$. Here, a bundle morphism from β: $E \to B$ to β': $E' \to B'$ is a pair of mappings Φ: $E \to E'$ and φ: $B \to B'$ such that $\Phi|\beta^{-1}(b)$ is a linear isomorphism with $\beta'^{-1}(\varphi(b))$.

Observe now that $G(k)$ is in a canonical manner the base space of a k-vector bundle

$$\gamma(k)\colon E(k) \to G(k)$$

called, of course, the Grassmann bundle. The fibre over $U \in G(k)$ is created by anamnesis: The k-dimensional subspace U recalls that it is formed by vectors. And the projection $\gamma(k)$ is amnesis: A vector in U forgets its identity and just falls back into the subspace U from which it was born.

The **Fundamental Theorem of Vector Bundles** now states: Every k-vector bundle β: $E \to B$ is induced by a bundle mapping into the Grassmann bundle $\gamma(k)$. I. e., there is a commutative diagram of mappings

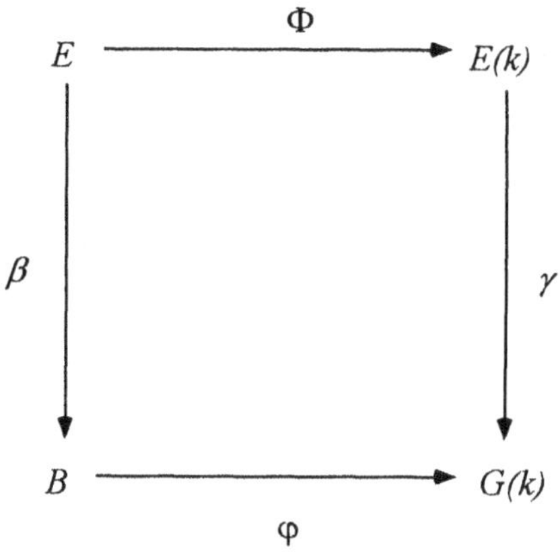

such that, for each $b \in B$, the restriction of Φ to the fibre over b is a linear isomorphism onto the fibre over $\varphi(b)$. φ is determined up to homotopy.

Even the non-expert will recognize: The Grassmann bundle is the universal bundle for all k-bundles; it comprises every imaginable type of k-bundle.

Only a truly great mathematician could have had the vision to create such a fundamental mathematical object at a time where there existed no precise idea of such a concept as the category of vector bundles.

Mathematisches Institut, Universität Bonn

REFERENCE

John W. Milnor and James D. Stasheff, *Characteristic Classes*. Annals of Mathematics Studies, no. 76. (Princeton Univ. Press: Princeton, N. J., 1974).

ARNO ZADDACH

REGRESSIVE PRODUCTS AND BOURBAKI

§ 1. EXTERIOR POWERS OF A VECTOR SPACE

The construction of the Grassmann Algebra (or Exterior Algebra) over a given real n-dimensional vector space $\mathcal{V}$, is executed by certain axioms.[1] To do this, a metric ("point product" between vectors) is **not needed.** For the present exposition, it is supposed that the readers are already well acquainted with Grassmann's "progressive multiplication" (ordinary wedge product $\wedge$), the multivectors (his "extensives") and their "levels" (or "steps") from 0 to n.

The modern concepts and notations can be found in Bourbaki's *Algèbre* (Ch. III: *Algèbres tensorielles, extérieures et symétriques*). We will call "cleavable" (or "decomposable") every multivector of the form $\vec{a}_1 \wedge \vec{a}_2 \wedge \cdots \vec{a}_p \neq 0$ ($\vec{a}_i \in \mathcal{V}$). The cleavable elements of level p ($1 \leq p \leq n$) generate the pth **exterior power** of $\mathcal{V}$, which is a vector space of dimension $\binom{n}{p} = \binom{n}{n-p}$; symbol: $\Lambda^p \mathcal{V}$. Besides, there exists a special element $E_\emptyset$, the neutral of the algebra. Its multiples (by real factors $\neq 0$) are of level 0 and form the one-dimensional space $\Lambda^0 \mathcal{V}$. The direct sum of all these vector spaces, endowed with the associative multiplication $\wedge$, gives us the Grassmann Algebra $\Lambda \mathcal{V}$ of dimension $\sum \binom{n}{p} = 2^n$.

Any two simple vectors (elements from $\mathcal{V} = \Lambda^1 \mathcal{V}$) have a wedge product that is anti-commutative: $\vec{b} \wedge \vec{a} = -\vec{a} \wedge \vec{b}$. For the multivectors A (of level p) and B (of level q), there holds $B \wedge A = (-1)^{pq} \cdot A \wedge B$. - A wedge product of simple vectors is equal to 0, iff the vectors are linearly dependent. No level is assigned to this zero element of the Grassmann Algebra.

In the progressive product $A \wedge B \neq 0$, the levels of the factors add up to the level of the result.

---

[1] For the details see Pickert 1961, § 11, and Zaddach 1994, § 43.

G. Schubring (ed.), Hermann Günther Graßmann (1809–1877): Visionary Mathematician, Scientist and Neohumanist Scholar, 285–295.

Remarks relating to **bases:** Let $(\vec{e}_1,\ldots,\vec{e}_n)$ be a basis for $\mathcal{V}$. We will write the products $\vec{e}_i \wedge \vec{e}_j = -\vec{e}_j \wedge \vec{e}_i$ with the indices in any order, but when we shorten the symbols to $E_{ij} = \vec{e}_i \wedge \vec{e}_j$, we will do so only under the restriction i<j. Then the $\binom{n}{2}$ bivectors E_{ij} will form a **basis** for the second power $\Lambda^2 \mathcal{V}$. All this holds analogously with the higher levels.

In the symbols E_H, each H is an "index block" $(h_1, h_2, \ldots, h_p)$ where $h_1 < h_2 < \cdots < h_p$. The index block N = (1, 2,..., n) will have a special importance. $\Lambda^p \mathcal{V}$ has, as one of its bases, the $\binom{n}{p}$ multivectors E_H, $H \subset N$ with all the index blocks H of the same cardinality p. (Within each block, the numerical indices remain always arranged in their natural order.)

The substructure $\Lambda^0 \mathcal{V}$ is isomorphic to the scalar field $\mathbb{R}$. But as soon as the "regressive multiplication" ($\vee$), our principal topic in this exposition, is put on the stage, the identification of $\Lambda^0 \mathcal{V}$ with $\mathbb{R}$ **should be avoided**. Therefore, in these pages, $E_\varnothing$ will be considered as distinct from the real number 1.

Let us emphasize the following detail: Although usually we may consider E_N as resulting from $\vec{e}_1 \wedge \vec{e}_2 \wedge \ldots \wedge \vec{e}_n$, this factorization (i.e. the distinction of a complete basis for $\mathcal{V}$) is **not necessary** for our purposes (see below: Determinants).

For the regressive multiplication, **one** multivector E_N must be selected as the neutral element for $\vee$. This requires an arbitrarily chosen distinction - not well emphasized in the works that were written in the "Old Grassmann Style." Therein, the highest power $\Lambda^n \mathcal{V}$ had been identified with both $\mathbb{R}$ and $\Lambda^0 \mathcal{V}$ (this last one, not even recognized as a structure with a separate neutral $E_\varnothing$). We find in those older books the forcible identification $E_N = E_\varnothing = 1$. Such a negligence opens the door to **contradictions** like this one:

$$\vec{e}_1 \wedge E_N = \vec{e}_1 \wedge E_\varnothing = \vec{e}_1, \text{ whereas}$$

$$\vec{e}_1 \wedge E_N = (\vec{e}_1 \wedge \vec{e}_1) \wedge \vec{e}_2 \wedge \cdots \wedge \vec{e}_n = 0 \cdot \vec{e}_2 \wedge \cdots \wedge \vec{e}_n = 0.$$

It was the great merit of Bourbaki to have given precise definitions to the exterior powers, with

$$\Lambda^n \mathcal{V} \neq \Lambda^0 \mathcal{V} \text{ and } E_N \neq E_\varnothing.$$ [2]

---

2 See also Flanders 1963, pp. 5-8.

Bourbaki created the symbol $\vee$, too (while $\wedge$ already appeared in a work composed by Burali-Forti and Marcolongo in 1909, p. 28).

§ 2. DETERMINANTS

Since the exterior product is nowadays denoted by the wedge $\wedge$, the "box" symbol [......] has become available to introduce determinants in Grassmann's Theory as follows:

$$\vec{a}_1 \wedge \vec{a}_2 \wedge \cdots \wedge \vec{a}_n = [\vec{a}_1, \vec{a}_2, \cdots, \vec{a}_n] \cdot E_N.$$

The real number $[\vec{a}_1, \cdots, \vec{a}_n]$ is the **determinant** of the n-tuple of vectors $\vec{a}_i$ - but with reference to E_N ! For this and other reasons, E_N plays a central role in our present exposition.

By distinguishing this multivector E_N of the highest level n, the ground vector space $\mathcal{V}$ is now equipped with an **orientation**: Any basis $(\vec{a}_1, \cdots, \vec{a}_n)$ of $\mathcal{V}$ is conceived as "positively oriented", iff the "box product" $[\vec{a}_1, \cdots, \vec{a}_n]$ is positive with reference to E_N.

Let *dim* $\mathcal{V} = 4$ (vectors of this dimension are the adequate tool for the algebraic treatment of the projective three-space). - Skew-symmetric determinants of 4 rows have certain agreeable qualities. Consider the bivectors $G = \vec{a} \wedge \vec{b}, H = \vec{c} \wedge \vec{d}$. Because these are cleavable, they can be put into a box symbol, each one to replace two rows, and we may write

$$[\vec{a}, \vec{b}, \vec{c}, \vec{d}] = [G, \vec{c}, \vec{d}] = [G, H] = [H, G]$$

If coordinates are employed, then to G, H there belong the skew-symmetric matrices (g^{ij}) and (h^{ij}). The well-known Laplace expansion $[G,H] = g^{12}h^{34} + g^{13}h^{42} + \cdots + g^{34}h^{12}$ (with 6 terms) is easily deduced from $G \wedge H = [G,H] \cdot E_N$, N = (1, 2, 3, 4). It is worth noting that the skew-symmetric determinant of 4 rows is equal to the square $(g^{12}g^{34} + g^{13}g^{42} + g^{14}g^{23})^2$. Its value is zero, iff the bivector is cleavable: $G = \vec{a} \wedge \vec{b}$.

The simplest example of a non-cleavable bivector is given by $\vec{e}_1 \wedge \vec{e}_2 + \vec{e}_3 \wedge \vec{e}_4$ ($\vec{e}_1, \cdots, \vec{e}_4$ being the basic vectors for the coordinate system). For this one, the aforementioned expression yields $g^{12}g^{34} + g^{13}g^{42} + g^{14}g^{23} = 1 + 0 + 0 \neq 0$.

dim $\mathcal{V} = 4$ implies that every trivector is cleavable.

§ 3. GRASSMANN'S SUPPLEMENTS FOR MULTIVECTORS

Already in the first *Ausdehnungslehre* "A_1" (1844), there existed strong reasons to avoid any mention of a metric, because the Exterior Algebra finds its most convincing applications in **Projective** Geometry. - In his "A_2" (1862), Grassmann creates for every multivector X its supplement ("Ergänzung") $|X$ by means of formulae which are reminiscent of orthogonality relations.

These supplements refer to an **orientation** of $\mathfrak{V}$ and require, therefore, utmost caution with the **signs**; they refer, too, to a complete basis $(\vec{e}_1,\cdots,\vec{e}_n)$ of $\mathfrak{V}$ and must be defined, separately, for each dimension and for each level of X. (This nuisance was removed, more than 80 years later, by Bourbaki.)

dim $\mathfrak{V} = 2$. Grassmann **defines**, taking tacitly as prototye the "positively directed" perpendicular:

$$|\vec{e}_1 = +\vec{e}_2,\ |\vec{e}_2 = -\vec{e}_1.$$

dim $\mathfrak{V} = 3$. Grassmann's definitions are:

$|E_{12} = +\vec{e}_3,\ |E_{13} = -\vec{e}_2,\ |E_{23} = +\vec{e}_1$ and vice versa.

As to *dim* $\mathfrak{V} = 4$, some details will be discussed later. - These formations are easily expanded, by a process of "linear prolongation", from the basic elements to the exterior powers, generated by these. But with increasing dimension, the description will require more and more display.

In brevity, without restriction of n, Grassmann gives the general formula for his supplements of the 2^n basic multivectors E_H in the Exterior Algebra over the vector space of dimension n:

$$|E_H = \pm E_K$$

Herein, the index blocks H and K are to be understood as ordered subsets of N = (1,2, , n), all the 2^n subsets with their indices in the natural order, and being K the complementary set to H with regard to the index block N. - The **sign** is determined in such a manner as the above special lines do suggest. But it is irritating that we must here mention the basic elements E_H.[3]

Nevertheless, the "supplements" in the "A_2" enable Grassmann to define, in a transparent manner, his regressive products (which depend on the dimension of $\mathfrak{V}$!):

3 We will see later, using Bourbaki's mapping φ, that there is no need to distinguish a complete basis.

$$A \vee B = |(|A \wedge |B) \qquad (1)$$

(and equal to 0 if there results, inside the bracket, the zero element of the algebra $\bigwedge \mathcal{V}$).

The vector space $\bigwedge \mathcal{V}$, now equipped with $\vee$, becomes an algebra that is isomorphic to ($\bigwedge \mathcal{V}$, $\wedge$). - The new multiplication $\vee$ operates in the opposite direction to $\wedge$: With p, q, s as the levels of A, B, $A \vee B$, the "complementing levels" n-p etc. are related by n-s = (n-p) + (n-q), so that we have

$$s = (p+q) - n \qquad (2)$$

(Note the reference to **n** !)

Easiest example (with *dim* $\mathcal{V} = 2$): $\vec{a} \vee \vec{b} = [\vec{a}, \vec{b}] \cdot E_{\varnothing}$.

Another example (here with *dim* $\mathcal{V} = 3$):

$$(\vec{a} \wedge \vec{b}) \vee (\vec{c} \wedge \vec{d}) = [\vec{c}, \vec{d}, \vec{a}]\, \vec{b} - [\vec{c}, \vec{d}, \vec{b}]\, \vec{a} \qquad (3)$$

(Proved by substituting the basic vectors $\vec{e}_1, \vec{e}_2, \vec{e}_3$ in their different combinations.)
The reference to the (arbitrarily distinguished) element E_N is hidden
--- on the left hand side, in $\vee$,
--- on the right hand side, in the box products.

Peano does **postulate** (3), in order to **define** the regressive product under the given circumstances (Peano 1888, p. 80). So he can avoid the basic vectors $\vec{e}_1, \vec{e}_2, \vec{e}_3$ and the supplements. But then, he imposes the entire burden of motivation upon the reader. As already said, with increasing dimensions, there are to be considered factors of higher levels in new combinations, and then Peano's procedure to introduce $\vee$ becomes unsatisfactory.

Geometric interpretation of (3): In the projective plane, the intersection (meet) of the two joins (lines) AB and CD is here well expressed as a point on the line AB (and may be treated, in the same fashion, as a point on the line CD).[4]

What must be qualified as a **serious error** in Grassmann's strategy (copied by his most eminent followers like Peano and Forder), is his goal to identify

$$\bigwedge^n \mathcal{V} = \bigwedge^0 \mathcal{V} = \mathbb{R}, \; E_N = E_{\varnothing} = 1 \;\; (!!)$$

---

4 As to meet and join of two lines in projective three-space, see Zaddach 1994, p. 227.

and to merge progressive and regressive multiplications in one sole operation for which the associative law cannot hold any longer.

§ 4. OTHER CONCEPTS OF BOURBAKI

On the set of all linear mappings $\mathcal{V} \rightarrow \mathbb{R}$, there is "imprinted", in a natural way, a vectorial structure, too. This aspect was recognized since 1920, beginning with H. Weyl's famous work *Space - Time - Matter*. Later on, these "linear functions" (or "linear forms") received the name of "co-vectors"; their structure is nowadays known as the "dual vector space" with reference to $\mathcal{V}$, of the same dimension n. Our notation here will be $\mathcal{V}^*$, and we will denote the **co-vectors** by symbols like $\underset{\rightarrow}{u}$.

Following these ideas of the present century, Bourbaki conceives the application of such a function $\mathcal{V} \rightarrow \mathbb{R}$ as a bilinear **composition**, co-vector with vector, the result being a scalar. The bracket notation $< \underset{\rightarrow}{u}, \vec{a} >$ emphasizes the dual roles of the two "factors."

In an analogous manner, co-multivectors are created, proceeding from each exterior power $\Lambda^p \mathcal{V}$, $(p = 0, \cdots, n)$. Bourbaki shows that it is justified to identify $(\Lambda^p \mathcal{V})^*$ with $\Lambda^p (\mathcal{V}^*)$; in words: There is no need to distinguish between co-multivectors and multi-covectors.

The composition $< \underset{\rightarrow}{u}, \vec{a} >$ should not be called an "inner product", because in Bourbaki's work, there appear "inner products" in a more general context, the factors being a co-multivector and a multivector, even if the levels do not coincide. The results are multivectors (incl. $E_\emptyset$) or co-multivectors (incl. $E^\emptyset$), as the case may be, whereas $< \underset{\rightarrow}{u}, \vec{a} >$ yields a scalar.

Let $(\vec{e}_1, \cdots, \vec{e}_n)$ be a basis for $\mathcal{V}$. Without any metric, there can be defined the corresponding "co-basis" $(\underset{\rightarrow}{e}^1, \ldots, \underset{\rightarrow}{e}^n)$ (i.e. a basis for $\mathcal{V}^*$) by the set of simple relations $< \underset{\rightarrow}{e}^i, \vec{e}_j > = \delta^i_j \quad (i, j = 1, \cdots, n)$.

This construction, too, can be extended to the exterior powers. We need it here only for the highest level n.

Without bothering to fix a basis $(\vec{e}_1, \cdots, \vec{e}_n)$ of $\mathcal{V}$, let us select a multivector E_N of level n as a basis for the highest exterior power $\Lambda^n \mathcal{V}$ (of dimension 1). The corresponding co-multivector E^N is declared as the linear function that maps every $aE_N \in \bigwedge^n \mathcal{V}$ upon its coefficient $a \in$IR, what may be condensed into the formula $< E^N, E_N >= 1$.

With reference to E^N, Bourbaki now creates a one-to-one mapping $\varphi: \bigwedge \mathcal{V} \to \bigwedge \mathcal{V}^*$ as follows. Given $A \in \bigwedge^p \mathcal{V}$, its image $\varphi(A)$ is that co-multivector from $\bigwedge^{n-p}\mathcal{V}^*$ that satisfies $< \varphi(A), X > \; = \; < E^N, A \wedge X >$ for all the $X \in \bigwedge^{n-p}\mathcal{V}$. (See Bourbaki 1970, A III, p. 158).

Here we have, in Bourbaki's language, $\varphi(A)$ expressed as the "right inner product" of E^N with A. The element $A \wedge X$ is a multivector of the highest level n, so that the composition $< E^N, A \wedge X >$ is well defined. Moreover, $\varphi(0)$ is defined as the zero element in the dual algebra $\bigwedge \mathcal{V}^*$.

Especially, we have $\varphi(E_\varnothing) = E^N$and $\varphi(E_N) = E^\varnothing$. We may consider φ as the **perfection** of Grassmann's formation of the supplements, because it only requires a reference to E^N, thus avoiding much ballast. But **when** we factorize $E_N = \vec{e}_1 \wedge \vec{e}_2 \wedge \cdots \wedge \vec{e}_n$, then we get nearly the same images as formerly seen by handling Grassmann's operator | , because by applying φ upon $\vec{e}_i$, $\vec{e}_i \wedge \vec{e}_j$ and so on, then the same expressions as before will result, only with the "little difference" that they are now members of the co-bases, figuring its indices in the upper position.

Examples with *dim* $\mathcal{V}$ = 4 (for the treatment of the projective **three-**space):

$$\varphi(\vec{e}_1) = +E^{234}, \; \varphi(\vec{e}_2) = -E^{134}, \text{etc.}$$
$$\varphi(E_{12}) = +E^{34}, \varphi(E_{13}) = -E^{24}, \text{etc.}$$
$$\varphi(E_{123}) = \underset{\to}{e}^4, \quad \varphi(E_{124}) = -\underset{\to}{e}^3, \text{etc.}$$

(Sign + in the cases of **even** permutations of 1, 2, 3, 4.)

As Grassmann did before (1), so Bourbaki tackles the problem to explain the regressive products - but now with the assistance of the dual algebra $\bigwedge \mathcal{V}^*$ and the progressive product for its co-elements (here denoted by the slightly modified $\wedge'$). As a mere exercise, he gives at the end of his *Algèbre, Chapitre III* (Bourbaki 1970, A III, 200) the definition $A \vee B = \varphi^{-1}(\varphi(A) \wedge' \varphi(B))$, so that

$$\varphi(A \vee B) = \varphi(A) \wedge' \varphi(B)$$

(and $A \vee B = 0$ if $\varphi(A) \wedge' \varphi(B)$ is the zero element of $\bigwedge \mathcal{V}^*$).

§ 5. PROJECTIVE THREE-SPACE

It was the treatment of Projective Geometry by Grassmann's methods for which Gibbs coined the name of "Point Analysis." Herein, the projective space (to be constructed) and the employed vector space (as foundation) have **not** the same dimension, but the latter is equal to the former plus one! In other terms: All the coordinates of points, straight lines etc. are **homogeneous** coordinates. When $\vec{p} \in \mathcal{V}$, $\vec{p} \neq \vec{0}$, then $\vec{p}$ is considered as **one** of the representatives of a certain point P, and $\lambda \vec{p}$ (with $\lambda \neq 0$) represents the same point.

We shall give here only some illustrations for the projective three-space (therefore, *dim* $\mathcal{V} = 4$). The very important second exterior power, $\bigwedge^2 \mathcal{V}$, has now the dimension 6. **Only** the cleavable bivectors can represent lines. This explains why these lines form here a variety of 4 parameters, in spite of the aforementioned dimension 6.[5]

The straight lines are here especially interesting because they play a self-dual role in the projective three-space. The join of the two points A, B has a representative of the form $\vec{a} \wedge \vec{b}$. But the same line may be treated, too, in a dual manner as the meet of two planes - expressed by the **regressive** product $U \vee V$ of the **trivectors** which represent these planes.

Any **plane** admits the representation by a trivector $U = \vec{a} \wedge \vec{b} \wedge \vec{c}$ (the join of 3 linearly independent points A, B, C) - or by means of a certain co-vector $\underset{\rightarrow}{u}$ which figures in the "equation" of the plane, $< \underset{\rightarrow}{u}, \vec{x} > \ = 0$, the condition for "incidence" of the plane with its points.

There holds the relation $\underset{\rightarrow}{u} = \varphi(U) = \varphi(\vec{a} \wedge \vec{b} \wedge \vec{c})$, perhaps with the help of a real factor $\lambda \neq 0$. Already here, one may appreciate the great flexibility of Grassmann's apparatus.

In the present conditions, the important regressive products are the following ones. In each example, the form with $\vee$ gives a clear idea of the geometric constellation.

(i) Regressive product, trivector T by vector $\vec{a}$.
Resulting level: s = 3 + 1 - 4 = 0, according to (2).
Reduction to determinants: $T \vee \vec{a} = [T, \vec{a}] \cdot E_{\varnothing}$.

(ii) Trivector T by bivector $\vec{a} \wedge \vec{b}$; s = 3 + 2 - 4 = 1.

$$T \vee (\vec{a} \wedge \vec{b}) = [T, \vec{b}]\, \vec{a} - [T, \vec{a}]\, \vec{b}.$$

---

5 As to non-cleavable bivectors and multivectors see Zaddach 1994, pp. 107, 133 and 231.

Proved by substituting vectors $\vec{e}_1,\cdots,\vec{e}_4$ such that $\vec{e}_1 \wedge \vec{e}_2 \wedge \vec{e}_3 \wedge \vec{e}_4 = E_N$ (distinguished neutral element for $\vee$).

Forder **postulates** (ii) in order to **define** this special regressive product.[6] He could not give (i), because in the "Old Style", there is no difference between the two expressions.

(iii) Regressive product of two trivectors, $\left(\vec{a} \wedge \vec{b} \wedge \vec{c}\right) \vee T$. Resulting level:
s = 3 + 3 - 4 = 2. - Reduction to determinants:

$$(\vec{a} \wedge \vec{b} \wedge \vec{c}) \vee T = [\vec{a},T]\, \vec{b} \wedge \vec{c} + \left[\vec{b},T\right] \vec{c} \wedge \vec{a} + [\vec{c},T]\, \vec{a} \wedge \vec{b}.$$

With a different notation, this identity, too, is already present in Grassmann's "A_2", formula 181 (p. 136 in the 1896 edition). Geometric interpretation: Intersection of two planes; representation of this line as a linear combination of 3 arbitrary lines that are situated in one of the planes.

(iv) A new look at $\left(\vec{a} \wedge \vec{b}\right) \vee \left(\vec{c} \wedge \vec{d}\right)$: There are two non-trivial possibilities. - Either the result is of level 0 $\left(\lambda E_{\varnothing} \neq 0\right)$. In this case, the represented points A, B, C, D are linearly **independent**, and the two joins AB and CD are **skew lines** (the numerical value of λ being without significance in Projective Geometry). --

Or we have $\left(\vec{a} \wedge \vec{b}\right) \vee \left(\vec{c} \wedge \vec{d}\right) = 0$, the zero element of the algebra. Then the two joins are coplanar, intersecting each other. (To determine their common plane and their intersection, a special technique is useful; see Zaddach 1994, p. 227).

§ 6. THE HYPERBOLOID OF ONE SHEET

Let F, G, H be the (cleavable) bivectorial representatives of three straight lines, belonging to the same family of generators (of the given ruled hyperboloid). At the beginning, we need, too, the factorization $G = \vec{a} \wedge \vec{b}$ (such that the corresponding points A, B are situated on this generator, but not on the two others). Let us now inspect the condition

$$[\vec{x},F,\vec{a}]\left[\vec{b},H,\vec{x}\right] = \left[\vec{x},F,\vec{b}\right][\vec{a},H,\vec{x}] \tag{4}$$

The presupposed skewness of the generators warrants that, in the determinants, each trivectorial coefficient of $\vec{x}$ is different from zero:

6 Forder 1941, p. 104. See commentary on (3).

$$F \wedge \vec{a},\ F \wedge \vec{b},\ H \wedge \vec{a},\ H \wedge \vec{b} \neq 0.$$

For this reason, the condition (4) is exactly of grade 2 with respect to $\vec{x}$. - Obviously, (4) is satisfied by all the points on the lines represented by F and H. As to the points P situated on the third generator, they can be represented in the form $\vec{p} = \alpha\vec{a} + \beta\vec{b}$, and it is easily verified that they do satisfy (4).

In this manner, (4) is confirmed as the equation of the ruled hyperboloid. After multiplying by $E_\varnothing$ and applying (i), we have

$$\left((\vec{x} \wedge F) \vee \vec{a}\right) \left[\vec{b}, H, \vec{x}\right] - \left((\vec{x} \wedge F) \vee \vec{b}\right) \left[\vec{a}, H, \vec{x}\right] = 0.$$

This can be transformed by the use of (ii):

$$(\vec{x} \wedge F) \vee \left(\left(\vec{a} \wedge \vec{b}\right) \vee (H \wedge \vec{x})\right) = 0,$$

and so we have derived already, in modern notation, Grassmann's splendid representation of the ruled hyperboloid (1855):

$$(\vec{x} \wedge F) \vee G \vee (H \wedge \vec{x}) = 0,$$

unfortunately published at a time when no one had yet succeeded in penetrating the mysteries of the *Ausdehnungslehre* (see Grassmann 1855). In words: A point X (**not** on the generators represented by F and H) belongs to the hyperboloid, iff the planes with trivectorial representatives, $\vec{x} \wedge F$ and $H \wedge \vec{x}$, have an intersection that meets with the generator AB (represented by $G = \vec{a} \wedge \vec{b}$).[7]

APPENDICES

(I) If the point P is situated on the generator represented by $G = \vec{a} \wedge \vec{b}$, then the bivector $(\vec{p} \wedge F) \vee (H \wedge \vec{p})$ represents the straight line through P that belongs to the **conjugate familiy** of generators.

(II) Grassmann's Algebra is the adequate tool to work with the **polarity**, too, that is defined by the hyperboloid. Exercise: Establish the formulae for the assignments

pole ---- polar plane and: line ---- polar line.

---

7 See also Forder 1941, p. 153.

(III) By $\vec{x} \mapsto G \vee (H \wedge \vec{x})$, there is induced a degenerated mapping of the projective three-space upon the straight line, represented by G, a "projection" by means of rays that are emitted from the other line.

Departamento de Matemática, Universidad de Tarapacá, Arica (Chile)

REFERENCES

N. Bourbaki, *Éléments de Mathématique. Algèbre, Chapitre 1 à 3* (Hermann: Paris 1970).

Cesare Burali-Forti, Roberto Marcolongo, *Elementi di calcolo vettoriale con numerose applicazioni* (Zanichelli: Bologna 1909).

Harley Flanders, *Differential Forms* (Academic Press: New York 1963).

Henry G. Forder, *The Calculus of Extension* (Cambridge: Cambridge University Press 1941). Reprint 1960 by Chelsea Publ.Comp., New York.

Hermann G. Grassmann, "Die stereometrische Gleichung zweiten Grades und die dadurch dargestellten Oberflächen" *Journal für reine und angewandte Mathematik*, 1855, *49*, 37-46. Also in: *Grassmanns gesammelte mathematische und physikalische Werke*, Bd. II, 1, (Teubner: Leipzig 1904), 170 - 179.

Giuseppe Peano, *Calcolo geometrico secondo l'Ausdehnungslehre di H. Grassmann* (Fratelli Bocca: Torino 1888).

Günter Pickert, *Analytische Geometrie*. Vierte Auflage (Akademische Verlagsgesellschaft Geest u. Portig: Leipzig 1961).

Arno Zaddach, *Grassmanns Algebra in der Geometrie* (B.I. Wissenschaftsverlag: Mannheim 1994).

GÜNTER BRAUNSS

THE GRASSMANN PRODUCT IN PHYSICS

I would like to begin my talk on alternating or Grassmann products in physics with a few remarks on the rôle of the name 'Grassmann' in mathematical physics, viewed from my personal experience in the past forty years .

REFERENCES TO GRASSMANN'S WORK IN THE PHYSICAL SCIENCES IN THE PAST 45 YEARS

When I began to study physics in 1952, textbooks in physics in which the name Grassmann appeared were a rarity. In fact, I remember one book only - Sommerfeld's lectures on mechanics - where this name appeared in a list of notations for the cross or vector product. Nothing, however, was mentioned in this book which explained Grassmann's contribution to the specific subject. Mathematical physics at that time had not yet gained the independence it has today. As a discipline on its own it was not wholly accepted by orthodox mathematicians and even less so by orthodox physicists. This began to change when in theoretical physics new mathematical problems came up, related to group theory, algebra, very sophisticated functional analysis and differential geometry. These problems were partly treated by theoretical physicists themselves; a prominent example is the work of Wigner on the representation of groups. But they were also taken up by mathematicians, like Gelfand and Mackey, to name only two. This helped mathematical physics to become more independent and self-assured.

In 1966, during a stay with the group of Bryce DeWitt at Chapel Hill, I came for the first time across the concepts of Grassmann algebras and Grassmann manifolds in two then just published books, *Foundations of Mechanics* by R. Abraham, and *Lie Groups for Physicists* by R. Hermann. Both books, however, gave no further reference as to Grassmann's work or to that of his contemporaries . The first time I found such a reference was in a book, *Mathematics Applied to Physics*, which appeared in 1970. It contained a contribution by G. Dechamps on exterior differential forms, in which a short review of the personal history and work of Grassmann was given. Since a few years there is

G. Schubring (ed.), Hermann Günther Graßmann (1809–1877): Visionary Mathematician, Scientist and Neohumanist Scholar, 297–301.

now a steadily increasing output of books and papers in the mathematical physical literature, in which the name Grassmann appears in the context of algebras, products, manifolds, bundles etc. named after him. To give an example: In one of the last issues of *Communications in Mathematical Physics* there appeared a paper titled "Quantum Grassmann Manifolds." As to books, I would like to mention the two excellent volumes, *Manifolds, Analysis and Physics*, written by coauthors Cècile DeWitt-Morette, Margaret Dillard-Bleick and Yvonne Choquet-Bruhat, as well as a recent monograph, *Geometry of Phase Spaces*, by Jan Sławianowski. It seems to me that while mathematical physics tries to pay due tribute to Herrmann Grassmann by naming concepts going originally back to him, mathematics has been rather reluctant in doing so. For example, some very popular textbooks on linear algebra, which deal extensively with exterior algebras, don't even once mention the name Grassmann. A notable exception is the book *Analytische Geometrie* by my esteemed colleague G. Pickert. In particular the name *Grassmann product* seems exclusively to be used in mathematical physics, whereas in the mathematical literature all kinds of fancy names are in use but not the name Grassmann product.

TWO EXAMPLES OF GRASSMANN ALGEBRAS IN PRESENT DAY PHYSICS

I am now coming to the proper subject of my talk, alternating *vulgo* Grassmann products in physics. Grassmann products, naturally, appear in the context of Grassmann algebras, which nowadays have such a wide range of applications in mechanics and physics, that it is impossible to give here a survey of all these applications. I therefore will restrict myself to two examples. The first example, which I only will sketch, is treated in detail in a recent very lucid monograph, *Infinite-Dimensional Dynamical Systems in Mechanics and Physics*, by Roger Temam.

To study the behavior of a dynamical system you need to know (among other things) how the volume, say, of the unit ball in the corresponding state space, changes under the influence of the given semi-group of time evolution. For a finite system in a finite-dimensional Euclidean space one can calculate changes of volume by using a finite-dimensional Grassmann algebra (together with the usual partial derivatives). This scheme can be extended to infinite systems acting in an infinite-dimensional Hilbert space by using an infinite- dimensional Grassmann algebra (together with Fréchet Derivatives). The Grassmann algebra can be constructed in a fairly easy way with the help of completely antisymmetric tensor products of a generating Hilbert space. In this way one can calculate changes of volume and, as a consequence, determine Lyapunow exponents and numbers, which in turn allow statements on the stability of the considered system.

As a second example I have chosen one in which the purely algebraic aspect of a Grassmann product is dominant- a point of view which, by the way, was emphasized by Grassmann himself in his *Lineale Ausdehnungslehre*. I am going to show that a mathematical formulation of what physicists call a "supersymmetric quantization" most naturally leads to the introduction of a Grassmann product.

One has in physics two types of quantization: Bose type and Fermi type. The first one is roughly characterized by commutators related with skew-symmetric viz. symplectic bilinear forms and by completely symmetric state space functions. It applies to particles with zero or integer spin. The second one can roughly be characterized by anticommutators related with symmetric bilinear forms and by completely antisymmetric state functions. It applies to particles with half-integer spin. Supersymmetric quantization tries to unify both concepts. An abstract mathematical formalism for a Bose type quantization can be constructed as follows. Let X be a linear topological space, e. g. $\mathbb{R}^{2N}$ in case of a system with N degrees of freedom or the space $S_r(\mathbb{R}^m)$ of real-valued rapidly decreasing functions in case of infinite-dimensional systems. Let further w_0: $X \rightarrow \mathbb{R}$ be a linear functional determined by the phase space coordinates of a Hamiltonian system, and let W_0: $X \rightarrow S^1 : f \rightarrow W_0(f) = \exp(iw_0(f))$. That is, W_0 is a character, which determines a (unitary) representation of the additive group X. Let $G_0(X,w_0)$ denote the group generated by the $W_0(f)$, $f \in X$, with respect to (ordinary) multiplication. The corresponding complex linear hull is a complex commutative algebra, which shall be denoted by $A_0(X,w_0)$. It can be considered as a generating algebra of the algebra of classical observables related with w_0 (meaning that we can construct the latter from $A_0(X,w_0)$ by taking appropriate limits). A Bose type quantization of $A_0(X,w_0)$ can now be constructed as a deformation which is induced by a projective representation in the following way. Let $\Gamma : X \times X \rightarrow \mathbb{R}$ be a skew-symmetric nondegenerate bilinear form (in general the symplectic form of the considered Hamiltonian system). Then the function

$$\gamma_{\hbar} : X \times X \rightarrow S^1 : \gamma_{\hbar}(f,g) = \exp(i\hbar\Gamma(f,g))$$

has the properties of a multiplier (in the sense of proj ective representation of a group). Hence

$$W_0(f) \circ_{\hbar} W_0(g) := \gamma_{\hbar}(f,g) W_0(f+g)$$

determines an associative (noncommutative) product "$\circ_\hbar$" (this product is also called *Moyal product* or *star product* and denoted by "$*_\hbar$"). Denote by $A_0(X,w_0,\hbar)$, the algebra generated by the $W_0(f)$, $f \in X$, with respect to this product. The algebra $A_0(X,w_0,\hbar)$ is then a generating algebra for a Bose type quantization (or rather a *weak* Bose type quantization since we have used a weak topology). The above mentioned commutator relation is then given by

$$w_0(f) \circ_\hbar w_0(g) - w_0(g) \circ_\hbar w_0(f) = 2i\hbar\Gamma(f,g) .$$

The symmetry properties of the system are reflected by the relation

$$w_0(f_1)\cdots w_0(f_n) = (1/n!)\sum_\pi w_0\left(f_{\pi(1)}\right)\circ_\hbar \cdots \circ_\hbar w_0\left(f_{\pi(n)}\right)$$

where $\pi:(1,\cdots,n) \to (\pi(1),\cdots,\pi(n))$ denotes a permutation. Trying to extend this quantization scheme to Fermi type systems simply by replacing the skew-symmetric by a symmetric bilinear form, however, fails. The crucial point is the commutative product. That is, in order to copy the above construction one also has to replace the hitherto commutative product by an alternating, that is, a Grassmann product. This means that the linear functional w_0 has to be replaced by a linear map w_1 from X into some (complex) Grassmann algebra with a unit element $\mathbf{1}$, such that

$$w_1(f)w_1(g) = -w_1(g)w_1(f) \quad \forall f,g \in X.$$

Let $G_1(X,w_1)$ be the group generated by the

$$W_1(f) = \exp(iw_1(f)) \quad (= \mathbf{1} + iw_1(f)), f \in X,$$

and let $A_1(X,w_1)$ denote the corresponding algebra. Deform $G_1(X,w_1)$ via a projective representation in the following manner. Let $S : X \times X \to \mathbb{R}$ be a nondegenerate *symmetric* bilinear form. Then

$$\sigma_\hbar : X \times X \to S^1 : \sigma_\hbar(f,g) = \exp(i\hbar S(f,g))$$

is a multiplier. Hence

$$(W_1(f_1)\cdots W_1(f_m)) \circ_\hbar (W_1(g_1)\cdots W_1(g_n)) :=$$
$$\left[\prod_{j=1}^m \prod_{k=1}^n \sigma_\hbar(f_j,g_k)\right] W_1(f_1)\cdots W_1(f_m) W_1(g_1)\cdots W_1(g_n)$$

determines again an associative multiplication "$\circ_\hbar$" (for which we use the same notation as in the Bose type case). Denote the resulting "deformed" algebra by $A_1(X, w_1, \hbar)$ and call it a "quantization" of the Grassmann algebra $A_1(X, w_1)$ This essentially yields then the desired Fermi type quantization. It is illustrated by the anticommutation relation

$$w_1(f) \circ_\hbar w_1(g) + w_1(g) \circ_\hbar w_1(f) = 2i\hbar\, S(f,g)\, \mathbf{1}$$

and a relation which reflects the inherent antisymmetry,

$$w_1(f)\cdots w_1(f_n) = (1/n!)\sum\nolimits_\pi sign(\pi) w_1\left(f_{\pi(1)}\right) \circ_\hbar \cdots \circ_\hbar w_1\left(f_{\pi(n)}\right),$$

where π: $(1,\ldots,n) \rightarrow (\pi(1),\ldots,\pi(n))$ again denotes a permutation and sign(π) stands for the signature of this permutation. In a final step one arrives then at a Z_2-graded algebra (the so-called "superalgebra" of a supersymmetric quantization)

$$A(X \oplus X, w, \hbar) := A_0(w_0, X, \hbar) \oplus A_1(w_1, X\hbar),$$

where $w(f \oplus g) \equiv w_0(f) + w_1(g)$ and - this is necessary condition -

$$w_0(f) \circ_\hbar w_1(g) = w_0(g) w_1(f) \quad \forall f, g \in X.$$ [1]

Mathematisches Institut, Universität Gießen

REFERENCES

Ralph Abraham, *Foundations of mechanics: a mathematical exposition of classical mechanics with an introduction to the qualitiative theory of dynamical systems and applications to the three-body problem* (New York: Benjamin, 1967).

Yvonne Choquet-Bruhat, Cécile DeWitt-Morette, Margaret Dillard-Bleick, *Analysis, manifolds and physics* (Amsterdam: North-Holland, 1977).

Robert Hermann, *Lie groups for physicists* (New York: Benjamin, 1966).

Günter Pickert, *Analytische Geometrie*. - 4. Auflage (Leipzig: Akad. Verlagsges., 1967).

Elie Roubine, *Mathematics applied to physics* (Berlin: Springer 1970).

Jan I. Sławianowski, *Geometry of Phase Spaces* (New York: J. Wiley, 1992).

Roger Teman, *Infinite-Dimensional Dynamical Systems in Mechanics and Physics* (New York: Springer, 1988).

[1] Proofs of the above relations will be given in a forthcoming paper, "Weak Supersymmetric Moyal Quantization," by the author. An introduction to Grassmann products, Grassmann algebras and Grassmann manifolds related with problems in mathematical physics can be found in the books mentioned above. An extensive and up-to-date survey of deformations in the context of quantum theory can be can be found in a forthcoming paper by M. Flato and D. Sternheimer: "Star Products, Quantum Groups, Cyclic Cohomology and Pseudodifferential Calculus." (Preprint, to appear in "Contemporary Mathematics," AMS).

OLE IMMANUEL FRANKSEN

INVARIANCE UNDER NESTING - AN ASPECT OF ARRAY-BASED LOGIC WITH RELATION TO GRASSMANN AND PEIRCE

1. TWO HISTORICAL THREADS INTERTWINED

Hermann G. Grassmann's *Die lineale Ausdehnungslehre* (1844) made no impact on logic, except perhaps for the indirect influence it exerted upon Ernst Schröder and a few others through the logical publications of Hermann's brother, Robert Grassmann (Grattan-Guinness 1996; Peckhaus 1996). For although Grassmann saw great generality in his remarkable algebraic-geometrical formulations, logic and its key question of inference were not his concern. Yet, in old age about 1909, Charles Sanders Peirce would remark that

"in my memoir of 1870 ... I made no reference to Grassmann ... which I am all but absolutely sure that I should have done had I been acquainted with either of Grassmann's volumes. So I infer that the too exclusive admiration of Hamilton in our household prevented my acquaintance with that great system" (Hartshorne *et al.* 1932, IV, §669, p. 566).

Peirce's 1870 memoir, entitled *Description of a notation for the logic of relatives, resulting from an amplification of the conceptions of Boole's calculus of logic* (ibid., III, §§45-149, pp. 27-98), is remarkable for the fact that it turned De Morgan's *logic of relations*, published in 1859, into an algebra of relations. In particular, it was in this paper Peirce introduced the first instance of a Boolean inner product, his relative product, to express the application of a relation such as *lw* for "whatever is a lover of a woman," in that he remarked that "this notation is the same as that used by Mr. De Morgan, although he appears not to have had multiplication in his mind" (ibid., III, §68, p. 389).

Referring to this paper in a public lecture in 1903, Peirce boasted that he had "made a contribution to this subject [i.e., logic] which nobody who masters the subject can deny was the most important excepting Boole's original work that has ever been made" (Eisele 1978, IV, p. 1519). These were prophetic words. No logician ever denied it. No logician ever paid any attention to it. Even Alonzo Church, that eminent

G. Schubring (ed.), Hermann Günther Graßmann (1809–1877): Visionary Mathematician, Scientist and Neohumanist Scholar, 303–335.

scholar of the history of logic, only mentioned it in passing (Church 1956, p. 57 & p. 156). In fact, the only one who seems to have studied it seriously was Hawkins (Hawkins 1971), and that was from the purely algebraic point of view of hypercomplex numbers. This was for good reason. For, however astounding it may seem, Peirce took as his starting point a purely logical analogy to Hamilton's quaternion, naming it a *logical quaternion*, with its units denoting mutually disjoint classes of individuals, his so-called *relatives*.

In this algebraic sense, the two historical threads from Grassmann and Peirce are intertwined. Both were concerned with some form of a linear combination: ax + by + cz + ..., amenable to various interpretations of the symbols. To put it in array-theoretic terms, they were concerned with *lists*, which are a very simple kind of array. But there is more to it. For in their exploration of this form, they shared a philosophical attitude which best perhaps could be described as an intuition - some might even say vision - about a structural extension in some hierarchical sense. Thus, in his 1870 memoir Peirce would explain:

"The conception of multiplication we have adopted is that of the application of one relation to another. So, ... the multiplication of quaternions is the application of one such relation to a second. Even ordinary numerical multiplication involves the same idea, for 2 ×3 is a pair of triplets, and 3×2 is a triplet of pairs, where "triplet of" and "pair of" are evidently relatives." (Hartshorne et al., 1932, III, §76, p. 44)

One interpretation of this "idea," and the one we shall adopt, is that we are concerned with *nested arrays* expressed in their simplest form of lists embedded within lists. For, as we intend to illustrate in the following representing Grassmann's "*Stufen*" as *levels of nesting*, the modern concept of a nested array provides a means of explanation of "the awkwardly expressed philosophical beliefs" (Nagel 1939, §37, p. 173) with which Grassmann interlined his *Ausdehnungslehre*, causing mathematicians and philosophers alike to ignore it as obscure and incomprehensible. In particular, to give perspective to Grassmann's gropings for a realization of his philosophical intuitions, we shall consider them in relation to the abstract concept of *invariance under nesting* that has been developed in connection with a novel technology known as array-based logic

The plan of the following discussion is therefore this. After a brief introduction to array-based logic and its underlying array-theoretic foundation, we shall consider Grassmann's definition of a pair as a means to implement the structural aspect of dialectic by nesting. Then, after a short explanation of some technical conventions for the computer display of arrays, we shall discuss the concept of invariance under nesting illustrated by Peirce's proof, in the context of quaternions, that "every associative algebra has a matrix representation." Finally, coming back again to Grassmann, we shall see that the common idea he was groping for in his discussions of the distributive law and the element-by-element addition, could have been the array-theoretic operation of interchanging two adjacent levels of nesting.

2. IN RELATION TO ARRAY-BASED LOGIC

Array-based logic is a novel technology founded on the algebraic tradition in the history of logic. The basic idea is to *explore the consequences of considering truth-values as physical measurements.* That is, the aim is to formalize logic in accordance with the theoretical structure of discrete systems and express this formalization algebraically in array-theoretic terms. In particular, by analogy to the modern approach to physics, invariance (or symmetries) under mathematical groups of transformations are examined to identify laws and introduce isomorphisms (or analogies) that unify various branches of logic (such as propositional algebra, predicate calculus, and relational algebra) as well as extensions to multi-valued logic. A major advantage of this approach is that a small set of geometrical operations will apply in general. For example, the fundamental rule of inference (besides substitution) is a geometrical interpretation of what Peirce, taking his terminology from William Whewell, called *colligation* (e.g., Hartshorne *et al.* 1932, II, §451, p. 276). Paraphrasing *Alice in Wonderland* by Lewis Carroll who also indulged in the 'game of logic', one might say that colligation is what remains of a tensor contraction when the addition is removed.

The identification of this set of geometrical operations is of fundamental importance from a technological point of view. For it admits the *simultaneous use of two isomorphic representations*, a *theoretical* one oriented towards the design and modelling of systems (Franksen 1978 & 1984) and a *technological* one to ensure maximum speed and minimum storage requirements in industrial computer implementations (Franksen 1977; Falster 1981). The early development of this new technology was undertaken jointly with the Danish hifi company Bang & Olufsen, whose audio-video products today are all logically controlled by software based on this technology and implemented in a small processor. Based on the experiences of applying this technology to their own products, the company marketed in 1990 the first software product, *Beologic*™, for applications in logic control of electronic apparatus.

Inspired by Gabriel Kron's (Kron 1939) application of tensors to electrical engineering problems, our interest was originally focused on *non-nested arrays* such as he had introduced to make the representation of tensor components operational in a numerical sense. However, as the potential of the technology was explored in an ever widening range of industrial applications, a need was recognized to express meaning and simplify the formulations by applying nesting of ordered arrays. Since further it had been recognized for quite some time that array-oriented programming languages such as APL, although well suited for industrial purposes, were not mathematically consistent in the boundary cases (Falster 1981), it was decided to adopt as 'algebraic platform' *the theory of nested arrays* developed by Trenchard More, Jr. at IBM (now retired) over a lifetime (More 1973, 1981, 1993).

A deciding factor, among others, for this decision was More's (More 1976a) use of the computer to generate proofs of the consequences of the *associative law of composition* of the operations in his theory for a table of some 64,000 compositions. Curiously, this contribution to the establishment of a new algebra, one of the more astounding feats in the area of automated theorem-proving in our generation, has been largely neglected by mathematicians and computer scientists alike. Another factor of decisive importance was that More had made his theory *one-sorted*, ensuring only one set of rules of logical quantification by virtue of the fact that all variables are of the same kind, namely nested arrays. A third factor was that, in the tradition initiated by a student of Hilbert, Moses Schönfinkel (Schönfinkel 1924), and fully developed by Haskell Curry (Curry 1958) and others, More based his theory exclusively on *monadic operations*. Thus, as proposed originally by Schönfinkel (Schönfinkel 1924), any left argument is "*curryed*" (after Curry) to the operation so as to form a new monadic operation characterized by a left-associated argument. This admits the possibility of invoking curryed operations at any level of nesting by means of an ingenious operator "EACH" invented by More (see More 1973). Though, from the viewpoint of logic, the true significance of founding array theory on monadic functions derives from the fact that, with quantifications substituted by conjunctions or disjunctions, it guarantees a solution by the truth-table technique to the decision problem ("Entscheidungsproblem"), whether or not an expression is *valid* (or tautological) (see Bernays and Schönfinkel, 1928).

3. GRASSMANN'S DEFINITION OF A PAIR

The philosophical influence of Schleiermacher's *Dialektik* upon Grassmann's conception of algebraic form in his *Ausdehnungslehre* is perhaps best illustrated by Grassmann's definition of a pair. For it is in this definition that his perception of the *structural form of dialectic* (with its transition from the two opposites of *thesis* and *antithesis* to a *synthesis)* found its algebraic expression in the abstract form of nested pairs:

"When two quantities or forms (which latter name we prefer as the general one, ...) are mutually connected, they are called the *terms* of the connection, [and] the form produced by the connection the *result* of the connection. Should the two terms need to be distinguished, then we name the one the *first term*, the other the *last term*. As a general designation of the connection we take the symbol $\frown$; now if a and b are such terms, and thus a the first term, b the last term, then we denote the result of the connection $(a\frown b)$; in that here the parentheses should express that the connection should be perceived not any more in the separation of its terms, but as a conceptual unity [*Footnote*: Now, in what way this unity took effect, and what was thereby given up each time of the perception of each of the interconnected terms, depend on the nature of each particular connection].

The result of the connection can again be connected with other forms, and thus one arrives at a connection of several terms, which first and foremost, however, always appears as a connection of two at a time. For convenience, we use the usual abbreviated parenthetical notation, namely in that we omit parentheses around symbols belonging together, if the opening parenthesis [(] either stands at the beginning of the whole expression, or would follow after another opening parenthesis; e.g., instead of $((a\frown b)\frown c)$ we write $a\frown b\frown c$." (Grassmann 1844, §2, 2)

In **Figure 1** we have illustrated the main points in Grassmann's definition of a pair from an array-theoretic point of view, representing his left-associative, so-called *synthetic connection* '$\frown$' by its corresponding left-associative operation '*pair*'. Thus, in the three columns one may compare step-by-step the parallel developments in the dialectic process, Grassmann's definition, and the building-up of a nested list of pairs. In particular, with respect to the latter, we observe how the two isolated elements a and b are paired to form a new conceptual unity, namely a list of two elements defined by axis p, say. This list is subsequently paired with a third element c to form yet a new conceptual unity, namely the nested list of two elements defined by axis q. Indeed, as we now construe this discovery, *pairing* is the structural operation that can nest arbitrary arrays as elements into more deeply nested arrays.

However, there is also a non-structural part to the recursive step of the dialectic process or argument derived from the underlying logical assumption. Namely that any thesis, since it expresses a truth, must have a logical negation (produced by application of the monadic operation "*not*") by the law of the excluded middle. That is, "a or not-a" is a tautology or always true. Equally important as pairing, therefore, is the determination of the *opposite* of a thesis, namely its antithesis. While it is generally accepted that Grassmann's "reflections upon the meaning of negative signs in geometry" (Nagel 1939, p. 168), and thus the notion of the opposite, is basic to his discussion of abstract forms, he did not succeed from our point of view to single out this idea in a form that is compatible with, yet independent of nesting. To solve this problem, the opposite must be defined as a monadic operation, say "*opp*," for changing sign. Since *opp* is its own inverse, it will act as a group generator with the defining relation: opp opp $a = a$.

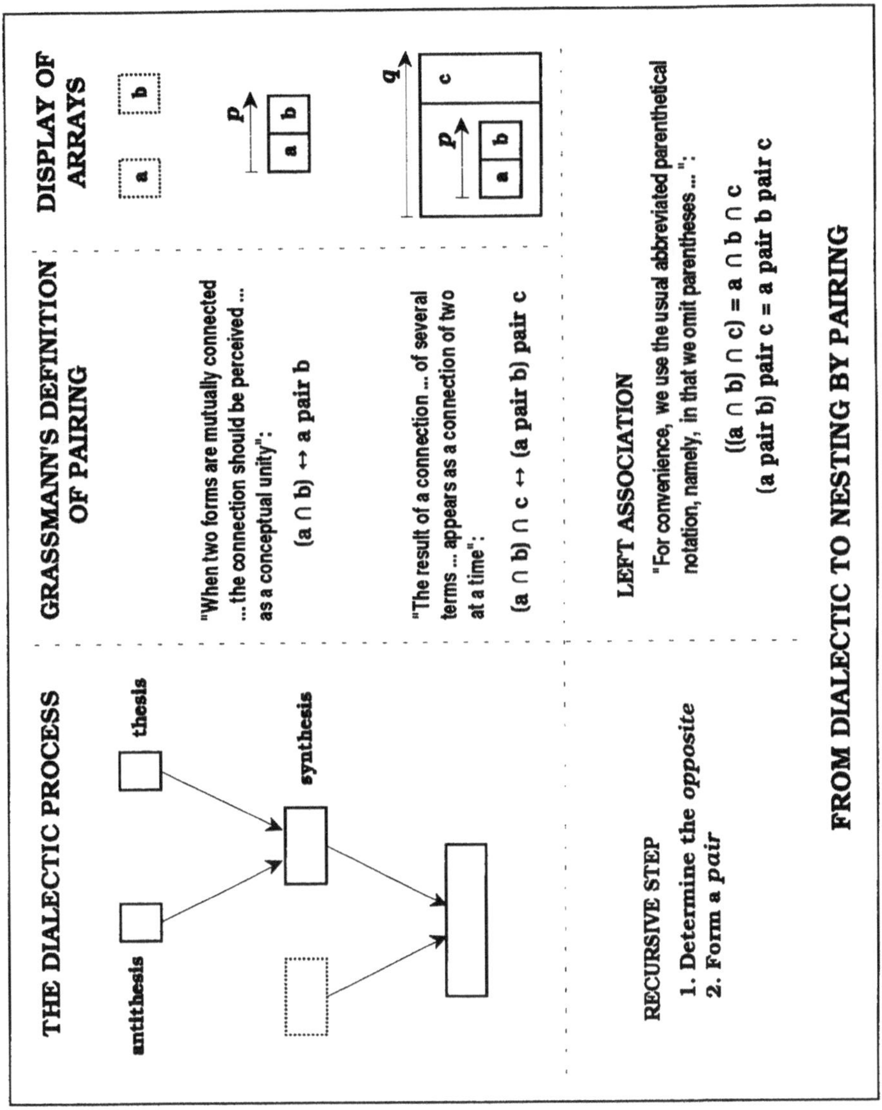

Figure 1.

This makes opposition isomorphic to logical negation, with the latter being defined by the law of double negation: not not $a = a$, discovered originally by the Stoics. That is, both operations are representations of the same abstract group, namely the cyclic group C_2 of second order.

The germinating idea in Grassmann's *Ausdehnungslehre* originated, as he pointed out in the preface (Grassmann 1844, p. V), in his observation that geometrical distances are directed lengths, with a negative sign indicating the opposite direction. His further study of this idea convinced him that the notion of opposite is founded on the inverse operations of addition and substraction. But that made it impossible for him to separate the structural properties from the non-structural ones in the above sense. In his subsequent discussion of an *analytical process* for decomposition of the result of the synthetic process, he was therefore forced to abandon the left-associative operation of pairing in favour of a more *einfach* or simple kind of connection, that he assumed to be both associative and commutative (Grassmann 1844, §5, p. 4). Under this assumption he defined the analytical process as follows (*ibid.*, §5, 5):

"Given the result of a connection and one of its terms, the analytical process consists in seeking the other term. Thus, two analytical processes belong to a connection, namely according to whether the first term or the last term is sought; and only then will both processes produce the same result, when the two terms of the original connection are interchangeable. Since the analytical process can be conceived also as a connection, we consequently distinguish between the original or synthetic connection and the dissolving or analytic connection".

Grassmann's intuition, that there had to be some structural operation opposite to that of pairing, was however correct. The monadic *Konstanzfunktion* or constancy function, proposed by Schönfinkel in (Schönfinkel 1924), enables us recover either of the terms of a pair, say the first, without requiring knowledge of the other term (as stipulated in Grassmann's definition above). In curryed form we may denote Schönfinkel's operation: *a first*, in the sense that: (*a* first) $b = a$. Grassmann's definition of a pair was therefore basically sound and could have opened up for a structural treatment of nested forms. Yet, as presented in the *Ausdehnungslehre* the question remains: Why set forward a definition only to admit it cannot be used?

In his introduction, Grassmann emphasized that "the formal sciences must not take their point of departure from axioms ["Grundsätzen"], as do the real sciences, but will take definitions instead as their foundation" (Grassmann 1844, p. XIX). If we compare this statement with the remark in his preface, that he had recast his work "several times in different forms" including the "Euclidean" one "in möglichster Strenge" or with utmost rigour (*ibid.*, p. XVI), it is hard to believe that Grassmann would introduce his definition of the nested pair without having some definite purpose in mind. The hypothesis, to be discussed in detail later on, is that he recognized, perhaps intuitively, that the distributive law and the process of element-by-element addition were in some structural manner founded on the nested form of pairing.

4. EVERY ASSOCIATIVE ALGEBRA HAS A MATRIX REPRESENTATION

Though, to add perspective to the discussion of this hypothesis, let us first introduce the idea of invariance under nesting. As a convenient and rather intuitive vehicle of explanation, we shall consider a nested interpretation of the key idea in Peirce's detailed proof in a letter, dated "1909 Feb 26," to his friend William James, on how to find a matrix representation of a hypercomplex number which, for simplicity, we shall take to be a quaternion (Eisele 1976, III/2, pp. 836-866). Prior to this, however, we briefly need to consider the conventions for the display of nested arrays on a two-dimensional medium such as a piece of paper or a computer screen.

Gabriel Kron, who in the 1930s pioneered the use of multi-dimensional arrays in the numerical solution of engineering problems, observed that "*by experience* the most practical procedure" was to "break up" the arrays in tables and represent them so on paper (Kron 1939, p. 12). In principle this is still the approach. What has been added with the advent of powerful, modern computer languages for handling nested arrays, are systematic conventions for display of these tables in nested box diagrams of varying designs. According to More's convention (More 1981, pp. 536-538), which we shall adopt, (1) the *levels* of a nested array are organized in order from the top level and down and so displayed from outside the box diagram and in, while (2) the *axes on each level* are displayed in an alternating pattern of horizontal and vertical axes in order from behind. That is, the last axis is displayed horizontally from left to right; the next-to-the-last axis vertically in front from top to bottom; the third-to-the last axis, if any, again horizontally from left to right above the lastmentioned horizontal axis, etc.

Normally, this presents no problem. But in the very special situation where we wish to compare the results of moving one or more axes, say, from the top level down to the level immediately below, we need to take into account the influence of the conventions of display.

This is illustrated in **Figure 2**, depicting in its three columns from left to right: An array with all three axes: *a b c*, on the top level; the same array with the last axis: *c*, moved down; and, again, the same array but now with the last two axes: *b c*, moved down. Comparison of the displays as well as the spatial visualizations of the array in the left and right columns, respectively, reveal that the orientations of the axes on the top level are preserved when the last two (in general, the last *even* number) of axes are moved down.

However, by contradistinction, as illustrated in the center column, the displays and hence our spatial visualizations of the array, appear with transposed axes on the top level, when solely the last axis (in general, the last *odd* number of axes) is moved down. Thus, to preserve the orienta-

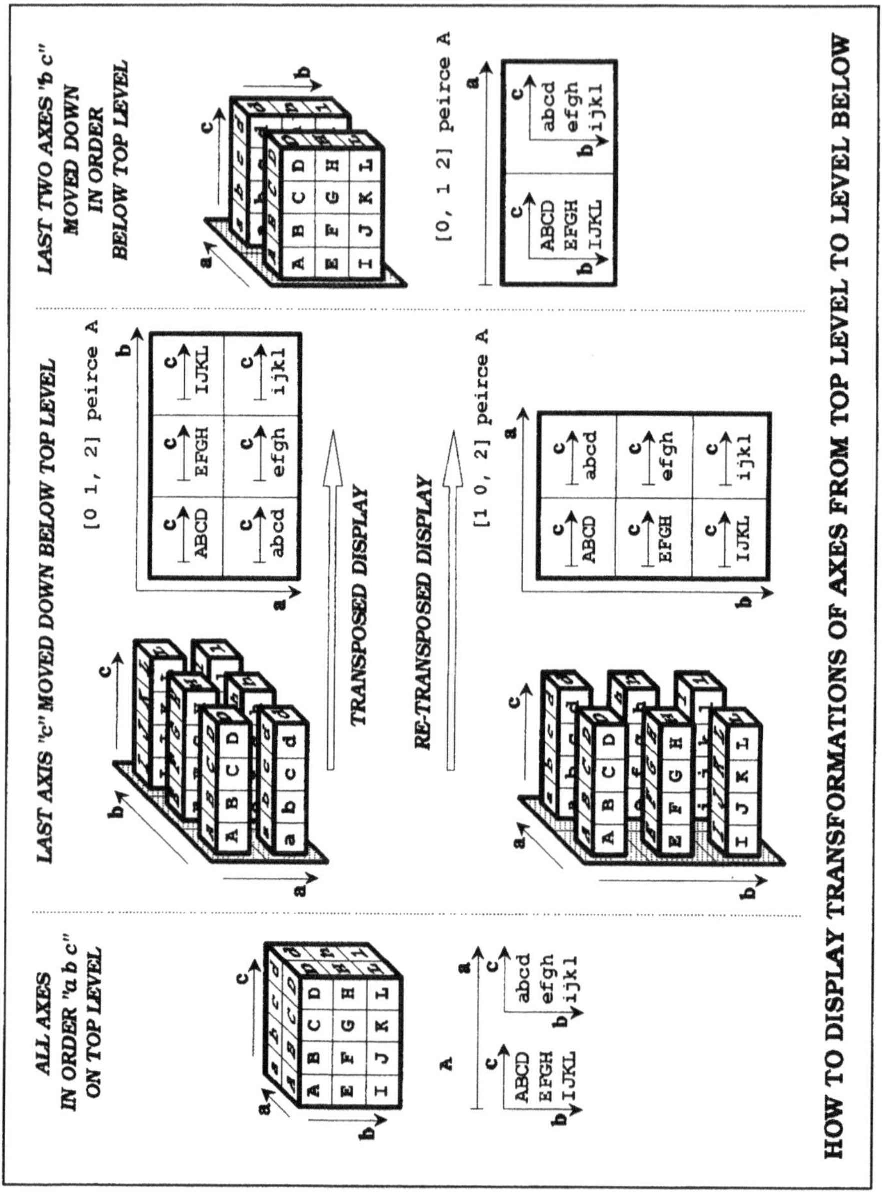

Figure 2.

tions of the axes on the top level when the last axis is moved down, it is necessary to *re-transpose* the display.

Since by convention the axes of a nested array are indexed: 0 1 2 etc., in order from the top level and down, these transformations of axes over two adjacent levels could be performed by a curryed operation[1], '*peirce*,' the curryed or left-associated argument of which is a pair, that here is specified in bracket notation with a comma to separate the axes on the top level from those on the level below. That is, with axes: *a b c*, of array *A* in the left-hand column of figure 2 indexed 0 1 2, respectively, then, to preserve the orientation of the first two axes 0 1 (or: *a b*) moving the last axis 2 (or: *c*) down, we should specify the curryed argument: [1 0, 2] and *not* [0 1, 2], as one perhaps might be inclined to think. This sets the stage for our discussion of the basic ideas underlying Peirce's proof, that every associative algebra has a matrix representation. In his 1870 memoir, Peirce had to "assert, upon reasonable inductive evidence," the general validity of his procedure (Hartshorne *et al.* 1932, III, § 130, p. 80). However, a few years later, in March 1873, he wrote his father, the astronomer and mathematician Benjamin Peirce, that he had found a proof (Eisele 1976, III/2, p. 1136). It appeared in print as an addenda to his father's remarkable *Linear Associative Algebra* published posthumously in (B. Peirce 1881).

Editing this work for publication, Charles added extensive footnotes with the matrix representation (in an index tuple notation he called "*the relative form*") of almost every algebra discussed by his father, though, as mentioned earlier, his most detailed description was in the letter to William James.

As we shall interpret Peirce, the first step is to establish a nested array representation of the multiplication table (*not* the group table) of the associative algebra in question. This is illustrated in **Figure 3** for the multiplication table of a quaternion, which is a hypercomplex number:

$$Q = \alpha\times\mathbf{1} + \beta\times\mathbf{i} + \gamma\times\mathbf{j} + \delta\times\mathbf{k}$$

---

[1] In 1880, arranging his *n*-adic "*relatives*" or logical relations as non-nested arrays in the form of "blocks [tables], cubes, and so forth", Peirce generalized the notion of the *converse* of a dyadic relation [i.e., the relation with its two arguments interchanged] to, as he expressed it, "*transpositions*" of relations (Hartshorne *et al.* 1932, III, §§ 220-225, pp. 140-144). In honour of this pioneering contribution, the author felt justified a century later, in 1988, to use Peirce's name (pronounced "purse") for the generalized operation of transposition or permutation of axes over two adjacent levels of nested arrays (Franksen 1991). In fact, as we defined this operation, it generalizes the idea of permutation of axes to *colligation* of axes, depending upon the specification of the curryed argument. The rationale behind this extension was the discovery of a generalized inverse for a so-called *multiset* (in the context of nested lists), that would encompass the ordinary notation of permutations as a special case. Called "*heap*" in 1886 by the originator, A.B. Kempe, a multiset is like a set, except that it may contain multiple occurrences of the same element.

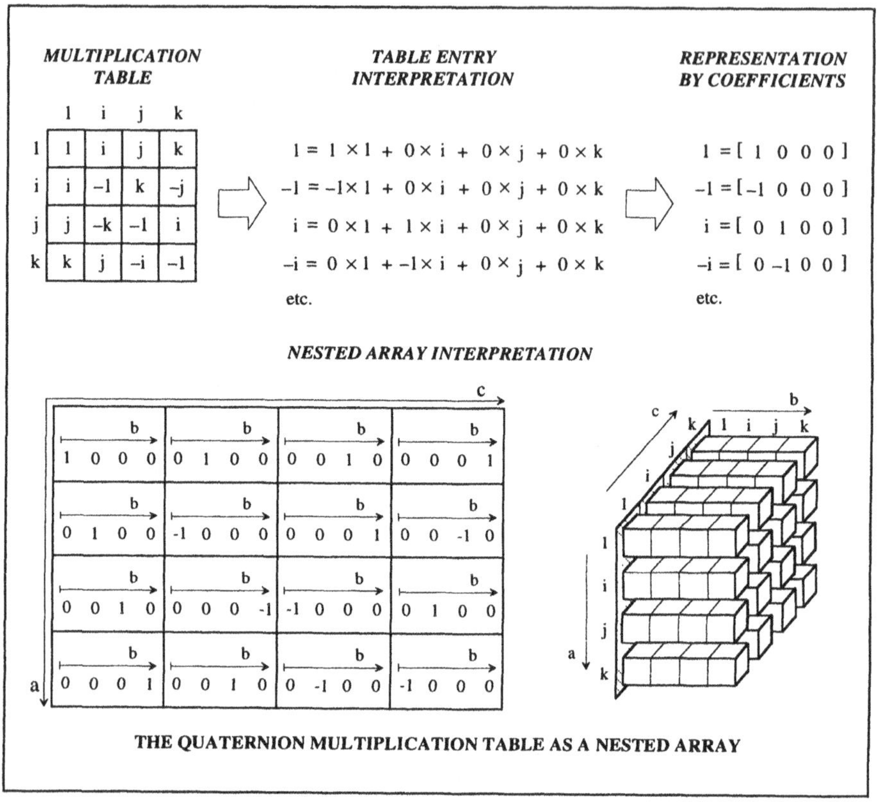

Figure 3.

Now, since the multiplication table is defined in terms of the hypercomplex units: **1 i j k**, whereas the matrix representation must be expressed by means of the real-number coefficients: α β γ δ, it is necessary to introduce a new representation of the multiplication table in terms of the units of the latter coefficients, as shown in the upper half of the figure. This gives rise to the nested array representation exhibited in the lower half of the figure, with the two-dimensional display to the left and its corresponding spatial visualization to the right.

The second step, illustrated for the spatial visualizations of the nested array in **Figure 4a** and the corresponding two-dimensional displays in **Figure 4b**, consists in moving one axis down, so as to transform the nested array from a table-of-lists to a list-of-tables. The underlying idea of this transformation, is (1) that the latter list-of-tables will have, on the top level, a list of four elements each defined by one of the hypercomplex units: **1 i j k** in that order; while (2), on the level below the top level, we obtain for each of these four hypercomplex units a corresponding table of real-number unit coefficients. Each of these tables will therefore appear as a matrix representation of the hypercomplex unit in question. That is, if we multiply each of these four tables by its corresponding real-number coefficient: α β γ or δ, and add the resulting four tables or matrices element-to-element, we have produced a matrix representation of the quaternion.

To see why this is true, let us consult Figure 4a. In the nested array representation of the multiplication table (at the extreme left), we have designated the two axes on the top level: *a c*, and the axis on the level below: *b*, assigning the order: *a b c* to the three axes. This is against our convention of display, so we have no notation for showing that the middle axis *b* is down in between the two axes *a* and *c* on the top level. But if we adopt Grassmann's conceptual use of parentheses in his definition of a pair, we could perhaps denote it: *a* (*b*) *c*. Now, preserving the order of these axes (since the quaternion multiplication is non-commutative), we can establish a table or matrix on the level below the top level in either of two ways. On one hand, as shown along the upper path towards the right of Figure 4a, we could move axis *a* down to form a list-of-tables: (*a b*) *c*. This turns out to establish, what one might call, the "modern matrix representation," the usual interpretation of which (due to Cayley) is that the quaternion is a complex number with complex coefficients. On the other hand, as shown in the lower path of the figure, we could move axis *c* down to create an alternative list-of-tables: *a (b c)*. This gives rise to the "classical matrix representation," discussed in the 19th century literature but now almost completely neglected.

As an intermediate step common to either of these processes, we could instead have moved axis b up on the top level, as shown in the center column of the figure, so as to produce an array in the form of a cube: a b c, with all three axes on the top level. From there we might proceed, either moving down the pair of axes: (a b) or the pair of axes: (b c). Now, as shown in the dotted frame to the right of Figure 4a, any nested form of this array represents all possiblecompositions by multiplication of the four hypercomplex units on each of its three axes. Therefore, these nested forms constitute but a geometrical representation of the associative law in the sense that the levels of nesting play the role of a structural interpretation of the association denoted by the parentheses.

That is, if we accept the algebraic law of associativity for the quaternion, then we must also accept the structural idea of equality of the three nested forms: $a\ b\ c = (a\ b)\ c = a\ (b\ c)$ in the figure; and even further, that they again must equal the original form: *a* (*b*) *c*, with the lower-level axis *b* in between axes *a* and c on the top level. This suggests the geometrical interpretation, that we are concerned with one and the same, abstract object, here rather loosely called "the array," which are depicted in different frames of reference. However, before we go into more details on this point, a brief remark is in order on how to denote the curryed or left-associated argument to operation *peirce* subject to our conventions of display.

For this purpose, let us consider Figure 4b which in its two columns illustrates the two alternatives, with the "classical matrix representation" represented by the form *a* (*b* *c*) to the left and the "modern matrix representation" represented by the form (*a* *b*) *c* to the right. According to our convention that the axes are numbered in order from the top level and down, the three axes of the original array: *a* (*b*) *c* must be denoted [*a* *c*, *b*] = [0 1, 2], since the two axes *a* *c* (or 0 1) on the top level must precede the axis *b* (or 2) on the level below. With the indices thus assigned to the axes, the "classical representation": *a* (*b* *c*) must be written [0, 2 1], while the "modern representation": (*a* *b*) *c* must be expressed [1, 0 2], since axis *c* (or 1) on the top level must precede the other two axes *a* *b* (or 0 2) below.

What facilitated this explanation was the fact that, to begin with, we arbitrarily placed the lower level axis *b* in the middle. But as we now recognize by comparison with Figure 2, this was simply a convenient means of performing the re-transposition prior to moving down the single axis. Of course, the reason it is necessary to nullify the transposition introduced implicitly by the conventions of display as the axis is moved down, is that otherwise this transposition of the tables would invoke the *commutative law* by virtue of the fact, that the axes, and hence the arguments of the product, are interchanged. But, as William Rowan Hamilton discovered in 1843 much to the surprise of contemporary mathematicians, quaternions form a self-consistent algebra even if they do *not* commute under multiplication.

Whether or not Peirce visualized his derivation of the two matrix representations from the associative law in terms of nested arrays, is not known. As mentioned earlier, he conceived relations rather early on as multi-dimensional, non-nested arrays. Since further, as often repeated in his writings, he maintained that "multiplication expresses relation" (e.g., Eisele 1976, III/1, p. 530), it would be natural to infer that he also might have introduced non-nested arrays, breaking them up into slabs, as a geometrical illustration of the associative law in this proof. His remark to William James in 1909 that "the proof of this is the simplest conceivable" (*ibid*., III/2, p. 856), would perhaps suggest that. Particularly intriguing in this connection is a description, published in 1911, of what Peirce himself called "the slightest possible sketch ... [of]

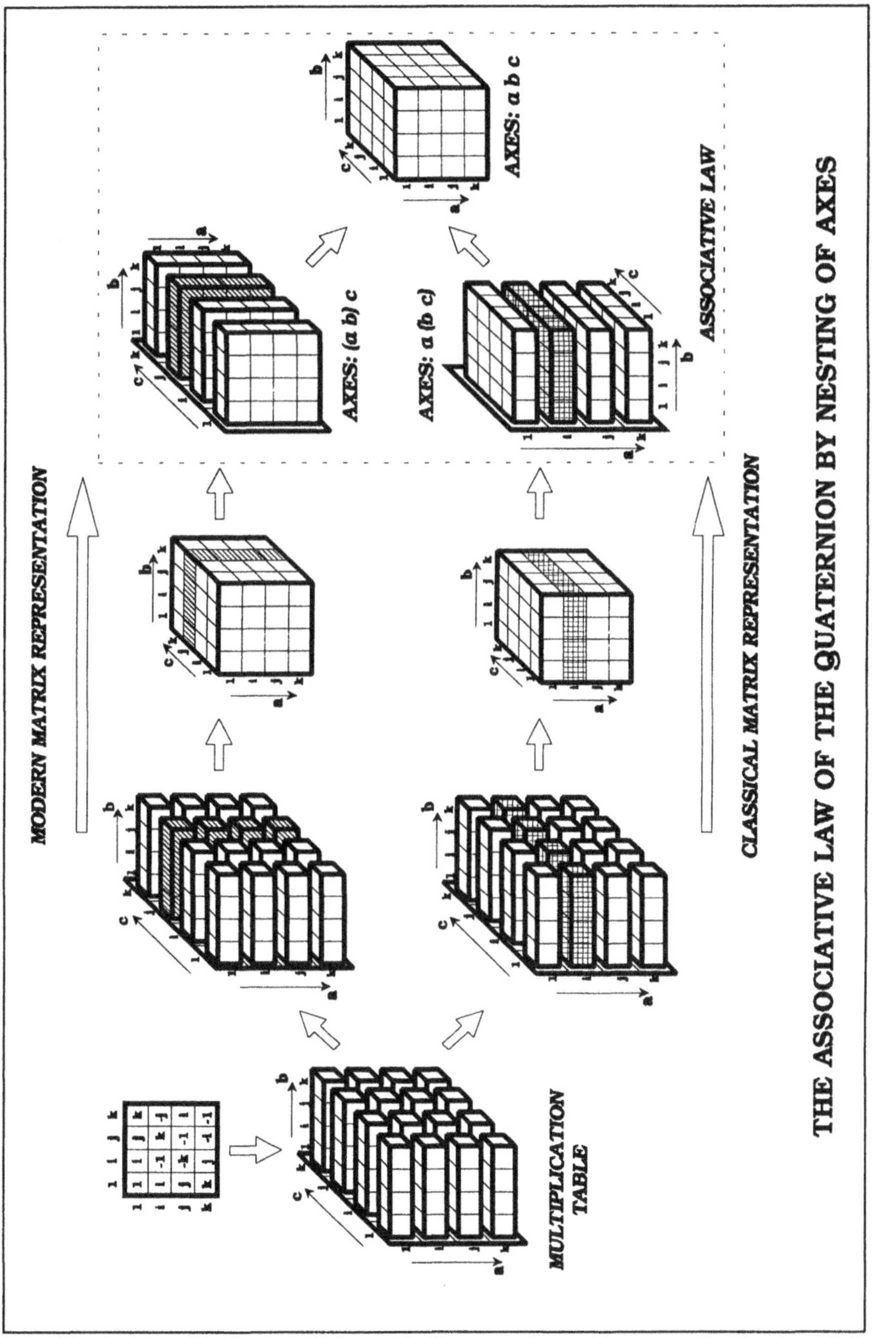
THE ASSOCIATIVE LAW OF THE QUATERNION BY NESTING OF AXES
MODERN MATRIX REPRESENTATION
CLASSICAL MATRIX REPRESENTATION
ASSOCIATIVE LAW
AXES: (a b) c
AXES: a (b c)
AXES: a b c
MULTIPLICATION TABLE

Figure 4a.

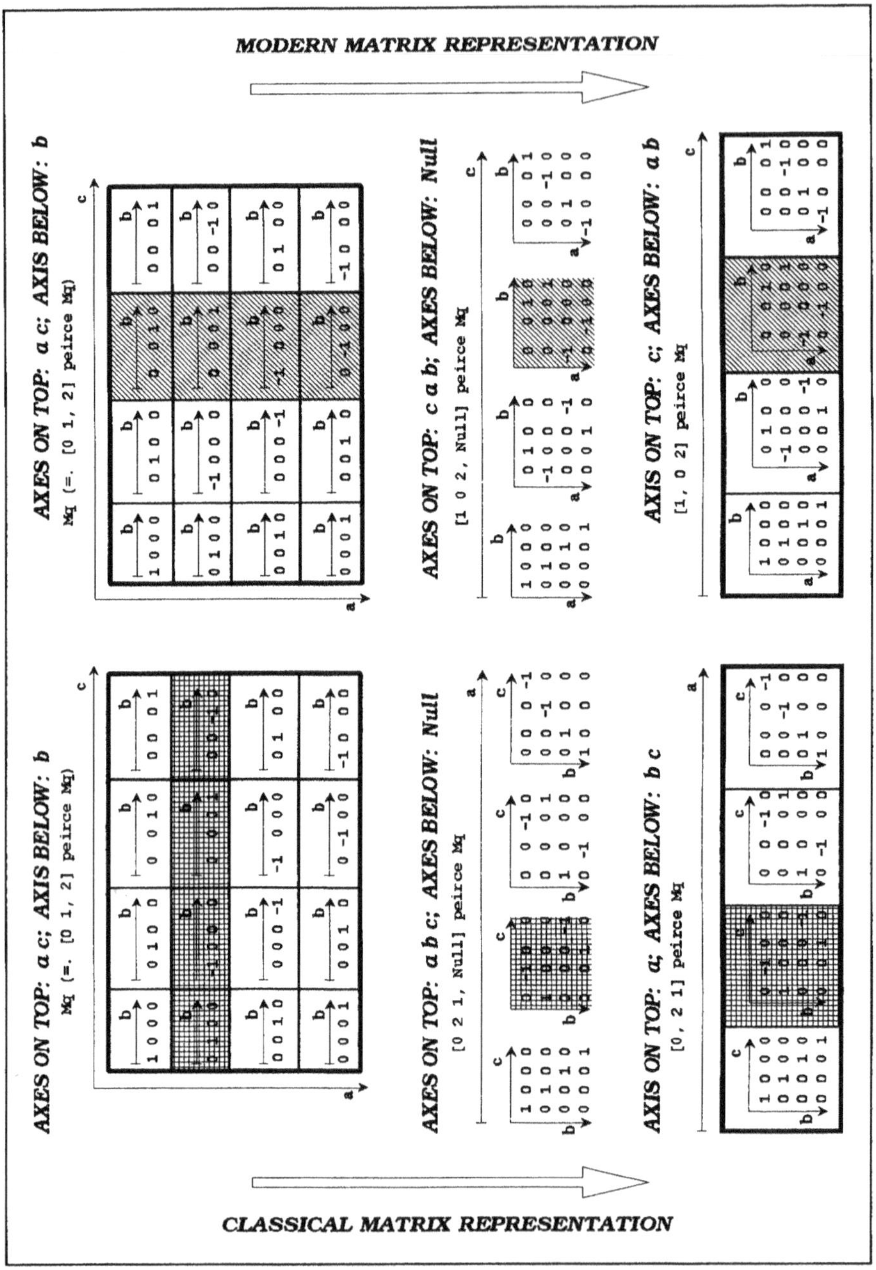
MODERN MATRIX REPRESENTATION
CLASSICAL MATRIX REPRESENTATION
AXES ON TOP: a c; AXIS BELOW: b
Mq (=. [0 1, 2] peirce Mq)
AXES ON TOP: c a b; AXES BELOW: Null
[1 0 2, Null] peirce Mq
AXIS ON TOP: c; AXES BELOW: a b
[1, 0 2] peirce Mq
AXES ON TOP: a b c; AXES BELOW: Null
[0 2 1, Null] peirce Mq
AXIS ON TOP: a; AXES BELOW: b c
[0, 2 1] peirce Mq

Figure 4b.

by far the best *general* system which has yet been devised" on how "to write a proposition without asserting it" (Hartshorne *et al.* 1932, IV, §§378-391, pp. 322-328). Because this description, which Peirce had "had under examination for five years with continually increasing satisfaction," concerns a "square bracket" notation explained geometrically in terms of a "multitude of enclosures ... composed of vertical and horizontal lines" in a kind of box-diagrammatic arrangement. Although not conclusive evidence, it seems to indicate that, during a narrow span of years before his death in 1914, Peirce was concerned with orthogonal arrangement and nesting simultanously.

5. INVARIANCE UNDER TRANSFORMATIONS OF NESTING

About 1970, when More formalized his notion of a nested array combining orthogonality with nesting, he founded its algebraic model on the concept of hierarchical nesting from set theory, in the sense that "the nesting of arrays introduces the relation of membership" (More 1973, p. 11). The associated insights of set theory stood him in good stead during the quarter of century he has devoted since then to the development of his remarkable *theory of nested arrays*. One example of pertinent interest later on concerns the concept of a *scalar* which, in vector or tensor algebra, is a geometrical object possessing magnitude but no direction. It was natural, therefore, to define its array-theoretic counterpart as a special kind of nested array, a so-called *single*, with no axes (and, hence, of no extension) holding only one element, though, the true difficulty was that the latter element too had to be a nested array to satisfy the basic assumption of one-sortedness. Namely, the elements of an array should again be arrays without exception. Traditionally, we conceive scalars as plain numbers. So the immediate generalization would be that singles may contain some nested array other than just a number. But what if the element of a single is in fact a number or character? To solve this problem, More introduced from set theory W.V. Quine's idea of "self-membered individuals" or *self-containment*, as it is now usually called. That is, the representation of an individual as a unit set containing itself. Thus, calling a number or a character an *atom* (since "atoms" in logic are the indivisible individuals), we can make an atom a nested array by defining it as a single that contains itself. Singles, therefore, provide a rather elegant array-theoretic generalization of the traditional notion of a scalar.

But the influence of set theory upon More's work went beyond this boundary case of *singular* arrays with an element but no axes as well as its *dual* boundary case of *empty* arrays with no element but axes. For example, consideration of the Zermelo-Fraenkel-Skolem axiom system of set theory inspired him to define some of the more important operations in array-theory (More 1981). His endeavours to deal with irregular arrays, in which the elements do not *conform* in the sense of having the same shape (i.e., are spanned by corresponding sets of axes of corresponding lengths), further sustained More in his set-theoretical interpretation of nesting. Namely, that *nesting is the intrinsic property of membership of any array itself.*

In array-based logic, however, the special case of conforming arrays in the above sense is usually predominant in the formulation of system models, as it is generally experienced whenever we deal with discrete systems in physics and engineering. This suggests that, *under the assumption of conforming arrays*, we can introduce, by analogy to physics, the powerful approach of invariance under transformations of reference frame, if we associate nesting with the reference frame itself instead of considering it a property of the array. Hence, whenever arrays with elements of conforming shapes can be assumed, we submit that it is possible to introduce the double interpretations of transformation, conventionally known as *alibi* and *alias* in algebra. In this algebraic sense, More's view, that nesting expresses the property of membership of an array itself, constitutes the alibi interpretation of transformations that *an element is moved elsewhere*. Contrariwise, our novel proposal of considering *nesting as an intrinsic property of the reference itself* establishes the alias interpretation of transformations, that *an element is renamed*. The intrinsic properties of a system, its "laws" if you will, cannot depend on the accidental circumstances of an arbitrary reference frame. They must hold generally in all reference systems. As supported by analogy to physics, the importance of the alias interpretation proposed here, is therefore that the intrinsic properties can now be determined as those that are preserved under transformations of the nested reference frame. Briefly expressed, this is what we will understand by *invariance under nesting*.

By the conventions of display, the orthogonal axes of an array are ordered by levels and within each level from the top level down. The peculiar property of a *nested reference frame* is therefore that it establishes not only the orthogonality of the axes, but also the associated hierarchical property that each axis belongs precisely to one distinct level of nesting. Accordingly, a simple representation might be, that this kind of reference frame is formed by an ordered list of axes *distributed* into the levels of nesting, with the latter assigned the order from the top level and down. In other words, the nested reference frame is a composition of, on one hand, a permutation of the axes and, on the other hand, a distribution[2] of the axes into the ordered levels of nesting.

Now, with respect to this novel kind of reference frame, the nested array is merely a representation of some abstract, geometrical object (henceforth simply named the "array"), the intrinsic properties of which must be independent of the arbitrary reference frame in which it happens to be measured. This conception, it will be recalled, was forced upon us earlier by the associative law of multiplication for quaternions. For the satisfaction of this law demanded equality between the differently-nested array-representations of the multiplication table. In general, to require equality between the different representations of the abstract object or "array," which are produced by transformations of the nested reference frame, is tantamount to say, that the transformations bringing the representations into one another must define a mathematical group. In other words, to establish the intrinsic properties of the abstract "array," we must determine whatever remains invariant under a mathematical group of transformations of the nested reference frame.

To illustrate this point, let us briefly reconsider the two matrix representations of the quaternion in Figure 4. What we observe here, is that the corresponding two array-representations can be produced from each other by a cyclic shift of one position of the axes. That is, from [*a*, *b* *c*] = [0, 2 1] to [*c*, *a* *b*] = [1, 0 2], or *vice versa* for the opposite shift, if we fix the distribution associated with these two permutations such that the first axis goes into the top level and the last two into the level below. Since the associative law requires preservation of the order of the three axes, *a b c*, it is readily appreciated that, in principle, the mathematical group of transformations of the reference frame must be the cyclic group of shifts of positions of the axes relatively to the distribution of the shifted axes into the top two levels. For example, shifting all axes into the top level leaving no axes on the level below, we obtain the two valid representations: [*a b c*, *Null*] = [0 2 1, *Null*] and [*c a b*, *Null*] = [1 0 2, *Null*]. Here, the array *Null* denotes the empty list of axes in agreement with the fact that all elements of the multiplication table are singles and hence of no extension, as defined in Figure 3. Further, we recognize that the representation: [*a c*, *b*] = [0 1, 2] must also belong to this group of cyclic shifts, since the transposition (or reversal) of the two axes on the top level was a purely notational aspect caused by the conventions

[2] In general, the reason we are considering a *distribution* of the axes into the levels of nesting rather than a *partition* of the axes (with the parts of the partition defined by the levels), is that a level may well violate the defining requirement of non-empty parts of a partition. Thus, if the only element on a level is a *single*, that level is characterized by having no axes whence this level can only be produced by a distribution of the axes.

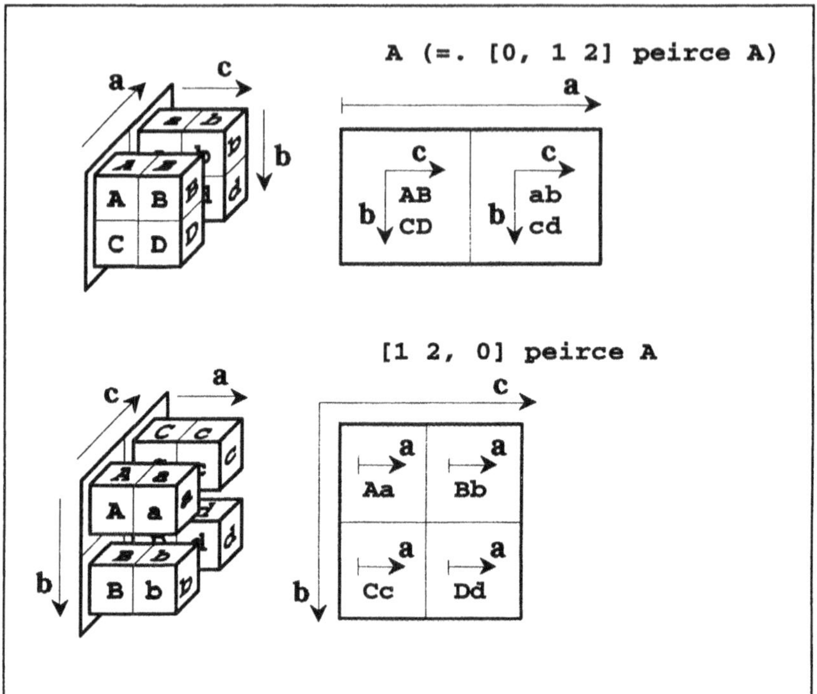

INTERCHANGING THE TOP TWO LEVELS

Figure 5.

of display. In summary, therefore, we see that the cyclic group of positional shifts of the axes with respect to a distribution of these axes on the top two levels, determines the intrinsic or structural properties of the *associative law* of multiplication. Obviously, this result is generally valid, since we made no use of the actual contents of the multiplication table, but merely considered the quaternion as an explanatory vehicle.

Earlier, the *commutative law* was briefly mentioned when we touched upon the fact that quaternions do not commute. Since this law is merely based on a transposition (or reversal) of a pair of axes in the multiplication table, this law too can be explained in terms of a group of transformations, namely the cyclic group of reversals. What this suggests, since the possible transformations of the nested reference frame are determined by the symmetric group of all permutations of the axes in combination with the distribution into the top two levels, is that subgroups of the latter group give meaning to the structural properties of laws such as those of association and commutation.

A third subgroup, of particular interest in relation to the following discussion of Grassmann's contribution, defines the transformations of reference frame that are produced by an *interchange of the top two levels of axes* of a nested array. The basic idea is illustrated in **Figure 5** for the simple transformation of a list-of-tables to a table-of-lists. Since the two tables are of the same shape in the list-of-tables shown in the upper half of the figure, our assumption of conforming arrays is satisfied. Further, purely as a matter of convenience, all three axes have been assigned the same length, namely length two. For this example, the axes on the top two levels of the list-of-tables in the upper part of the figure may be denoted by the pair: [*a*, *b c*] = [0, 1 2]. If we introduce an operation "*reverse*" that returns the elements of a list in reverse order, then, applying this operation to the pair, we find: *reverse* [*a*, *b c*] = [*b c*, *a*] or, tantamount to this: *reverse* [0, 1 2] = [1 2, 0], which defines the axes on the top two levels of the resulting table-of-lists shown in the lower part of the figure. Since operation *reverse* is its own inverse (i.e., performing *reverse* twice is the neutral operation), this transformation is invariant under the cyclic group of order two, whence it forms yet an isomorphism to logical negation.

To explain the significance of this transformation, let us imagine that we wish to add the elements on corresponding positions of the said two tables or, as it is often expressed, to add the two tables *element-by-element*. By the conventional rule of matrix addition, the result will be a table of the same shape in which we obtain the elements: *A*+*a*, *B*+*b*, etc. Now, if we consider the outcome of the transformation in the lower half of the figure, we see that the resulting table-of-lists is but the the structural part of this element-by-element addition, since in the corresponding positions of the table: *A* is paired with *a*, *B* is paired with *b*, etc. That is, a subsequent addition performed on each of these pairs will produce the desired result. Therefore, when More (More 1973) introduced the transformation of interchanging the top two levels, later to be known as operation "*pack*," he perceived that, in combination with his operator "*EACH*" (introduced simultaneously), it would provide the fundamental means of generalizing arithmetical or logical operations to conforming arrays. Thus, if "*f*" is any such operation, then its transformed operation: *EACH f* will cause *f* to act on each element on the level below the top level, whence the composite operation: *EACH f pack* (excuted in the usual order: first *pack* then *EACH f*) will define the generalization of operation *f* to arrays. Further, since the interchange of the top two levels applies to conforming arrays of arbitrary shapes and depths of nesting, he realized, very perceptively, that any such case can be dealt with in a uniform manner simply by applying the operation:

EACH f pack recursively so as to finally invoke *f* on the atoms (numbers or truth-values) at the bottom of the array.

This significant idea, that *an interchange of the top two levels will produce the structural part of an element-by-element operation*, provides yet an example (in addition to our previous discussion of associativity and commutativity) of how nesting in combination with orthogonality introduce a new degree of freedom into algebra with respect to the explicit separation of form and content. Further, it illustrates the very flexible manner in which the interplay between these two properties enable us to deal with structural forms in isolation as well as the kind of meaning or interpretation that can be ascribed to these forms. Simultaneously, however, it also demonstrates the dependence of this approach upon the assumption of conforming arrays. In fact, as More discovered, the only feasible way of extending a structural operation like "*pack*" to *all* arrays, was to establish some - let us call it - "generally acceptable" procedure, that would convert non-conforming arrays into conforming arrays *prior* to the interchange of the top two levels.

This conversion process, as More construed it, should increase the dimensionality (or number of axes) of all elements to match the element of the greatest dimensionality, and then *either* truncate each axis to the length of the shortest corresponding axis *or* extend each axis to the length of the longest corresponding axis replicating some arbitrary "fill item." Rather fortuitously More (More 1973, 1981) chose the former, calling the conversion operation "*trim*," because several years later he was able to prove that the latter would lead to contradictions within the context of his array theory. By necessity, however, any such procedure will appear quite arbitrary, since there is no precedence in algebra except for one well-established fact. Namely, that *scalars replicate* to form a component array of corresponding size in the products of a scalar by a vector, a scalar by a matrix, or a scalar (tensor of order zero) by a tensor. Though, in the context of tensor algebra, the last-mentioned case is not explained as a replication, but as an *outer product* in the *cartesian* sense. That is, the result is obtained by multiplication of each component in the first tensor by each component in the second tensor. But this explanation apart, the simplest way to incorporate these special cases into the conversion operation *trim* was undoubtedly by means of the rule proposed by More (More 1973), that *a single, which need not just be atomic, is replicated to conforming size*.

With the conversion of non-conforming arrays into conforming arrays defined explicitly in this manner, More (More 1973) decided to generalize the concept of interchanging the top two levels from conforming arrays to non-conforming arrays implicitly extending operation *pack* by the trimming procedure. Clearly, this is consistent with his fundamental viewpoint based on the *alibi* interpretation, discussed previously. Namely, that transformations relatively to the levels of nesting will produce *different arrays in one-to-one correspondence with each other*. Incorporating the trimming conversion into *pack* would extend the content of meaning of this operation but at a cost.

That is, for non-conforming arrays we lose the one-to-one correspondence yet gain, for example, that replications of a single becomes an intrinsic property of the transformation producing the new array.

Conversely, as we see it, adopting an *alias* interpretation of the transformations with respect to the levels of nesting, the basic assumption of conforming arrays precludes any consideration of the trimming conversion in general, and the replication of singles in particular, within the purview of invariance under nesting. Yet, since the concept of a single is generalized from that of a scalar (which is the fundamental invariant under transformation of coordinates by virtue of the fact that it has no axes), we need to identify a structural operation which, with the same effect as replication of singles, is meaningful from our point of view. In this respect, the abovementioned tensorial explanation is noteworthy, that singles are replicated as a consequence of defining multiplication as an outer product in the cartesian sense. In logic, we find the same idea proposed originally by Boole whose so-called *development formula* may be interpreted as an index tuple representation of the outer product. In array-based logic, accordingly, we define all logical and arithmetic operations, say *f*, as outer products, adopting More's definition of the operator (More 1973): *OUTER f A* = *EACH f cart A*, with the nested array *A* as argument. The importance of this definition, is that the cartesian product is introduced explicitly by the operation *cart* which, like *pack*, only affects the top two levels, since it merely establishes all possible combinations in structured form of the elements of its argument. In other words, the cartesian product will produce all those combinations with a single, which otherwise would have required the single to be replicated.

By way of illustration, these two viewpoints (that singles are either replicated to conforming size or established in the appropriate combinations by the cartesian product) may be compared by relating them to the conventional formulations of the *left and right distributive laws* of paramount importance in Grassmann's *Ausdehnungslehre*. In fact, it will suffice to consider only one of these laws, say: $(a + b) \times c = (a \times c) + (b \times c)$, with a b c interpreted as scalars in the conventional sense. Stripping this law of the addition and multiplication operations, we obtain the structural counterpart: $(a\ b)\ c = (a\ c)(b\ c)$, which, to emphasize the nested pairing, we may denote in the bracket notation introduced earlier: [[*a*, *b*], *c*] = [[*a*, *c*], [*b*, *c*]]. Thus, on the left hand side we have a pair only the first element of which is a pair (cf. Figure 1), while on the right hand side we have a pair of pairs. Now, the point is that, with *a b c* interpreted as arbitrary arrays, the application of *cart* to the left hand side: *cart* [[*a*, *b*], *c*] will produce the right hand side: [[*a*, *c*], [*b*, *c*]]. Alternatively, the same result obtains if *cart* is substituted by More's generalized version of *pack*, provided that *c* in this case is but a single. This suggests that it is *cart*, rather than *pack*, which provides the the fundamental explanation[3] of the underlying geometrical idea which, as pointed out previously, originates in the tensorial interpretation of multiplication as an outer product. In a historical perspective, however, it is illuminating to compare these modern insights with that of Cayley who in 1860 said, explaining the "recent terminology" of the matrix algebra he had founded a few years earlier, that "as a particular case of the multiplication or composition of matrices [i.e., the conventional matrix or inner product] we have the multiplication of a matrix by a single quantity, effected by multiplying by such quantity each term of the matrix" (Cayley 1891, IV, p. 602).

6. GRASSMANN'S "STUFEN" INTERPRETED AS LEVELS OF NESTING

Lewis has expertly dealt with the influence of Schleiermacher's *Dialektik* upon the first edition of 1844 of Grassmann's *Ausdehnungslehre* (Lewis 1977) . In the preface to the second edition, published seventeen years after the first, Grassmann pointed out that to give his presentation in "the most rigorous mathematical form, that we know at all, the Euclidean, ..., [it] was a necessary consequence of such a change of plans, that all the results of the first part [edition] ... to be included in the new version had to be derived anew according to the

3 The observation that the geometric essence of the distributive laws is captured by the array-theoretic cartesian product (and *not* by the one-to-one correspondence manifest in the operation *pack*), is implicit in More's [1976b] original, albeit brief statement of the array-theoretic form of these laws: *prod EACH sum A* = *sum EACH prod cart A*. Here, the operations *prod* and *sum* are multiplication and addition, respectively, while the argument *A* can be interpreted at will as either of the nested pairs: (*a b*) *c* or *a* (*b c*), so as to encompass the conventional formulations of both the right and left distributive laws. Though, what makes More's statement particularly elegant is that it directly applies to the more general formulations of the distributive law. For example, the argument *A* may be interpreted as the nested list: (*a b c*)(*d e*) = [[*a*, *b*, *c*] , [*d*, *e*]] derived from the formulation of the left hand side of the distributive law: $(a + b + c) \times (d + e)$. Here, *cart* will correctly produce a nested table for the right hand side which, denoted as a list: (*a d*)(*a e*)(*b d*)(*b e*)(*c d*)(*c e*) determines the expression: $(a{\times}d) + (a{\times}e) + (b{\times}d) + (b{\times}e) + (c{\times}d) + (c{\times}e)$. This result cannot be obtained by *pack*, since implicitly the argument is trimmed to the conforming shapes: (*a b*)(*d e*) which only yields: (*a d*)(*b e*) with the top two levels interchanged.

altered plans" (Grassmann 1862, p. iv). The implication of this remark, readily verified by comparison of the two editions, is that, in reworking his material for the second edition, Grassmann completely removed any trace of his attempt in the first edition to present his mathematical system as the embodiment of the philosophy described in Schleiermacher's *Dialektik*. For example, although he evidently felt compelled to preserve his terminology of "Stufen" in the second edition by virtue of the fundamental role it played in the first, he appeared rather disinclined to actually explain, let alone to justify this term (Grassmann 1862, §5, p. 2; §16, p. 6). For this reason, it is exclusively with reference to the first edition, that we now shall endeavour to bring to light in a modern perspective Grassmann's attempt to take advantage of his definition of nested pairing.

A main theme in the first edition of Grassmann's *Ausdehnungslehre* is that, in contradistinction to the one and only kind of multiplication in arithmetic, it is possible to develop different kinds of multiplication for hypercomplex numbers, his "extensive magnitudes," each kind with a distinct geometrical meaning and, as he also attempted to demonstrate by many examples, an associated variety of useful applications. As he tells us in his preface, referring to his examination of A.F. Möbius's *Der barycentrische Calcul* (1827), or calculus based on the concept of the center of mass:

"I was not a little pleased to find there the same concept of summation of points [i.e., the interpretation of the center of gravity as a sum] to which I had been led. ... However, since the concept of a product of points does not occur at all in this work, although it is first with this concept, as it enters into combination with that of a sum, that the unfolding of the new analysis begins, I could not expect from it any further advancement of my purpose." (Grassmann 1844, pp. viii-ix)

With this remark, Grassmann introduced the general idea of his approach now considered almost commonplace in mathematics. That is, *if we have an operation identified as addition, then any operation satisfying the distributive laws in terms of this addition may be called multiplication.* Grassmann did not use the term "distributive law" introduced by F.J. Servois in 1814, but merely referred to it as the "*Grundgesetz der Multiplikation*" or the basic law of multiplication (Grassmann 1844, §32, p. 53), in that he simply introduced the explicit statement of the left and right distributive laws as the definition of the concept of multiplication (ibid., §10, p. 11). But let us go directly to his introductory *Uebersicht der allgemeinen Formenlehre*, or survey of the general theory of forms, to consider Grassmann's own formulation of the abstract form of the distributive laws:

"So far we have considered only one kind of synthetic connection in itself and in its relation to the corresponding analytic one. What now matters, is to add to that the relation between two different kinds of synthetic connections. To this end the one must be defined through the other in accordance with its concept. This definition of concept depends upon the manner in which an expression, containing both kinds of connections, can be transformed without changing the total result.

The simplest way in which both connections can appear in an expression, is that the result of one connection is submitted to the other; thus, if $\frown$ and $\wedge$ are the symbols for the two connections, then the relation between the two depends on the transformations that may be carried out on the expression $(a \frown b) \wedge c$.

If the second connection should relate uniformly to both terms of the first, then the simplest transformation would be to submit each term of the first connection to the second, and then take these individual results as terms of the first kind of connection.

If this transformation can be undertaken without changing the total result, i.e., if $(a \frown b) \wedge c = (a \wedge c) \frown (b \wedge c)$, then the second connection is called the next higher step corresponding to that of the first connection." (ibid., §9, p. 10)

Evidently, the first paragraph illustrates the kind of sweeping generalizations that made H.R. Baltzer, a contemporary mathematician and subsequent editor of Möbius's collected works, refuse to review Grassmann's *Ausdehnungslehre* with the remark: "It is impossible for me to enter into his ideas - my head becomes dizzy and everything appears sky-blue when I try reading him" (Nagel, 1939, p. 174). Even Eduard Study, in his commentaries on Grassmann's *Gesammelte Werke* in 1894, had to forego any explanation:

"It is not clear why one method of connection must be specified through the other; in any case the sentence contributes nothing to the clarity of the whole and can be omitted without harm" (Kannenberg 1995, note 5, p. 302).

However, the meaning of this statement becomes perfectly clear if rendered concrete by saying that "multiplication" must be defined through "addition" in accordance with the concept of "multiplication," namely distribution through addition. In fact, this is how Grassmann himself explained it when, in the introduction to his definition of the *exterior product,*[4] he starts out from a very general formulation of the distributive laws, such as: $(a + b) \frown c = (a \frown c) + (b \frown c)$, telling us that this relation determines the general sign $\frown$ of connection as multiplication (Grassmann 1844, §29, p. 49). A further illustration in almost the same words, is provided by Grassmann's definition of a scalar type product (ibid., §§ 96-97, p. 136) aptly denoted "weight multiplication" in Swimmer's delightful introduction to Grassmann algebra (Swimmer 1989).

---

[4] Grassmann's own term was "*äussere Produkt*" or outer product. But since he understood it in the sense of the *vector cross product*, as we now call it, his term is in terminological conflict with the tensorial (and array-theoretic) definition of an outer product in the cartesian sense. The alternative term, *exterior product*, has therefore enjoyed a long tradition in the mathematical literature. On the other hand, since Grassmann defined his "*innere Produkt*" or inner product in the sense of the *vector dot product*, his use of this term agrees with our understanding as exemplified by the inner product in matrix algebra.

Equally significant in the first paragraph of the above quotation, however, is the last sentence, emphasizing the property of invariance under transformation from one side of the distributive law to the other. What inspired this idea of preservation "of the total result" might well have been Grassmann's intensive study of Lagrange's *Mécanique Analytique* (1811-1815), evidenced by the many references to this work in his *Ausdehnungslehre* of 1844. For in England about the same time, it was Boole's early reading of Lagrange that inspired him to his discovery of the algebraic theory of invariants, subsequently developed into a proper branch of mathematics by, in particular, Arthur Cayley (MacHale 1985, pp. 54-55). In analytical mechanics, the fundamental idea is the preservation of energy of a closed system, originally proposed by Leibniz solely in terms of kinetic energy. Therefore, it is the measure of energy being conserved under transformations of the reference system that gave rise to the abstract concept of a scalar as we now define it. But in 1844 all this was not well understood. Still, with the insight Grassmann had acquired from his study of Lagrange, we may perhaps assume that he had developed an intuitive notion of a scalar not too different from the modern concept.

The array-theoretic interpretation of the remainder part of this quotation is summarized under four headings in **Figure 6**. Beginning at the top, we stipulate that Grassmann conceived the abstract form of the left hand side of the distributive law to be in agreement with his dialectical idea of pairing. That is, the left hand side is established by "connection of two at a time" (cf. Figure 1), but now revised to be composed of two *different* kinds of connection. In principle, this is in accordance with the statement on the form of presentation in Grassmann's introduction. Namely, that in the philosophical method "the unity of idea is the primitive" whence "the overview of the whole predominates," as "in essence it proceeds by oppositions progressing from the general to the particular" (Grassmann 1844, §13, p. xxix). Pairing is such a primitive idea expressing the structural property of the repeated oppositions. So by detailing two rather than one kind of connection Grassmann went from the general case to a particular one. His terminological idea of classifying the last connection as a "next higher step" relatively to the first, emphasizes the overview of the whole. This is illustrated by the nested array at the top right, since, with the two kinds of connection represented by axes p and q respectively, the "next higher step" is geometrically interpreted in the sense of the *next higher level*.

The actual transformation from the left hand side to the right hand side of the distributive law is dealt with under the last two headings of the

GRASSMANN'S FORMULATION | **DISPLAY OF ARRAYS**

A CONNECTION OF TWO AT A TIME

"The relation between the two depends on the transformation that may be carried out on the expression (a ∩ b) ⩀ c".

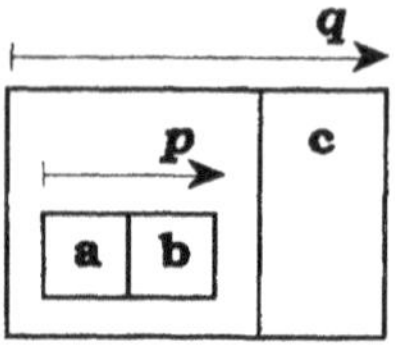

TERMINOLOGY OF "NEXT HIGHER"

"If this transformation can be undertaken without changing the total result, i.e., if

(a ∩ b) ⩀ c = (a ⩀ c) ∩ (b ⩀ c)

then the second connection is called the next higher step corresponding to that of the first connection."

Axis *q* "next higher" of axis *p*, with *q* and *p* as representations of the second ⩀ respectively the first ∩ connection.

UNIFORM RELATION TO BOTH

"If the second connection should relate uniformly to both terms of the first,"

(a pair b) pair single c

INTERCHANGE TOP TWO LEVELS

"... submit each term of the first connection to the second, and then take these individual results as terms of the first kind of connection":

(a ⩀ c) ∩ (b ⩀ c)

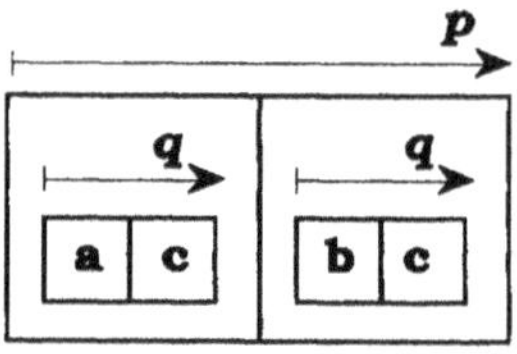

(a pair c) pair (b pair c)

THE DISTRIBUTIVE LAW

Figure 6.

figure. Exactly what Grassmann meant by a "uniform relation" of the second connection to both terms of the first connection, is an open question. Possibly, he attempted to express some intuitive understanding of the modern idea of multiplication by a scalar. If this was indeed the case, we can make it explicit by turning the last term *c* on the left hand side into a single, simply by submitting it to the operation *single.*[5] That is, we substitute *c* by *single c* with the result that the second connection now appears as *pair single*. This admits the interesting interpretation in the alibi sense of More, that the transformation from the left hand side to the right hand side of the distributive law may be brought about by an interchange of the top two levels, if we apply More's operation *pack* to the left hand side. In other words, the transformation is established: pack ((*a* pair *b*) pair single *c*) = ((*a* pair *c*) pair (*b* pair *c*)). Here, on the left-hand side, if *c* is atomic (i.e., a scalar in the conventional sense), then by virtue of the property of self-containment mentioned earlier: *c* = single *c*.

Did Grassmann have an intuitive understanding of the transformation along these lines? Perhaps he did. For in a later discussion he appears to come back to the interchange of the top two levels but now in connection with the addition of two nested lists. Thus he said:

"We have already shown above ... that if several stretches are generated in a corresponding way, then not only to each element and each part of one of them corresponds an element and a part in each of the others, but also the sum is correspondingly generated in the same way, namely so that the sum of corresponding parts always corresponds to these parts.

Now, if one has two arbitrary stretches of the system, say p_1 and p_2, both of which are formulated as sums of stretches belonging to the original kind of changes for the entire system, namely

$$p_1 = a_1 + b_1 + \ldots$$

$$p_2 = a_2 + b_2 + \ldots$$

then one has

$$p_1 + p_2 = (a_1 + a_2) + (b_1 + b_2) + \ldots,$$

and if further α_1, α_2, β_1, β_2,... are corresponding parts of the stretches a_1, a_2, b_1, b_2..., whence also $(\alpha_1+\alpha_2)$, $(\beta_1+\beta_2)$,... are in the same sense corresponding parts of (a_1+a_2), (b_1+b_2), then ... each part of the sum (p_1+p_2) is obtained as the sum of corresponding parts; that is, such a sum is always equal to

$$(\alpha_1+\alpha_2)+(\beta_1+\beta_2)+\ldots.$$

$$\text{i.e.} = (\alpha_1+\beta_1+\ldots)+(\alpha_2+\beta_2+\ldots)$$

where the first term represents a part of p_1, the second the corresponding one of p_2. Thus, each element of the sum (p_1+p_2) is generated by that means that one changes its initial

[5] The author is indebted to Trenchard More for this suggestion which, incidentally, generalizes the array-theoretic formulation, such that the last argument *c* may now be an arbitrary array just as *a* and *b*.

element for each arbitrary part of p_1 and then for the corresponding one of p_2." (Grassmann 1844, § 19, pp. 29-30)

Our principal interest in this quotation is due to the fact that Grassmann is concerned with addition of geometrical line-elements, his so-called *stretches* ("Strecken"), which consecutively over three levels are built up by addition of still smaller stretches. Thus, on the top level we have the stretches p_i, formed by addition of the stretches a_i b_i ... on the level below the top level, each of which again is formed by addition of the stretches α_i ... respectively β_i ... one level further down. In an array-theoretic sense, this very general approach of adding lengths end-to-end may be interpreted in terms of operation *pack* for interchanging the top two levels of a conforming array. This is illustrated in **Figure 7** where, in the three columns, we have excerpts from Grassmann's explanation (left hand column), an interpretation of the appropriate excerpts in terms of an interchange of the top two levels by means of *pack* (center column), respectively an interpretation in terms of an interchange of the two levels immediately below the top level by means of *EACH pack* (right hand column).

As documented by this quotation on addition over three levels, Grassmann clearly mastered the problem of dealing with nested levels. In fact, it would appear that he saw the concept of nested levels as an abstract form which, adopted from dialectic in the sense of a philosophical "primitive," he attempted to impose on algebra as a "unity of idea." For example, in the course of his discussion of the distributive law Grassmann interpreted addition as a connection of the first level ("erster Stufe"), multiplication as a connection of the second level ("zweiter Stufe"), commenting in a footnote that as a third level ("dritte Stufe") one might add the raising to a power or exponentiation (Grassmann 1844, §9, pp. 10-11). Another interpretation of his, is the purely geometrical one, that stretches make up the first level, planes the second level, and volumes the third level (ibid., §16, pp. 21-22). In short, therefore, we would venture the opinion that *Grassmann perceived his "Stufen" in the abstract sense of nested levels, which were given different meanings depending upon the application.*

Did Grassmann also have some intuitive understanding of the idea of interchanging the top two levels of his nested list? The nearest we can come to an answer to this question by consideration of his *Ausdehnungslehre*, is the following typographical layout of his statement in the above quotation, that the sum (p_1+p_2) is always equal to:

$$(\alpha_1+\alpha_2)+(\beta_1+\beta_2)+....$$

$$\text{i.e.} = (\alpha_1+\beta_1+...)+(\alpha_2+\beta_2+...)$$

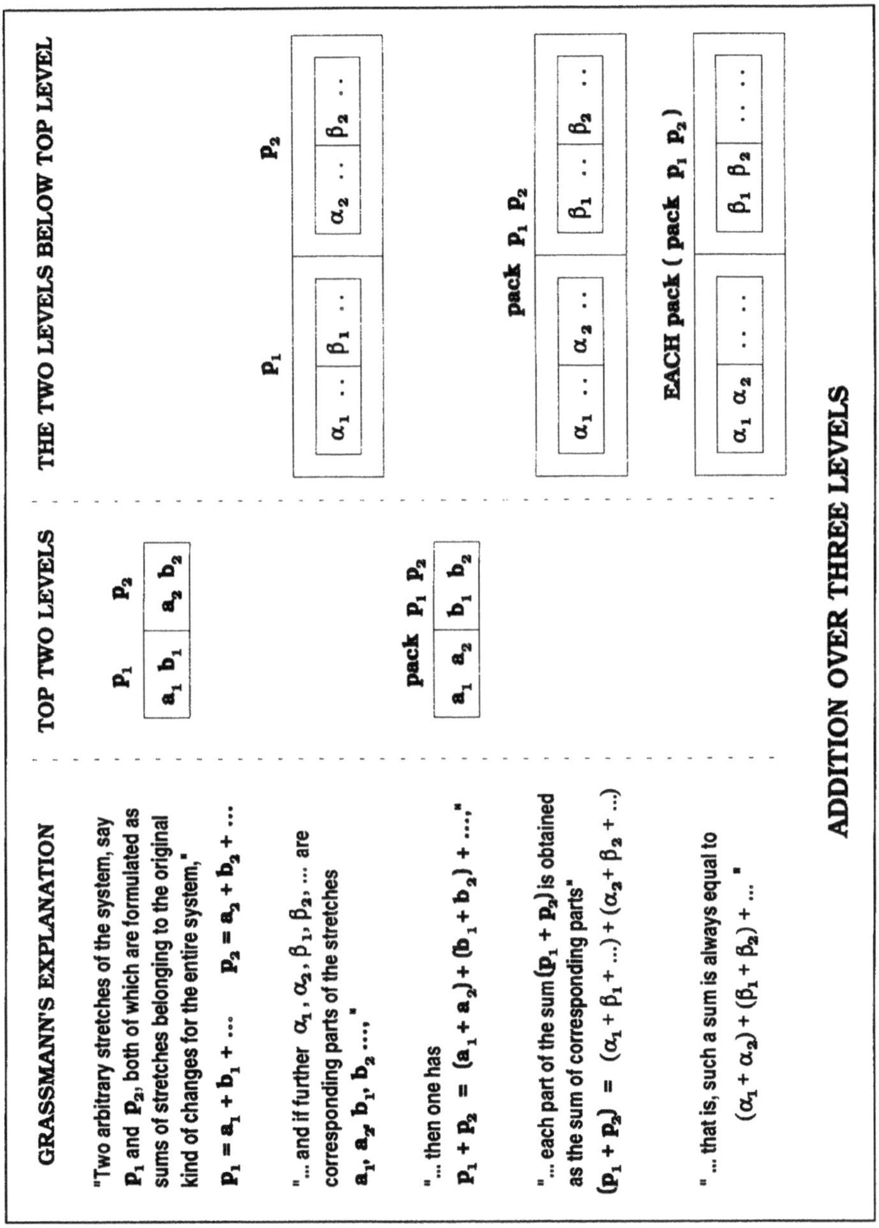
GRASSMANN'S EXPLANATION
"Two arbitrary stretches of the system, say p1 and p2, both of which are formulated as sums of stretches belonging to the original kind of changes for the entire system,"
p1 = a1 + b1 + ... p2 = a2 + b2 + ...
"... and if further α1, α2, β1, β2, ... are corresponding parts of the stretches a1, a2, b1, b2 ...,"
"... then one has
p1 + p2 = (a1 + a2) + (b1 + b2) + ...,"
"... each part of the sum (p1 + p2) is obtained as the sum of corresponding parts"
(p1 + p2) = (α1 + β1 + ...) + (α2 + β2 + ...)
" ... that is, such a sum is always equal to
(α1 + α2) + (β1 + β2) + ... "
TOP TWO LEVELS
p1 p2
a1 b1
a2 b2
pack p1 p2
a1 a2
b1 b2
THE TWO LEVELS BELOW TOP LEVEL
p1 p2
α1 .. β1 ..
α2 .. β2 ..
pack p1 p2
α1 .. α2 ..
β1 .. β2 ..
EACH pack (pack p1 p2)
α1 α2 .. ..
β1 β2 .. ..
ADDITION OVER THREE LEVELS

Figure 7.

because it gives the impression of an interchange of levels. Other clues of a more indirect nature relate to the very few times Grassmann found it appropriate to introduce a general symbol of connection in the two main parts of his work. For in contrast to his introductory "Survey of the General Theory of Forms," with its frequent use of general symbols of connection to represent a variety of different situations, it would appear that operation *pack* applies to the handful of situations formulated in the two main parts by means of a general symbol of connection. For example, Grassmann's formulation of the matrix equation: $Ax = B$ could be interpreted as the form: (*pack A x*) *B* (Grassmann 1844, §45, pp. 70-72). Whether these observations should be considered tantalizing glimpses or mere coincidences, that is all we can offer by way of an answer to this question.

7. MATHEMATICS AS THE EMBODIMENT OF A PHILOSOPHY

Imbued with the spirit of Friedrich Schleiermacher's lectures on dialectics at Berlin University, Grassmann was inspired to give mathematics a new vision as the embodiment of this philosophy. This radical departure from the traditional view is usually considered, to use the parlance of today, as "philosophical noise" which, due to Grassmann's formal training in theology rather than in mathematics, must be filtered away in order to assess the true contribution of his *Ausdehnungslehre* of 1844. But, as apparently only Peirce perceived because of his thorough study of Kant and other German philosophers, it is the vision, whether fulfilled or not, that made Grassmann's work a "great system." For it is in the perspective of new visions that radically new discoveries or inventions are made.

As we have endeavoured to show here, Grassmann was the first to recognize in algebra the importance of combining order or orthogonality with hierarchy or nesting. That he failed to carry through this program with regard to nesting is not what counts in retrospect. Rather, what counts is that he had the vision to go against all odds as far as did.

ACKNOWLEDGEMENTS

For indispensable help in solving the difficult problem of translating the quotations from Grassmann into English, the author is particularly indebted to Gert Schubring, Günther Braunss, and Ivor Grattan-Guinness. Other sources of invaluable importance in this respect have been the translations by Albert C. Lewis (1977) and Lloyd C. Kannenberg (1995). For very useful comments and suggestions on earlier versions of this paper, grateful thanks are due to Grattan-Guinness,

Trenchard More, Jr., and Alvin Swimmer. Finally, the author wishes to express his appreciation of some inspiring discussions with his close colleague, Peter Falster, during the conception of the paper.

Department of Electric Power Engineering, Technical University of Denmark

REFERENCES

Paul Bernays, Moses Schönfinkel, "Zum Entscheidungsproblem der mathematischen Logik," *Mathematische Annalen*, 1928, *99*: 342-372.

Arthur Cayley, *The Collected Mathematical Papers*, Vols. I-XIII (Cambridge: At the University Press, 1889-1898. New York: Johnson Reprint Corp., 1963).

Alonzo Church, *Introduction to Mathematical Logic*, Vol. I (Princeton, N.J.: Princeton University Press, 1956).

Haskell B. Curry, Robert Feys, *Combinatory Logic*, Vol. I (Amsterdam: North-Holland, 1958).

Carolyn Eisele (ed.), *The New Elements of Mathematics by Charles S. Peirce*, Vols. I-IV (The Hague-Paris: Moulton Publishers, 1976).

Peter Falster, "Notational Aspects of APL. Illustrated by Relational Database Applications," *Structures and Operations in Engineering and Management Systems*, eds. Ø. Bjørke & O.I. Franksen (Trondheim, Norway: Tapir Publishers, 1981), 399-445.

Ole I. Franksen, "An Operational Index Notation: (1) Trema Functions for Indexing & (2) Relational Database Illustrations," Proposal for a new research program on "Base Operations in Industry," *First Lerchendal Conference* at NTH, Trondheim, Norway, 1977.

Ole I. Franksen, "Group Representation of Finite Polyvalent Logic. A Case Study Using APL Notation," *IFAC VII World Congress*, Helsinki, June, 1978. *A Link Between Science and Applications of Automatic Control*, ed. A. Niemi (Oxford and New York: Pergamon Press, 1979) 875-887.

Ole I. Franksen, *H.C. Ørsted - a man of the two cultures* (Birkerød, Denmark: Strandbergs Forlag, 1981).

Ole I. Franksen, "Are Data-Structures Geometrical Objects?," *Systems, Analysis, Modelling, Simulation*, 1984. "Part I: Invoking the Erlanger Program," *1* (2):113-130; "Part II: Invariant Forms in APL and Beyond," *1* (2):131-150; "Part III: Linear Differential Operators," *1* (3): 251-260; "Part IV: Logic Invariants by Finite Truth-Tables," *1* (4):339-350.

Ole I. Franksen, "Comments on operation '*peirce*' and its family of related functions," Research Note, 1991-10-16, Elec. Power Eng. Dept., Tech. Univ. Denmark, 14 pp.

Hermann G. Grassmann, *Die lineale Ausdehnungslehre. Ein neuer Zweig der Mathematik*, Part 1 of *Die Wissenschaft der extensiven Grösse oder die Ausdehnungslehre, eine neue mathematische Disciplin* (Leipzig: Verlag von Otto Wigand, 1844).

Hermann G. Grassmann, *Die Ausdehnungslehre. Vollständig und in strenger Form bearbeitet* (Berlin: Verlag von Th. Chr. Fr. Enslin, 1862).

Ivor Grattan-Guinness, "Where does Grassmann fit in the History of Logic?," 1996 (this volume).

Charles Hartshorne, Paul Weiss (eds.), *Collected Papers of Charles Sanders Peirce*, Vols. I-IV (Cambridge, Mass.: The Belknap Press of Harvard University Press, 1932. Reprint 1960).

Thomas Hawkins, "Hypercomplex Numbers, Lie Groups, and the Creation of Group Representation Theory," *Archive for History of Exact Sciences*, 1971, *8* (1/2):243-287.

Lloyd C. Kannenberg (Translator): Hermann G. Grassmann, *A New Branch of Mathematics. The* Ausdehnungslehre *of 1844 and Other Works* (Chicago, La Salle: Open Court 1995).

Gabriel Kron, *Tensor Analysis of Networks* (New York: John Wiley & Sons, 1939).

Albert C. Lewis, "H. Grassmann's 1844 *Ausdehnungslehre* and Schleiermacher's *Dialektik*," *Annals of Science*, 1977, *34*:103-162.

Desmond MacHale, *George Boole. His Life and Work* (Dublin: Boole Press Ltd., 1985).

Trenchard More, "Axioms and Theorems for a Theory of Arrays," *IBM Journal of Research and Development*, 1973, *17* (2):135-175.

Trenchard More, "On the Composition of Array-theoretic Operations, *IBM Cambridge Scientific Center Tech. Rep. G320-2113*, April 1976a, iii+62 pp. An appendix: "A Table of Compositions of Array-theoretic Operations," i+108 pp., was privately circulated.

Trenchard More, "Types and Prototypes in a Theory of Arrays," *IBM Cambridge Scientific Center Tech. Rep. G320-2112*, May 1976b, 65 pp.

Trenchard More, "Notes on the Diagrams, Logic, and Operations of Array Theory," *Structures and Operations in Engineering and Management Systems*, eds. Ø. Bjørke & O.I. Franksen (Trondheim, Norway: Tapir Publishers, 1981), 497-666.

Trenchard More, "Transfinite Nesting in Array-theoretic Figures, Changes, Rigs, and Arms - Part I," *APL Quote Quad*, 1993, *24* (1):170-184.

Ernest Nagel, "The formation of modern conceptions of formal logic in the development of geometry," *Osiris*, 1939, *VII*:142-224.

Volker Peckhaus, "The influence of Hermann Günther Grassmann and Robert Grassmann on Ernst Schröder's Algebra of Logic," 1996 (this volume).

Benjamin Peirce: "Linear Associative Algebra. With Notes and Addenda by C.S. Peirce, Son of the Author," *American Journal of Mathematics*, 1881, *4*: 99-229

Moses I. Schönfinkel, "Über die Bausteine der mathematischen Logik." *Mathematische Annalen*, 1924, *92*:305-316.

James Joseph Sylvester, *The Collected Mathematical Papers*, Vols. I-IV (New York: Chelsea Publ. Co., 1973).

Alvin Swimmer, *Grassmann Algebra. Theory and Applications* (Tempe, Arizona: Math. Dept., Arizona State University, 1989).

JOCHEN PFALZGRAF

AN APPLICATION OF GRASSMANN GEOMETRY TO A PROBLEM IN ROBOTICS

In this short note, we shall give a brief description of an application of methods from Grassmann's geometry to a problem arising in robotics. We profit from this opportunity to give a short presentation of some basic notions which are used in the mathematical modeling of a robot manipulator arm - more details can be found in (Pfalzgraf 1995) and the literature quoted there.

We shall begin with some motivational remarks, with the objective of showing how a mathematical model for a robot can be established naturally, leading to the notion of a robot map in its general form which involves well-known mathematical manifolds. This is the reason why this modeling approach opens the possibility to apply various results and methods from geometry and (algebraic) topology in a very self-evident way. First we need some basic facts about the mechanical structure of the robot types which we want to model. A robot as we are considering it can be briefly described as a system of rigid parts (sub-arms, links), where adjoining parts are linked by joints and which, in many cases, is an open kinematic chain (tree-like structure). The motion of such systems is studied in the classical field of kinematics (classical mechanics, geometry). In kinematics, the motion of rigid bodies is studied without considering forces and momentums (the latter are considered in robot dynamics). Of basic importance for such investigations is a fundamental theorem established by L. Euler (1776),[1] which states that the motion of a rigid body where one point remains fixed is a rotation around a (fixed) axis going through that point. Thus, a robot model will be described as a kinematic chain consisting of links (sub-arms) and joints. Technically, the most common types of joints are rotational (revolute) and translational (prismatic) joints, respectively. The base of a robot is that part where a robot is fixed to the ground or where the wheels are attached, for example. The first sub-arm is then linked with the base. The

[1] Cf. e.g. Angeles 1988.

G. Schubring (ed.), Hermann Günther Graßmann (1809–1877): Visionary Mathematician, Scientist and Neohumanist Scholar, 337–343.

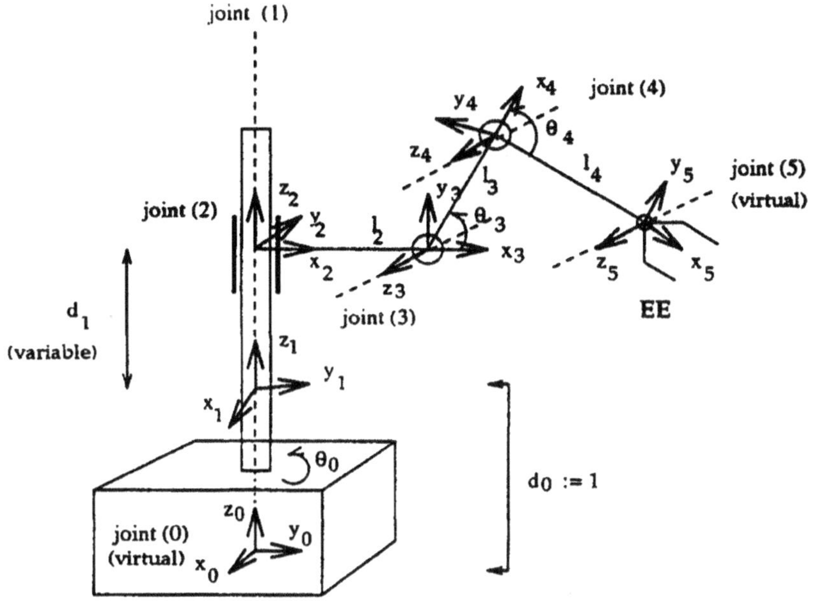

Figure 1: rtrr-Robot

the corresponding tools like screw driver, welding device, etc., it is called the end effector of the manipulator, abbreviated by EE. As an example, in the figure below we display a 4-joint robot arm having a rotational joint, followed by a translational joint and two rotational joints, this will be abbreviated by the notation rtrr-robot arm (cf. Pfalzgraf 1996 for a more detailed discussion of the kinematics model of that sample robot). To establish the robot map, let t denote the number of translational joints and r the number of rotational joints, respectively. The objective is to model the robot as a mathematical map from joint space (configuration space) $\mathcal{C}$ to work space $\mathcal{W}$, this is the space where the movements of the end effector (EE) are described

$$\mathrm{R}: \mathcal{C} \rightarrow \mathcal{W}$$

The degree of freedom (DOF) of the robot is defined by DOF = n = r+t (this is the total number of joints $n = r+t$), i.e. the “joint space manifold” $\mathcal{C}$ has dimension r+t. Thus, in the previous example we have DOF = 4. In this approach, we can ask what would be the largest possible joint space mathematically (where all smaller joint spaces can be embedded). The parameter of a translational joint runs through a real interval, we model this, more generally, by the whole real line R .

Rotational joint parameters can analogously be interpreted as elements of the unit circle S^1 embedded in the real plane. We recall that $S^1 = \{z = e^{i\phi} \mid 0 < \phi \le 2\pi\}$[2] and a rotation through angle ϕ corresponds to a multiplication of a 2-dimensional vector by $e^{i\phi}$ in the corresponding "rotation plane" (the complex numbers represented as 2-dimensional vectors with real and imaginary parts as x- and y-components in the real plane). Thus, in the general case, the configuration space (joint space) manifold C can be represented by

$$\underbrace{(R \times \cdots \times R)}_{t} \times \underbrace{(S^1 \times \cdots \times S^1)}_{r} = R^t \times (S^1)^r$$

where the first cartesian product has t factors and the second one r factors. The cartesian product $(S^1)^r$ is also called *r-dimensional torus* T^r with $T = S^1$, the 1-dimensional torus. From a mathematical point of view, this configuration space is a nice differentiable manifold, actually it is an Abelian Lie group.

Concerning the *work space*, we can ask analogously what is the complete set of information which is necessary to describe the location of the end effector (EE) of a robot? The complete information about the EE movement is described by a position vector $a = (a_1, a_2, a_3) \in \mathbb{R}^3$ and the orientation of the EE. Orientation in Euclidean 3-space can be uniquely determined by an orthonormal frame. The set of all frames can be described by the group of all proper rotations, the *special orthogonal group*: $SO_3(\mathbb{R}) = \{A \mid A$ is an orthogonal 3×3 matrix with $\det A = +1\}$. Hence the whole work space manifold, describing all possible EE locations in the most general case, is $\mathbb{R}^3 \times SO_3(\mathbb{R})$ the *Euclidean motion group*. Again, this is a nice differentiable manifold. It is of dimension 6 and this is why 6-DOF robots are so widespread in practical applications. More precisely, a manipulator with the capability that the EE can reach every point in the complete work space must at least have six DOF, i.e. it needs six joints and at least three of them have to be rotational (to provide all possible orientations). Summarizing, in the most general case, a robot can be mathematically described by a (differentiable) map, robot map

$$R: \mathbb{R}^t \times (S^1)^r \rightarrow \mathbb{R}^3 \times SO_3(\mathbb{R})$$

2 For the algebraic description we prefer here the complex number representation of points of the real plane.

This general mathematical model of a robot then forms the basis for further study of robotics problems. Obviously, the mechanical structure of a robot determines its workspace structure (geometry and topology).

Using the previously introduced notation, let R: $C \rightarrow \mathcal{W}$ be the robot map associated to a selected robot model with joint space C and work space $\mathcal{W}$. Let $q \in C$ denote the joint coordinate vector (n-dimensional) and $x \in \mathcal{W}$ the corresponding coordinate vector (m-dimensional) in the work space, respectively. Let $q \in C$ be a point in the joint space manifold and $x = \mathsf{R}(q)$ then $d(\mathsf{R})(q)$: $T_qC \rightarrow T_x\mathcal{W}$ denotes the derivative (the tangent map) of R in q, where T_qC and $T_x\mathcal{W}$ are the corresponding linear tangent spaces, respectively. If local coordinate systems are chosen then $d(\mathsf{R})(q)$ is represented by the corresponding Jacobi matrix. Based on that, we can define now: $q \in C$ is a singular (joint) configuration (singular point), if the derivative $d(\mathsf{R})(q)$ (the Jacobian matrix in that point) does not have maximal rank. Interpreting the derivative and the tangent spaces as a linear approximation of the robot map on the manifolds, one can say that in a small neighborhood of a singular configuration the derivative cannot guarantee a full linear approximation of the space near the point $\mathsf{R}(q)$ since there are dimensions missing (not the full rank for the linear approximation of the tangent space in $\mathsf{R}(q)$ is available). Mathematically, this is the reason why in such configurations there exist non reachable points. Or the velocity can have very high values near a singular point (cf. e.g. J.J. Craig 1986). Technically, such configurations cause problems. The manipulator can move in a locked position, then singularity means that the EE cannot be moved along a small path segment with only small changes of the joint parameters (this would be possible in non singular points). It is therefore desirable to have a complete overview of all singular configurations of a robot. The two main classes of singularities of different characteristics are the work space boundary singularities (e.g. when the robot arm is fully stretched) and the work space interior singularities (cf. J. J. Craig 1986). Now we touch such a singularity problem.

We briefly discuss a very interesting application of methods and results from (classical projective) Grassmann geometry to problems arising in the classification of singular configurations of so-called parallel manipulators (so-called Stewart platforms). Our exposition here is a kind of a very concise report of work presented in (J.D. Boissonat/Laumond 1989; cf. also J.P. Merlet 1985). We consider it as an impressive example showing how classical results from geometry can help to clarify difficult problems where other approaches did not succeed to solve them completely or where serious difficulties remained. We refer to J. P. Merlet's interesting work (in J. D. Boissonat/Laumond

1989) for further details - here we shall give only a rather brief sketch of the essential ideas.

A parallel manipulator consists of a basis platform and an end effector (EE) platform and in our case both are linked by 6 (hydraulic driven) rods (the links of this robot). The rod lengths are the link parameters. Varying the lengths of the rods causes changes of the EE location. (Such platforms come to application, for example, in flight simulators). An EE location corresponds to a unique rod lengths configuration, in general. As already mentioned previously, from a technical point of view it is desirable to know all singular configurations of such a manipulator. Therefore a basic question is: how can a singular configuration be characterized? Mechanically, a singular configuration is an unstable constellation of the whole system. If we interpret (idealize) the parallel manipulator skeleton as a polyhedron, then a theorem by Cauchy (1831) states that a convex polyhedron is always stable (rigid). Hence, an unstable configuration of a parallel manipulator can never belong to a convex rod configuration. The analytical determination of the singularities on basis of the Jacobian determinant approach leads to the task of finding all roots of the determinant. It turned out to be rather difficult and did not result in a complete solution of the problem (we refer to corresponding remarks in Boissonat/Laumond 1989). Now, it is shown how Grassmann geometric methods can help to make the problem more transparent; they lead some steps forward towards a general solution of the singularity problem. In this sense the question is how can geometry enter the discussion? Geometrically, a rod is replaced (in an abstract meaning) by an associated straight line in 3-space. Hence a rod constellation of the robot corresponds to a configuration of the associated 6 lines, respectively. The idea is then to translate the whole problem of classifying singular configurations of the parallel manipulator to a classification problem of corresponding ("singular") line configurations. And then this problem can be treated resorting to the help of results from projective (Grassmann) geometry. In this case here it amounts to the classification of varieties generated by such (singular) line configurations. Algebraically one can work in 4-dimensional affine real space (we denote it here by $\mathcal{V}^4$) expressing points by 4 homogeneous coordinates -- or, equivalently, in projective 3-space. We apply the Grassmann representation of a (straight) line via the binary exterior product in the Grassmann algebra (exterior algebra) $\Lambda^2 \mathcal{V}^4$ i.e. for $x, y \in \mathcal{V}^4$, the line joining x and y is expressed by the product $x \wedge y \in \Lambda^2 \mathcal{V}^4$.

We note that the dimension of this linear variety of all lines, $\Lambda^2 \mathcal{V}^4$, is $\binom{4}{2} = 6$. With respect to a fixed basis in $\mathcal{V}^4$ a straight line in $\mathcal{V}^4$ now can be expressed by the corresponding 6 Plücker coordinates, denoted by $(p_{ij})_{1\leq i<j\leq 6}$, in $\Lambda^2 \mathcal{V}^4$ (the corresponding projective space is 5-dimensional, of course).

Now one can exploit the following correspondences:

singular configuration ↔ unstable rod configuration (non convex polyhedron)

and on the other hand

non singular (stable) configuration ↔ the 6 lines (6 Plücker coordinate vectors) are linearly independent

consequently, the classification can be done as follows:

singular configuration ↔ line configuration with rank < 6

Summarizing, the classification of all singular rod configurations of the parallel manipulator corresponds to the classification of all varieties (generated by corresponding line configurations) having *rank* < 6. And to work on this, finally, one can apply existing results and methods from classical geometry -- cf. J. P. Merlet (Merlet 1988). This geometric method has been successfully applied by Merlet to a special parallel manipulator type called TSSM (triangular simplified symmetric manipulator). In the most general case of a parallel manipulator, there remain still difficult problems, but this geometric approach helped to make things more transparent and to make progress.

There are further interesting examples showing how methods and results from geometry and topology can be usefully applied to support work on typical robotics problems. Some material has been collected in (J. Pfalzgraf 1995). We would also like to point here to the very interesting article by J. E. Hopcroft (c.f. J. E. Hopcroft 1986) where various aspects of mathematical applications to problems arising in applied fields like the area of robotics are discussed. So, for example, one can read there a motivation how to benefit from the theorem of Riemann-Roch in such contextes.

Research Institute for Symbolic Computation, Kepler Universität Linz

REFERENCES

J. Angeles, *Rational Kinematics* (New York-Berlin: Springer, 1988).

J.-D. Boissonat, J.-P. Laumond (eds.), *Geometry and Robotics*, 391: *Lecture Notes in Computer Science* (Springer Verlag 1989).

J. J. Craig, *Introduction to Robotics* (Addison: Wesley Publ. Co. 1986).

J. E. Hopcroft, "The Impact of Robotics on Computer Science," *Communications of the ACM*, 1986, *29*: 486-498.

J.-P. Merlet, *Parallel Manipulators. Singular Configurations and Grassmann Geometry*, 791: *Technical Report, INRIA* (Le Chesnay, France: Rocquencourt 1988).

J. Pfalzgraf, "On Mathematical Modeling in Robotics," *Artificial Intelligence and Symbolic Mathematical Computing*, ed. J. Calmet and J. A. Campbell, Proceedings, Lecture Notes in Computer Science 737 (Berlin: Springer Verlag 1993), 116-132.

J. Pfalzgraf, "On geometric and topological reasoning in robotics," to appear in *Annals of Mathematics and Artificial Intelligence*, 1996.

NOTES ON CONTRIBUTORS

Albrecht Beutelspacher served, after his studies, from 1986 to 1988 at the Siemens research laboratories in Munich, and has been Professor of Mathematics at the University of Gießen (Germany) since 1988. He has written numerous research articles and several books in the fields of geometry, discrete mathematics and cryptology. He devotes a great deal of his time to making mathematics better understandable, both for students and for the man in the street.

Günter Braunss is Professor at the Mathematics Department, University of Gießen (Germany). He has published numerous papers on aerodynamics, differential equations, functional analysis and various topics in mathematical physics, and he is the co-author of a book on Boolean and switching algebras. His present research interests are quantum dynamics and applications of computer algebra.

Aldo Brigaglia is Associate Professor of *Matematiche Complementari* at the Mathematics Department of the Palermo University (Italy). His main research interests are history of mathematics, and above all the history of interrelations between algebraic versus geometrical methods from the late sixteenth to the twentieth centuries. Presently, he is also moving towards educational interests.

Andrea Brini is Professor of Algebra at the University of Bologna (Italy). His main research interests are Umbral Calculus, Invariant Theory and Algebraic Combinatorics.

Jean-Luc Dorier is a Maître de Conférences at the Université Pierre Mendès France in Grenoble (France). His research on the teaching of vector space theory at French Science universities led him to investigate the historical aspects of this theory.

Ole Immanuel Franksen is Professor at the Department of Electric Power Engineering, Technical University of Denmark, Lyngby (Denmark). He has been establishing the new field of systems science at the Technical University of Denmark since 1968, exploring the potential of the modern computer in engineering design and production. The novel technology, *Array-based Logic*, developed since 1978 by him and his associates in cooperation with Danish industry, illustrates the enormous perspectives of this new tool. In this work, the "historical method" has proved to be a powerful means for linking the separate disciplines. Outside his proper field of research, the author cultivates his own historical interests as Vice Chairman on the boards of the Technical

Museum of Denmark in Elsinore and the Ole Rømer Museum south of Copenhagen.

Ivor Grattan-Guinness is Professor of the History of Mathematics and Logic at Middlesex University (England). His latest individual book is *Convolutions in French Mathematics, 1800-1840* (1990). He edited a substantial *Companion encyclopedia of the history and philosophy of the mathematical sciences* (1994). His projects include a general history of mathematics for the Fontana History of Science series, and a history of mathematical logic and the foundations of mathematics from 1870 to 1930.

David Hestenes is Professor of Physics at Arizona State University (USA). He is the principal architect of *Geometric Calculus*, a unified mathematical language with historical roots in Hermann Grassmann's work. He has written several books and many papers developing the language and applying it to all domains of physics. He is currently concerned with pedagogical and software applications of the system.

Marie-Luise Heuser is affiliated to the Düsseldorf University and to the Stuttgart University (Germany). Her main research interests are *Naturphilosophie*, history of science and philosophy of mathematics, in particular in the nineteenth century, and philosophy of self-organization as applied to recent developments in the sciences. Several publications on F.W.J. Schelling and romanticist *Naturtheorien*.

Erika Hültenschmidt, Bielefeld (Germany), is engaged in research on the history of the language sciences, especially in France and Germany during the eighteenth and nineteenth centuries. At present, she is finishing a book on *Sprachwissenschaften in Deutschland und Frankreich, 1800-1870: Entwicklung, Differenzierung, Transfer*.

Wilhelm P.A. Klingenberg is Professor Emeritus of Mathematics at the University of Bonn (Germany). Besides numerous papers, he has written various textbooks on geometry, some of which have been translated into several languages. Besides on Riemannian geometry, he works on the foundations of geometry and geometric algebra.

F. William Lawvere is a Professor of Mathematics at the State University of New York at Buffalo. He is currently working on the application of Synthetic Differential Geometry to Categorical Dynamics, and also on the application of the Topos Theory to a new philosophy of mathematics.

Albert Lewis is an historian of mathematics with special interests, besides on Hermann Grassmann, in U.S. mathematics, centering mainly on G.B. Halsted, Norbert Wiener, and R.L. Moore. He has recently helped prepare *The Collected Papers* of Bertrand Russell at McMaster University in Canada.

Zbyněk Nádeník is Professor of Mathematics at the Technical University Prague and works in geometry.

Jochen Pfalzgraf is an Assistant Professor at the RISC: Research Institute for Symbolic Computation at the Universität Linz. His current research interests are: toward a general semantics for relational structures (including sheaf semantics); mathematical modeling with fibered structures (especially the concept of "logical fiberings") and applications to cooperating agents (robots) scenarios. Categorical and geometric methods applied to connectionist networks modeling. Industrial applications and cooperations in the "area of tension: theory and practice."

Volker Peckhaus is presently Wissenschaftlicher Assistent in the *Institut für Philosophie*, University Erlangen-Nürnberg, at the chair of Christian Thiel. He is member of the editorial boards of *Historia Mathematica* and *Modern Logic*. Beside numerous articles, he wrote *Hilbertprogramm und Kritische Philosophie* (1990), and *Hermann Ulrici (1806-1884)* (1995). His research interests are concerned with the history of formal logic and foundations as well as methodological problems in the history of science.

Karin Reich is Professor of History of Science at the Institut für Geschichte der Naturwissenschaften, Hamburg University (Germany). Her special interests are history of vector and of tensor calculus.

David E. Rowe is Professor of History of Mathematics and Exact Sciences at the Johannes Gutenberg Universität Mainz (Germany). His principal research interests are the history of geometry and the overall development of mathematics in Europe and North America during the nineteenth and twentieth centuries. He is a former editor of *Historia Mathematica*.

Karl-Heinz Schlote is head of the *Arbeitsgruppe für Geschichte der Naturwissenschaften und Mathematik* at the *Sächsische Akademie der Wissenschaften* (Leipzig, Germany). His research interests are history of mathematics in the nineteenth and twentieth centuries, emergence of abstract algebraical thinking, Emmy and Fritz Noether.

Erhard Scholz works as an historian of mathematics in the Mathematics Department of the University of Wuppertal (Germany) and is particularly involved in teacher education. His professional interests focus on the mathematics of the nineteenth and twentieth centuries and in particular on the exchange between mathematics and other cultural fields. His main interests at present centers on H. Weyl and F. Hausdorff.

Gert Schubring is Department Head at the research institute *Institut für Didaktik der Mathematik*, University of Bielefeld (Bielefeld, Germany). Besides research on the history of mathematics education, his research interests focus on the history of mathematics and the sciences in the eighteenth and nineteenth centuries and their broader cultural context. Recent publications include the edition of *'Einsamkeit und Freiheit' neu besichtigt. Universitätsreformen und Disziplinenbildung in Preußen als Modell für Wissenschaftspolitik im Europa des 19. Jahrhunderts* (1991).

Heinz Schwartze was a student at the *Marienstifts Gymnasium* in Stettin and passed the *Abitur* there. He is Professor for Didactics of Mathematics with special emphasis on geometry at the Gießen university (Germany).

Alvin Swimmer first heard of Grassmann Algebra in 1952 from Professor Lipman Bers when he was a first year graduate student at New York University. One year later, he wrote his Master's Thesis on Grassmann Algebra under his direction. Since that time, his primary mathematical interest has been to learn as much about the subject and its applications as possible. Since 1963 he has been on the faculty at Arizona State University, Tempe (USA).

Antonio G.B. Teolis is the director of the information systems department of the Municipality of Bologna (Italy). His main research interests are Theoretical Computer Science and Computational Invariant Theory.

Renate Tobies, Mathematics Department at the Kaiserslautern University (Germany), is involved in research on the history of mathematics and the sciences in the nineteenth and twentieth centuries, on development of communication and instruction, and on women in mathematics and the sciences. She is the managing editor of the *International Journal of History and Ethics of Natural Sciences, Technology and Medicine*.

R. Steven Turner teaches the history of science in the Department of History of the University of New Brunswick in Fredericton, N. B., Canada. He is the author of *In the Eye's Mind: Vision and the Helmholtz-Hering Controversy* (1994).

Arno Zaddach, born 1924 in Germany, has held the position of a Professor of Mathematics at the University of Chile since 1958, first in Santiago and then in Arica. Now retired, he lives in Arica (Chile). His research interest is the intertwining of algebraic structures with geometry.

NOTES AND CREDITS TO THE ILLUSTRATIONS

1. (p. ii, frontispiece) Portrait of Hermann G. Graßmann (date and artist unknown). The original hung in the teachers' conference room of the *Marienstiftsgymnasium*, Stettin, and was most probably destroyed during World War II. The first reproduction was published by F. Engel in volume 1.1 of Grassmann's *Gesammelte mathematische und physikalische Werke* in 1894.

2. (p. xxx) Photograph of Participants of the International Graßmann Conference, 23 - 28 May 1994, during their excursion on the Rügen Island, in the *Schloßpark* of the island's capital Putbus. The statue shows Fürst Malte I. (1783-1854), after 1815 the first Prussian *Kurator* (Regierungsbevollmächtigter) of Greifswald University. Courtesy of Ivor Grattan-Guinness.

3. (p. 2) Drawing of the Friedrich-Wilhelms-Schule Stettin, front façade, during Grassmann's time as a teacher there - simultaneously the period of his greatest mathematical production. The drawing is a part of the front page for the school's 1890 *Schulprogramm*: its *Festschrift* for the celebration of the school's fiftieth anniversary:

Festschrift zum 50jährigen Jubiläum der Friedrich-Wilhelms-Schule zu Stettin 1890 (Stettin: Druck von R. Grassmann, 1890).

The *Festschrift* contains a history of the school: "Geschichte der Friedrich-Wilhelms-Schule während der ersten 50 Jahre ihres Bestehens 1840-1890", by its director Hermann Fritsche.

4. (pp. 17-18) Reproduction of the first two pages in Grassmann's personal file in the provincial school administration for the Pomeranian province in Stettin, the *Provinzialschulkollegium*. In 1866, this *Provinzialschulkollegium* established for the first time individual personal files for the *Gymnasium* teachers in its administration. These files begin with two pages of a form which the teachers had to fill in themselves by answering twenty (also handwritten) questions. The forms had been copied by a contemporary, relatively primitive copying technique so that the form entries are not always quite legible.

Location of the personal file: Vorpommersches Landesarchiv Greifswald, Rep. 62, Nr. 1050, Die Personalien des Lehrers Hermann G. Graßmann.

Reproduction by courtesy of the Vorpommersches Landesarchiv Greifswald.

Transscription of the form and of Grassmann's items:

Zu den Personalacten des Königlichen Provinzial-Schul-Collegiums in Stettin
Ausgestellt den 1ten Juni 1866

1. Zuname	Graßmann
2. Vornamen /:mit Unterstreichung der Rufnamen:/	Hermann Günther
3. Stand des Vaters	Professor
4. Tag und Jahr der Geburt	15. April 1809
5. Geburtsort	Stetting
6. Konfession	evangelisch
7. An welchem Gymnasium und wann er das Zeugniß der Reife empfangen?	am Gymnasium zu Stettin Michaelis 1827
8. Was und auf welcher Universität er studiert hat?	in Berlin Theologie und Philologie
9. Von welcher wissenschaftlichen Prüfungs-Kommission und unter welchem Datum er das Zeugniß pro facultate docendi empfangen? evtl. von welchem Seminar und wann das Zeugniß für das Elementarschulamt?	von der wissenschaftlichen Prüfungskommission in Berlin, das Zeugniß pro fac. doc. datirt vom 31. Dec. 1831; das Zeugniß über eine zweite Prüfung für Mathematik, Physik, Mineralogie und Chemie bei derselben Kommission, datirt vom 1 Mai 1840
10. Ob er sonstige Prüfungen, etwa die pro schola oder theologische bestanden, eventuell wann und wo?	hat das erste theologische Examen (pro venia concionandi) in Stettin am 22 Mai 1834 und das zweite (pro ministerio) ebendaselbst am 10 Juli 1839 gemacht
11. Verhältniß zum Militairdienst	als unbrauchbar zurückgestellt
12. Wann und wo er das Probejahr abgeliefert?	am Gymnasium in Stettin Ostern 1831 bis 32
13. Etwanige sonstige Beschäftigung an Schulen vor der definitiven Anstellung	Von da ab als Hülfslehrer am Gymnasium in Stettin
14. Zeit und Ort der ersten definitiven Anstellung	Neujahr 1836 als erster Lehrer an der Otto Schule hier
15. Tag der ersten eidlichen Verpflichtung für den preußischen Schuldienst	im Januar oder Februar 1836, den Tag kann ich nicht angeben
16. spätere Dienstverhältniße	Michaelis 1842 wurde ich als ordentlicher Lehrer ans Gymnasium zu Stettin versetzt, Ostern 1843 an die Friedrich-Wilhelms-Schule, Johannis 1852 an das Gymnasium als erster Lehrer der Mathematik
17. Orden und Ehrenzeichen mit Angabe des Jahres der Verleihung	

18. Familienverhältnisse - ob unverheirathet, oder verheirathet, oder Wittwer? wie viel Kinder?	verheirathet, 8 Kinder
19. Einkommen aus dem Hauptamt	1424 Thaler, 11 Groschen, 3 Pfennige
20. Nebenämter und Einkommen aus denselben	keine

5. (p. 26) Photograph of the official residences (*Amtswohnungen*) of the professors of the *Marienstiftsgymnasium*. The buildings at the former Königsplatz, now plac Zołnierza Polskiego, were destroyed during World War II and have been reconstructed in the old style. Courtesy of Peter Bergau.

6. (p. 28) Drawing of the front façade of the *Marienstiftsgymnasium* as it looked during the period 1832-1884. In 1884, the building was enlarged. Destroyed during World War II, it was reconstructed after the war and serves again as a secondary school.

Source: Max Wehrmann, "Geschichte des Marienstiftsgymnasiums 1544-1894", *Festschrift zum dreihundertfünfzigjährigen Jubiläum des Königlichen Marienstifts-Gymnasiums zu Stettin* (Stettin: Herrcke und Lebeling, 1894), 1-165.

7. (p. 36) Portrait of Hermann's father Justus Günther Graßmann (1779-1852), date and artist unknown. Published in:

Karl Scheibert, *Geschichte des Geschlechts Graßmann und seiner Nebenlinien* (Görlitz: C. A. Starke, 1937)

8. (p. 86) The diploma of the *Ehrendoktor* degree which was conferred to Graßmann in 1876 by the Philosophical Faculty of Tübingen University in honor of his Sanskrit studies. Copies are preserved in the Universitätsarchiv Tübingen and in Graßmann's personal file in the Vorpommersches Landesarchiv (see note to illustr. 4).

The Latin laudatio (printed in the smallest letters!) reads:

"HERMANNUM GRASSMANN MARIANI GYMNASII STETTINENSIS PROFESSOREM QUI RARO HAC AETATE EXEMPLO MATHEMATICAE PERITIAM CONIUNXIT CUM SCIENTIA RERUM PHILOLOGICARUM ET IN UTROQUE STUDIORUM GENERE SCRIPTOR EXTITIT CLARISSIMUS MAXIME VERO ACUTISSIMA VEDICORUM CARMINUM INTERPRETATIONE NOMEN SUUM REDDIDIT ILLUSTRISSIMUM."

Karin Reich has described the initiatives taken and the eventual decisions in Tübingen in her article:

"Die Ehrenpromotion Hermann Graßmanns an der Universität Tübingen im Jahre 1876", *Hermann Graßmann - Werk und Wirkung. Internationale Fachtagung anläßlich des 150. Jahrestages des erste Erscheinens der "linealen Ausdehnungslehre"*, ed. Peter Schreiber

(Greifswald: Ernst-Moritz-Arndt-Universität, Fachrichtungen Mathematik/Informatik, 1995), 59-63.

Reproduction in this volume by the kind permission of the Universitätsarchiv Tübingen.

9. (p. 230) Memorial table for Hermann Graßmann inaugurated on 28 May 1994 by representatives of the Szczecin municipality and of Szczecin University, in the presence of participants of the Rügen Conference. The table was conjointly sponsored by the German and the Polish Mathematical Societies. It is not only the first monument for Graßmann at all, it is also the first monument for a German citizen in the now Polish town after the war.

The bilingual text, in Polish and in German, reads:

HERMANN GÜNTHER GRASSMANN
(15 IV 1809 - 26 IX 1877)
MATHEMATICIAN AND PHILOLOGIST
FAMOUS SON OF THIS TOWN
AT THE OCCASION OF 150 YEARS OF PUBLICATION OF HIS FIRST
"AUSDEHNUNGSLEHRE"
DEUTSCHE MATHEMATIKER VEREINIGUNG
POLSKIE TOWAZYSTWO MATEMATYCZNE
MAY 1994

The memorial table is situated at the entrance to the building of the Faculty for Mathematics and Physics of Szczecin University, ulica Wielkopolskej 15.

Courtesy of Gert Schubring.

INDEX OF NAMES

Boston Studies in the Philosophy of Science

Editor: Robert S. Cohen, *Boston University*

1. M.W. Wartofsky (ed.): *Proceedings of the Boston Colloquium for the Philosophy of Science, 1961/1962*. [Synthese Library 6] 1963 ISBN 90-277-0021-4
2. R.S. Cohen and M.W. Wartofsky (eds.): *Proceedings of the Boston Colloquium for the Philosophy of Science, 1962/1964*. In Honor of P. Frank. [Synthese Library 10] 1965 ISBN 90-277-9004-0
3. R.S. Cohen and M.W. Wartofsky (eds.): *Proceedings of the Boston Colloquium for the Philosophy of Science, 1964/1966*. In Memory of Norwood Russell Hanson. [Synthese Library 14] 1967 ISBN 90-277-0013-3
4. R.S. Cohen and M.W. Wartofsky (eds.): *Proceedings of the Boston Colloquium for the Philosophy of Science, 1966/1968*. [Synthese Library 18] 1969 ISBN 90-277-0014-1
5. R.S. Cohen and M.W. Wartofsky (eds.): *Proceedings of the Boston Colloquium for the Philosophy of Science, 1966/1968*. [Synthese Library 19] 1969 ISBN 90-277-0015-X
6. R.S. Cohen and R.J. Seeger (eds.): *Ernst Mach, Physicist and Philosopher*. [Synthese Library 27] 1970 ISBN 90-277-0016-8
7. M. Čapek: *Bergson and Modern Physics*. A Reinterpretation and Re-evaluation. [Synthese Library 37] 1971 ISBN 90-277-0186-5
8. R.C. Buck and R.S. Cohen (eds.): *PSA 1970*. Proceedings of the 2nd Biennial Meeting of the Philosophy and Science Association (Boston, Fall 1970). In Memory of Rudolf Carnap. [Synthese Library 39] 1971 ISBN 90-277-0187-3; Pb 90-277-0309-4
9. A.A. Zinov'ev: *Foundations of the Logical Theory of Scientific Knowledge (Complex Logic)*. Translated from Russian. Revised and enlarged English Edition, with an Appendix by G.A. Smirnov, E.A. Sidorenko, A.M. Fedina and L.A. Bobrova. [Synthese Library 46] 1973 ISBN 90-277-0193-8; Pb 90-277-0324-8
10. L. Tondl: *Scientific Procedures*. A Contribution Concerning the Methodological Problems of Scientific Concepts and Scientific Explanation.Translated from Czech. [Synthese Library 47] 1973 ISBN 90-277-0147-4; Pb 90-277-0323-X
11. R.J. Seeger and R.S. Cohen (eds.): *Philosophical Foundations of Science*. Proceedings of Section L, 1969, American Association for the Advancement of Science. [Synthese Library 58] 1974 ISBN 90-277-0390-6; Pb 90-277-0376-0
12. A. Grünbaum: *Philosophical Problems of Space and Times*. 2nd enlarged ed. [Synthese Library 55] 1973 ISBN 90-277-0357-4; Pb 90-277-0358-2
13. R.S. Cohen and M.W. Wartofsky (eds.): *Logical and Epistemological Studies in Contemporary Physics*. Proceedings of the Boston Colloquium for the Philosophy of Science, 1969/72, Part I. [Synthese Library 59] 1974 ISBN 90-277-0391-4; Pb 90-277-0377-9
14. R.S. Cohen and M.W. Wartofsky (eds.): *Methodological and Historical Essays in the Natural and Social Sciences*. Proceedings of the Boston Colloquium for the Philosophy of Science, 1969/72, Part II. [Synthese Library 60] 1974 ISBN 90-277-0392-2; Pb 90-277-0378-7
15. R.S. Cohen, J.J. Stachel and M.W. Wartofsky (eds.): *For Dirk Struik*. Scientific, Historical and Political Essays in Honor of Dirk J. Struik. [Synthese Library 61] 1974 ISBN 90-277-0393-0; Pb 90-277-0379-5
16. N. Geschwind: *Selected Papers on Language and the Brains*. [Synthese Library 68] 1974 ISBN 90-277-0262-4; Pb 90-277-0263-2
17. B.G. Kuznetsov: *Reason and Being*. Translated from Russian. Edited by C.R. Fawcett and R.S. Cohen. 1987 ISBN 90-277-2181-5

18. P. Mittelstaedt: *Philosophical Problems of Modern Physics*. Translated from the revised 4th German edition by W. Riemer and edited by R.S. Cohen. [Synthese Library 95] 1976
ISBN 90-277-0285-3; Pb 90-277-0506-2
19. H. Mehlberg: *Time, Causality, and the Quantum Theory*. Studies in the Philosophy of Science. Vol. I: *Essay on the Causal Theory of Time*. Vol. II: *Time in a Quantized Universe*. Translated from French. Edited by R.S. Cohen. 1980
Vol. I: ISBN 90-277-0721-9; Pb 90-277-1074-0
Vol. II: ISBN 90-277-1075-9; Pb 90-277-1076-7
20. K.F. Schaffner and R.S. Cohen (eds.): *PSA 1972*. Proceedings of the 3rd Biennial Meeting of the Philosophy of Science Association (Lansing, Michigan, Fall 1972). [Synthese Library 64] 1974 ISBN 90-277-0408-2; Pb 90-277-0409-0
21. R.S. Cohen and J.J. Stachel (eds.): *Selected Papers of Léon Rosenfeld*. [Synthese Library 100] 1979 ISBN 90-277-0651-4; Pb 90-277-0652-2
22. M. Čapek (ed.): *The Concepts of Space and Time*. Their Structure and Their Development. [Synthese Library 74] 1976 ISBN 90-277-0355-8; Pb 90-277-0375-2
23. M. Grene: *The Understanding of Nature*. Essays in the Philosophy of Biology. [Synthese Library 66] 1974 ISBN 90-277-0462-7; Pb 90-277-0463-5
24. D. Ihde: *Technics and Praxis*. A Philosophy of Technology. [Synthese Library 130] 1979
ISBN 90-277-0953-X; Pb 90-277-0954-8
25. J. Hintikka and U. Remes: *The Method of Analysis*. Its Geometrical Origin and Its General Significance. [Synthese Library 75] 1974 ISBN 90-277-0532-1; Pb 90-277-0543-7
26. J.E. Murdoch and E.D. Sylla (eds.): *The Cultural Context of Medieval Learning*. Proceedings of the First International Colloquium on Philosophy, Science, and Theology in the Middle Ages, 1973. [Synthese Library 76] 1975 ISBN 90-277-0560-7; Pb 90-277-0587-9
27. M. Grene and E. Mendelsohn (eds.): *Topics in the Philosophy of Biology*. [Synthese Library 84] 1976 ISBN 90-277-0595-X; Pb 90-277-0596-8
28. J. Agassi: *Science in Flux*. [Synthese Library 80] 1975
ISBN 90-277-0584-4; Pb 90-277-0612-3
29. J.J. Wiatr (ed.): *Polish Essays in the Methodology of the Social Sciences*. [Synthese Library 131] 1979 ISBN 90-277-0723-5; Pb 90-277-0956-4
30. P. Janich: *Protophysics of Time*. Constructive Foundation and History of Time Measurement. Translated from German. 1985 ISBN 90-277-0724-3
31. R.S. Cohen and M.W. Wartofsky (eds.): *Language, Logic, and Method*. 1983
ISBN 90-277-0725-1
32. R.S. Cohen, C.A. Hooker, A.C. Michalos and J.W. van Evra (eds.): *PSA 1974*. Proceedings of the 4th Biennial Meeting of the Philosophy of Science Association. [Synthese Library 101] 1976 ISBN 90-277-0647-6; Pb 90-277-0648-4
33. G. Holton and W.A. Blanpied (eds.): *Science and Its Public*. The Changing Relationship. [Synthese Library 96] 1976 ISBN 90-277-0657-3; Pb 90-277-0658-1
34. M.D. Grmek, R.S. Cohen and G. Cimino (eds.): *On Scientific Discovery*. The 1977 Erice Lectures. 1981 ISBN 90-277-1122-4; Pb 90-277-1123-2
35. S. Amsterdamski: *Between Experience and Metaphysics*. Philosophical Problems of the Evolution of Science. Translated from Polish. [Synthese Library 77] 1975
ISBN 90-277-0568-2; Pb 90-277-0580-1
36. M. Marković and G. Petrović (eds.): *Praxis*. Yugoslav Essays in the Philosophy and Methodology of the Social Sciences. [Synthese Library 134] 1979
ISBN 90-277-0727-8; Pb 90-277-0968-8

37. H. von Helmholtz: *Epistemological Writings*. The Paul Hertz / Moritz Schlick Centenary Edition of 1921. Translated from German by M.F. Lowe. Edited with an Introduction and Bibliography by R.S. Cohen and Y. Elkana. [Synthese Library 79] 1977 ISBN 90-277-0290-X; Pb 90-277-0582-8
38. R.M. Martin: *Pragmatics, Truth and Language*. 1979 ISBN 90-277-0992-0; Pb 90-277-0993-9
39. R.S. Cohen, P.K. Feyerabend and M.W. Wartofsky (eds.): *Essays in Memory of Imre Lakatos*. [Synthese Library 99] 1976 ISBN 90-277-0654-9; Pb 90-277-0655-7
40. Not published.
41. Not published.
42. H.R. Maturana and F.J. Varela: *Autopoiesis and Cognition*. The Realization of the Living. With a Preface to 'Autopoiesis' by S. Beer. 1980 ISBN 90-277-1015-5; Pb 90-277-1016-3
43. A. Kasher (ed.): *Language in Focus: Foundations, Methods and Systems*. Essays in Memory of Yehoshua Bar-Hillel. [Synthese Library 89] 1976 ISBN 90-277-0644-1; Pb 90-277-0645-X
44. T.D. Thao: *Investigations into the Origin of Language and Consciousness*. 1984 ISBN 90-277-0827-4
45. Not published.
46. P.L. Kapitza: *Experiment, Theory, Practice*. Articles and Addresses. Edited by R.S. Cohen. 1980 ISBN 90-277-1061-9; Pb 90-277-1062-7
47. M.L. Dalla Chiara (ed.): *Italian Studies in the Philosophy of Science*. 1981 ISBN 90-277-0735-9; Pb 90-277-1073-2
48. M.W. Wartofsky: *Models*. Representation and the Scientific Understanding. [Synthese Library 129] 1979 ISBN 90-277-0736-7; Pb 90-277-0947-5
49. T.D. Thao: *Phenomenology and Dialectical Materialism*. Edited by R.S. Cohen. 1986 ISBN 90-277-0737-5
50. Y. Fried and J. Agassi: *Paranoia*. A Study in Diagnosis. [Synthese Library 102] 1976 ISBN 90-277-0704-9; Pb 90-277-0705-7
51. K.H. Wolff: *Surrender and Cath*. Experience and Inquiry Today. [Synthese Library 105] 1976 ISBN 90-277-0758-8; Pb 90-277-0765-0
52. K. Kosík: *Dialectics of the Concrete*. A Study on Problems of Man and World. 1976 ISBN 90-277-0761-8; Pb 90-277-0764-2
53. N. Goodman: *The Structure of Appearance*. [Synthese Library 107] 1977 ISBN 90-277-0773-1; Pb 90-277-0774-X
54. H.A. Simon: *Models of Discovery* and Other Topics in the Methods of Science. [Synthese Library 114] 1977 ISBN 90-277-0812-6; Pb 90-277-0858-4
55. M. Lazerowitz: *The Language of Philosophy*. Freud and Wittgenstein. [Synthese Library 117] 1977 ISBN 90-277-0826-6; Pb 90-277-0862-2
56. T. Nickles (ed.): *Scientific Discovery, Logic, and Rationality*. 1980 ISBN 90-277-1069-4; Pb 90-277-1070-8
57. J. Margolis: *Persons and Mind*. The Prospects of Nonreductive Materialism. [Synthese Library 121] 1978 ISBN 90-277-0854-1; Pb 90-277-0863-0
58. G. Radnitzky and G. Andersson (eds.): *Progress and Rationality in Science*. [Synthese Library 125] 1978 ISBN 90-277-0921-1; Pb 90-277-0922-X
59. G. Radnitzky and G. Andersson (eds.): *The Structure and Development of Science*. [Synthese Library 136] 1979 ISBN 90-277-0994-7; Pb 90-277-0995-5

60. T. Nickles (ed.): *Scientific Discovery.* Case Studies. 1980
ISBN 90-277-1092-9; Pb 90-277-1093-7
61. M.A. Finocchiaro: *Galileo and the Art of Reasoning.* Rhetorical Foundation of Logic and Scientific Method. 1980 ISBN 90-277-1094-5; Pb 90-277-1095-3
62. W.A. Wallace: *Prelude to Galileo.* Essays on Medieval and 16th-Century Sources of Galileo's Thought. 1981 ISBN 90-277-1215-8; Pb 90-277-1216-6
63. F. Rapp: *Analytical Philosophy of Technology.* Translated from German. 1981
ISBN 90-277-1221-2; Pb 90-277-1222-0
64. R.S. Cohen and M.W. Wartofsky (eds.): *Hegel and the Sciences.* 1984
ISBN 90-277-0726-X
65. J. Agassi: *Science and Society.* Studies in the Sociology of Science. 1981
ISBN 90-277-1244-1; Pb 90-277-1245-X
66. L. Tondl: *Problems of Semantics.* A Contribution to the Analysis of the Language of Science. Translated from Czech. 1981 ISBN 90-277-0148-2; Pb 90-277-0316-7
67. J. Agassi and R.S. Cohen (eds.): *Scientific Philosophy Today.* Essays in Honor of Mario Bunge. 1982 ISBN 90-277-1262-X; Pb 90-277-1263-8
68. W. Krajewski (ed.): *Polish Essays in the Philosophy of the Natural Sciences.* Translated from Polish and edited by R.S. Cohen and C.R. Fawcett. 1982
ISBN 90-277-1286-7; Pb 90-277-1287-5
69. J.H. Fetzer: *Scientific Knowledge.* Causation, Explanation and Corroboration. 1981
ISBN 90-277-1335-9; Pb 90-277-1336-7
70. S. Grossberg: *Studies of Mind and Brain.* Neural Principles of Learning, Perception, Development, Cognition, and Motor Control. 1982
ISBN 90-277-1359-6; Pb 90-277-1360-X
71. R.S. Cohen and M.W. Wartofsky (eds.): *Epistemology, Methodology, and the Social Sciences.* 1983. ISBN 90-277-1454-1
72. K. Berka: *Measurement.* Its Concepts, Theories and Problems. Translated from Czech. 1983
ISBN 90-277-1416-9
73. G.L. Pandit: *The Structure and Growth of Scientific Knowledge.* A Study in the Methodology of Epistemic Appraisal. 1983 ISBN 90-277-1434-7
74. A.A. Zinov'ev: *Logical Physics.* Translated from Russian. Edited by R.S. Cohen. 1983 [*see also* Volume 9] ISBN 90-277-0734-0
75. G-G. Granger: *Formal Thought and the Sciences of Man.* Translated from French. With and Introduction by A. Rosenberg. 1983 ISBN 90-277-1524-6
76. R.S. Cohen and L. Laudan (eds.): *Physics, Philosophy and Psychoanalysis.* Essays in Honor of Adolf Grünbaum. 1983 ISBN 90-277-1533-5
77. G. Böhme, W. van den Daele, R. Hohlfeld, W. Krohn and W. Schäfer: *Finalization in Science.* The Social Orientation of Scientific Progress. Translated from German. Edited by W. Schäfer. 1983 ISBN 90-277-1549-1
78. D. Shapere: *Reason and the Search for Knowledge.* Investigations in the Philosophy of Science. 1984 ISBN 90-277-1551-3; Pb 90-277-1641-2
79. G. Andersson (ed.): *Rationality in Science and Politics.* Translated from German. 1984
ISBN 90-277-1575-0; Pb 90-277-1953-5
80. P.T. Durbin and F. Rapp (eds.): *Philosophy and Technology.* [*Also* Philosophy and Technology Series, Vol. 1] 1983 ISBN 90-277-1576-9
81. M. Marković: *Dialectical Theory of Meaning.* Translated from Serbo-Croat. 1984
ISBN 90-277-1596-3

Boston Studies in the Philosophy of Science

82. R.S. Cohen and M.W. Wartofsky (eds.): *Physical Sciences and History of Physics*. 1984. ISBN 90-277-1615-3
83. É. Meyerson: *The Relativistic Deduction*. Epistemological Implications of the Theory of Relativity. Translated from French. With a Review by Albert Einstein and an Introduction by Milić Čapek. 1985 ISBN 90-277-1699-4
84. R.S. Cohen and M.W. Wartofsky (eds.): *Methodology, Metaphysics and the History of Science*. In Memory of Benjamin Nelson. 1984 ISBN 90-277-1711-7
85. G. Tamás: *The Logic of Categories*. Translated from Hungarian. Edited by R.S. Cohen. 1986 ISBN 90-277-1742-7
86. S.L. de C. Fernandes: *Foundations of Objective Knowledge*. The Relations of Popper's Theory of Knowledge to That of Kant. 1985 ISBN 90-277-1809-1
87. R.S. Cohen and T. Schnelle (eds.): *Cognition and Fact*. Materials on Ludwik Fleck. 1986 ISBN 90-277-1902-0
88. G. Freudenthal: *Atom and Individual in the Age of Newton*. On the Genesis of the Mechanistic World View. Translated from German. 1986 ISBN 90-277-1905-5
89. A. Donagan, A.N. Perovich Jr and M.V. Wedin (eds.): *Human Nature and Natural Knowledge*. Essays presented to Marjorie Grene on the Occasion of Her 75th Birthday. 1986 ISBN 90-277-1974-8
90. C. Mitcham and A. Hunning (eds.): *Philosophy and Technology II*. Information Technology and Computers in Theory and Practice. [*Also* Philosophy and Technology Series, Vol. 2] 1986 ISBN 90-277-1975-6
91. M. Grene and D. Nails (eds.): *Spinoza and the Sciences*. 1986 ISBN 90-277-1976-4
92. S.P. Turner: *The Search for a Methodology of Social Science*. Durkheim, Weber, and the 19th-Century Problem of Cause, Probability, and Action. 1986. ISBN 90-277-2067-3
93. I.C. Jarvie: *Thinking about Society*. Theory and Practice. 1986 ISBN 90-277-2068-1
94. E. Ullmann-Margalit (ed.): *The Kaleidoscope of Science*. The Israel Colloquium: Studies in History, Philosophy, and Sociology of Science, Vol. 1. 1986 ISBN 90-277-2158-0; Pb 90-277-2159-9
95. E. Ullmann-Margalit (ed.): *The Prism of Science*. The Israel Colloquium: Studies in History, Philosophy, and Sociology of Science, Vol. 2. 1986 ISBN 90-277-2160-2; Pb 90-277-2161-0
96. G. Márkus: *Language and Production*. A Critique of the Paradigms. Translated from French. 1986 ISBN 90-277-2169-6
97. F. Amrine, F.J. Zucker and H. Wheeler (eds.): *Goethe and the Sciences: A Reappraisal*. 1987 ISBN 90-277-2265-X; Pb 90-277-2400-8
98. J.C. Pitt and M. Pera (eds.): *Rational Changes in Science*. Essays on Scientific Reasoning. Translated from Italian. 1987 ISBN 90-277-2417-2
99. O. Costa de Beauregard: *Time, the Physical Magnitude*. 1987 ISBN 90-277-2444-X
100. A. Shimony and D. Nails (eds.): *Naturalistic Epistemology*. A Symposium of Two Decades. 1987 ISBN 90-277-2337-0
101. N. Rotenstreich: *Time and Meaning in History*. 1987 ISBN 90-277-2467-9
102. D.B. Zilberman: *The Birth of Meaning in Hindu Thought*. Edited by R.S. Cohen. 1988 ISBN 90-277-2497-0
103. T.F. Glick (ed.): *The Comparative Reception of Relativity*. 1987 ISBN 90-277-2498-9
104. Z. Harris, M. Gottfried, T. Ryckman, P. Mattick Jr, A. Daladier, T.N. Harris and S. Harris: *The Form of Information in Science*. Analysis of an Immunology Sublanguage. With a Preface by Hilary Putnam. 1989 ISBN 90-277-2516-0

Boston Studies in the Philosophy of Science

105. F. Burwick (ed.): *Approaches to Organic Form.* Permutations in Science and Culture. 1987 ISBN 90-277-2541-1
106. M. Almási: *The Philosophy of Appearances.* Translated from Hungarian. 1989 ISBN 90-277-2150-5
107. S. Hook, W.L. O'Neill and R. O'Toole (eds.): *Philosophy, History and Social Action.* Essays in Honor of Lewis Feuer. With an Autobiographical Essay by L. Feuer. 1988 ISBN 90-277-2644-2
108. I. Hronszky, M. Fehér and B. Dajka: *Scientific Knowledge Socialized.* Selected Proceedings of the 5th Joint International Conference on the History and Philosophy of Science organized by the IUHPS (Veszprém, Hungary, 1984). 1988 ISBN 90-277-2284-6
109. P. Tillers and E.D. Green (eds.): *Probability and Inference in the Law of Evidence.* The Uses and Limits of Bayesianism. 1988 ISBN 90-277-2689-2
110. E. Ullmann-Margalit (ed.): *Science in Reflection.* The Israel Colloquium: Studies in History, Philosophy, and Sociology of Science, Vol. 3. 1988 ISBN 90-277-2712-0; Pb 90-277-2713-9
111. K. Gavroglu, Y. Goudaroulis and P. Nicolacopoulos (eds.): *Imre Lakatos and Theories of Scientific Change.* 1989 ISBN 90-277-2766-X
112. B. Glassner and J.D. Moreno (eds.): *The Qualitative-Quantitative Distinction in the Social Sciences.* 1989 ISBN 90-277-2829-1
113. K. Arens: *Structures of Knowing.* Psychologies of the 19th Century. 1989 ISBN 0-7923-0009-2
114. A. Janik: *Style, Politics and the Future of Philosophy.* 1989 ISBN 0-7923-0056-4
115. F. Amrine (ed.): *Literature and Science as Modes of Expression.* With an Introduction by S. Weininger. 1989 ISBN 0-7923-0133-1
116. J.R. Brown and J. Mittelstrass (eds.): *An Intimate Relation.* Studies in the History and Philosophy of Science. Presented to Robert E. Butts on His 60th Birthday. 1989 ISBN 0-7923-0169-2
117. F. D'Agostino and I.C. Jarvie (eds.): *Freedom and Rationality.* Essays in Honor of John Watkins. 1989 ISBN 0-7923-0264-8
118. D. Zolo: *Reflexive Epistemology.* The Philosophical Legacy of Otto Neurath. 1989 ISBN 0-7923-0320-2
119. M. Kearn, B.S. Philips and R.S. Cohen (eds.): *Georg Simmel and Contemporary Sociology.* 1989 ISBN 0-7923-0407-1
120. T.H. Levere and W.R. Shea (eds.): *Nature, Experiment and the Science.* Essays on Galileo and the Nature of Science. In Honour of Stillman Drake. 1989 ISBN 0-7923-0420-9
121. P. Nicolacopoulos (ed.): *Greek Studies in the Philosophy and History of Science.* 1990 ISBN 0-7923-0717-8
122. R. Cooke and D. Costantini (eds.): *Statistics in Science.* The Foundations of Statistical Methods in Biology, Physics and Economics. 1990 ISBN 0-7923-0797-6
123. P. Duhem: *The Origins of Statics.* Translated from French by G.F. Leneaux, V.N. Vagliente and G.H. Wagner. With an Introduction by S.L. Jaki. 1991 ISBN 0-7923-0898-0
124. H. Kamerlingh Onnes: *Through Measurement to Knowledge.* The Selected Papers, 1853-1926. Edited and with an Introduction by K. Gavroglu and Y. Goudaroulis. 1991 ISBN 0-7923-0825-5
125. M. Čapek: *The New Aspects of Time: Its Continuity and Novelties.* Selected Papers in the Philosophy of Science. 1991 ISBN 0-7923-0911-1

126. S. Unguru (ed.): *Physics, Cosmology and Astronomy, 1300-1700*. Tension and Accommodation. 1991 ISBN 0-7923-1022-5
127. Z. Bechler: *Newton's Physics on the Conceptual Structure of the Scientific Revolution*. 1991 ISBN 0-7923-1054-3
128. É. Meyerson: *Explanation in the Sciences*. Translated from French by M-A. Siple and D.A. Siple. 1991 ISBN 0-7923-1129-9
129. A.I. Tauber (ed.): *Organism and the Origins of Self*. 1991 ISBN 0-7923-1185-X
130. F.J. Varela and J-P. Dupuy (eds.): *Understanding Origins*. Contemporary Views on the Origin of Life, Mind and Society. 1992 ISBN 0-7923-1251-1
131. G.L. Pandit: *Methodological Variance*. Essays in Epistemological Ontology and the Methodology of Science. 1991 ISBN 0-7923-1263-5
132. G. Munévar (ed.): *Beyond Reason*. Essays on the Philosophy of Paul Feyerabend. 1991 ISBN 0-7923-1272-4
133. T.E. Uebel (ed.): *Rediscovering the Forgotten Vienna Circle*. Austrian Studies on Otto Neurath and the Vienna Circle. Partly translated from German. 1991 ISBN 0-7923-1276-7
134. W.R. Woodward and R.S. Cohen (eds.): *World Views and Scientific Discipline Formation*. Science Studies in the [former] German Democratic Republic. Partly translated from German by W.R. Woodward. 1991 ISBN 0-7923-1286-4
135. P. Zambelli: *The Speculum Astronomiae and Its Enigma*. Astrology, Theology and Science in Albertus Magnus and His Contemporaries. 1992 ISBN 0-7923-1380-1
136. P. Petitjean, C. Jami and A.M. Moulin (eds.): *Science and Empires*. Historical Studies about Scientific Development and European Expansion. ISBN 0-7923-1518-9
137. W.A. Wallace: *Galileo's Logic of Discovery and Proof*. The Background, Content, and Use of His Appropriated Treatises on Aristotle's *Posterior Analytics*. 1992 ISBN 0-7923-1577-4
138. W.A. Wallace: *Galileo's Logical Treatises*. A Translation, with Notes and Commentary, of His Appropriated Latin Questions on Aristotle's *Posterior Analytics*. 1992 ISBN 0-7923-1578-2
Set (137 + 138) ISBN 0-7923-1579-0
139. M.J. Nye, J.L. Richards and R.H. Stuewer (eds.): *The Invention of Physical Science*. Intersections of Mathematics, Theology and Natural Philosophy since the Seventeenth Century. Essays in Honor of Erwin N. Hiebert. 1992 ISBN 0-7923-1753-X
140. G. Corsi, M.L. dalla Chiara and G.C. Ghirardi (eds.): *Bridging the Gap: Philosophy, Mathematics and Physics*. Lectures on the Foundations of Science. 1992 ISBN 0-7923-1761-0
141. C.-H. Lin and D. Fu (eds.): *Philosophy and Conceptual History of Science in Taiwan*. 1992 ISBN 0-7923-1766-1
142. S. Sarkar (ed.): *The Founders of Evolutionary Genetics*. A Centenary Reappraisal. 1992 ISBN 0-7923-1777-7
143. J. Blackmore (ed.): *Ernst Mach – A Deeper Look*. Documents and New Perspectives. 1992 ISBN 0-7923-1853-6
144. P. Kroes and M. Bakker (eds.): *Technological Development and Science in the Industrial Age*. New Perspectives on the Science–Technology Relationship. 1992 ISBN 0-7923-1898-6
145. S. Amsterdamski: *Between History and Method*. Disputes about the Rationality of Science. 1992 ISBN 0-7923-1941-9
146. E. Ullmann-Margalit (ed.): *The Scientific Enterprise*. The Bar-Hillel Colloquium: Studies in History, Philosophy, and Sociology of Science, Volume 4. 1992 ISBN 0-7923-1992-3

147. L. Embree (ed.): *Metaarchaeology*. Reflections by Archaeologists and Philosophers. 1992 ISBN 0-7923-2023-9
148. S. French and H. Kamminga (eds.): *Correspondence, Invariance and Heuristics*. Essays in Honour of Heinz Post. 1993 ISBN 0-7923-2085-9
149. M. Bunzl: *The Context of Explanation*. 1993 ISBN 0-7923-2153-7
150. I.B. Cohen (ed.): *The Natural Sciences and the Social Sciences*. Some Critical and Historical Perspectives. 1994 ISBN 0-7923-2223-1
151. K. Gavroglu, Y. Christianidis and E. Nicolaidis (eds.): *Trends in the Historiography of Science*. 1994 ISBN 0-7923-2255-X
152. S. Poggi and M. Bossi (eds.): *Romanticism in Science*. Science in Europe, 1790–1840. 1994 ISBN 0-7923-2336-X
153. J. Faye and H.J. Folse (eds.): *Niels Bohr and Contemporary Philosophy*. 1994 ISBN 0-7923-2378-5
154. C.C. Gould and R.S. Cohen (eds.): *Artifacts, Representations, and Social Practice*. Essays for Marx W. Wartofsky. 1994 ISBN 0-7923-2481-1
155. R.E. Butts: *Historical Pragmatics*. Philosophical Essays. 1993 ISBN 0-7923-2498-6
156. R. Rashed: *The Development of Arabic Mathematics: Between Arithmetic and Algebra*. Translated from French by A.F.W. Armstrong. 1994 ISBN 0-7923-2565-6
157. I. Szumilewicz-Lachman (ed.): *Zygmunt Zawirski: His Life and Work*. With Selected Writings on Time, Logic and the Methodology of Science. Translations by Feliks Lachman. Ed. by R.S. Cohen, with the assistance of B. Bergo. 1994 ISBN 0-7923-2566-4
158. S.N. Haq: *Names, Natures and Things*. The Alchemist Jābir ibn Ḥayyān and His *Kitāb al-Aḥjār* (Book of Stones). 1994 ISBN 0-7923-2587-7
159. P. Plaass: *Kant's Theory of Natural Science*. Translation, Analytic Introduction and Commentary by Alfred E. and Maria G. Miller. 1994 ISBN 0-7923-2750-0
160. J. Misiek (ed.): *The Problem of Rationality in Science and its Philosophy*. On Popper vs. Polanyi. The Polish Conferences 1988–89. 1995 ISBN 0-7923-2925-2
161. I.C. Jarvie and N. Laor (eds.): *Critical Rationalism, Metaphysics and Science*. Essays for Joseph Agassi, Volume I. 1995 ISBN 0-7923-2960-0
162. I.C. Jarvie and N. Laor (eds.): *Critical Rationalism, the Social Sciences and the Humanities*. Essays for Joseph Agassi, Volume II. 1995 ISBN 0-7923-2961-9
     Set (161–162) ISBN 0-7923-2962-7
163. K. Gavroglu, J. Stachel and M.W. Wartofsky (eds.): *Physics, Philosophy, and the Scientific Community*. Essays in the Philosophy and History of the Natural Sciences and Mathematics. In Honor of Robert S. Cohen. 1995 ISBN 0-7923-2988-0
164. K. Gavroglu, J. Stachel and M.W. Wartofsky (eds.): *Science, Politics and Social Practice*. Essays on Marxism and Science, Philosophy of Culture and the Social Sciences. In Honor of Robert S. Cohen. 1995 ISBN 0-7923-2989-9
165. K. Gavroglu, J. Stachel and M.W. Wartofsky (eds.): *Science, Mind and Art*. Essays on Science and the Humanistic Understanding in Art, Epistemology, Religion and Ethics. Essays in Honor of Robert S. Cohen. 1995 ISBN 0-7923-2990-2
     Set (163–165) ISBN 0-7923-2991-0
166. K.H. Wolff: *Transformation in the Writing*. A Case of Surrender-and-Catch. 1995 ISBN 0-7923-3178-8
167. A.J. Kox and D.M. Siegel (eds.): *No Truth Except in the Details*. Essays in Honor of Martin J. Klein. 1995 ISBN 0-7923-3195-8

Boston Studies in the Philosophy of Science

168. J. Blackmore: *Ludwig Boltzmann, His Later Life and Philosophy, 1900–1906.* Book One: A Documentary History. 1995 ISBN 0-7923-3231-8
169. R.S. Cohen, R. Hilpinen and Q. Renzong (eds.): *Realism and Anti-Realism in the Philosophy of Science.* Beijing International Conference, 1992. 1995 ISBN 0-7923-3233-4
170. I. Kuçuradi and R.S. Cohen (eds.): *The Concept of Knowledge.* The Ankara Seminar. 1995 ISBN 0-7923-3241-5
171. M.A. Grodin (ed.): *Meta Medical Ethics*: The Philosophical Foundations of Bioethics. 1995 ISBN 0-7923-3344-6
172. S. Ramirez and R.S. Cohen (eds.): *Mexican Studies in the History and Philosophy of Science.* 1995 ISBN 0-7923-3462-0
173. C. Dilworth: *The Metaphysics of Science.* An Account of Modern Science in Terms of Principles, Laws and Theories. 1995 ISBN 0-7923-3693-3
174. J. Blackmore: *Ludwig Boltzmann, His Later Life and Philosophy, 1900–1906* Book Two: The Philosopher. 1995 ISBN 0-7923-3464-7
175. P. Damerow: *Abstraction and Representation.* Essays on the Cultural Evolution of Thinking. 1996 ISBN 0-7923-3816-2
176. G. Tarozzi (ed.): *Karl Popper, Philosopher of Science.* (in prep.)
177. M. Marion and R.S. Cohen (eds.): *Québec Studies in the Philosophy of Science.* Part I: Logic, Mathematics, Physics and History of Science. Essays in Honor of Hugues Leblanc. 1995 ISBN 0-7923-3559-7
178. M. Marion and R.S. Cohen (eds.): *Québec Studies in the Philosophy of Science.* Part II: Biology, Psychology, Cognitive Science and Economics. Essays in Honor of Hugues Leblanc. 1996 ISBN 0-7923-3560-0
Set (177–178) ISBN 0-7923-3561-9
179. Fan Dainian and R.S. Cohen (eds.): *Chinese Studies in the History and Philosophy of Science and Technology.* 1995 ISBN 0-7923-3463-9
180. P. Forman and J.M. Sánchez-Ron (eds.): *National Military Establishments and the Advancement of Science and Technology.* Studies in 20th Century History. 1996 ISBN 0-7923-3541-4
181. E.J. Post: *Quantum Reprogramming.* Ensembles and Single Systems: A Two-Tier Approach to Quantum Mechanics. 1995 ISBN 0-7923-3565-1
182. A.I. Tauber (ed.): *The Elusive Synthesis: Aesthetics and Science.* 1996 ISBN 0-7923-3904-5
183. S. Sarkar (ed.): *The Philosophy and History of Molecular Biology: New Perspectives.* 1996 ISBN 0-7923-3947-9
184. J.T. Cushing, A. Fine and S. Goldstein (eds.): *Bohemian Mechanics and Quantum Theory: An Appraisal.* 1996 ISBN 0-7923-4028-0
185. K. Michalski: *Logic and Time.* An Essay on Husserl's Theory of Meaning. 1996 ISBN 0-7923-4082-5
186. G. Munévar (ed.): *Spanish Studies in the Philosophy of Science.* 1996 ISBN 0-7923-4147-3
187. G. Schubring (ed.): *Hermann Günther Graßmann (1809–1877): Visionary Mathematician, Scientist and Neohumanist Scholar.* Papers from a Sesquicentennial Conference. 1996 ISBN 0-7923-4261-5

Also of interest:

R.S. Cohen and M.W. Wartofsky (eds.): *A Portrait of Twenty-Five Years Boston Colloquia for the Philosophy of Science, 1960-1985.* 1985 ISBN Pb 90-277-1971-3

Previous volumes are still available.

KLUWER ACADEMIC PUBLISHERS – DORDRECHT / BOSTON / LONDON

GPSR Compliance

The European Union's (EU) General Product Safety Regulation (GPSR) is a set of rules that requires consumer products to be safe and our obligations to ensure this.

If you have any concerns about our products, you can contact us on ProductSafety@springernature.com

In case Publisher is established outside the EU, the EU authorized representative is:

Springer Nature Customer Service Center GmbH
Europaplatz 3
69115 Heidelberg, Germany

Batch number: 10761599

Printed by Printforce, the Netherlands